Neurosciences for
allied health therapies

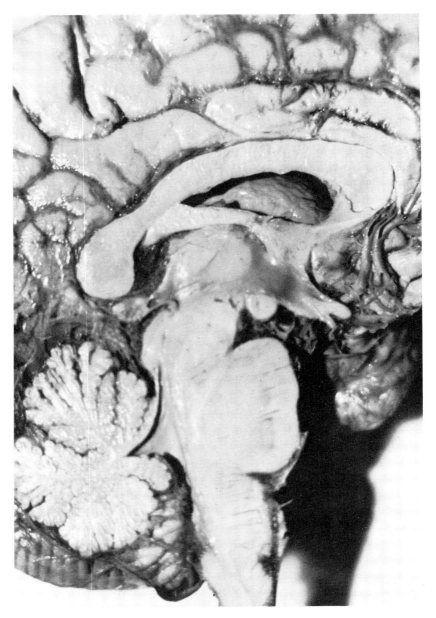

Cover design by Frank T. Adamo from a photograph of an *almost* midsagittal section of an adult human brain. Note the absence of the septum pellucidum, which allows the cavity of the left lateral ventricle and head of the caudate nucleus to be seen. This brain section was prepared by Donald R. Brown.

Neurosciences for allied health therapies

DONALD R. BROWN

Assistant Director School of Physical Therapy
Childrens Hospital of Los Angeles
Los Angeles, California

with 335 illustrations

The C. V. Mosby Company

ST. LOUIS • TORONTO • LONDON   1980

The C. V. Mosby Company
11830 Westline Industrial Drive, St. Louis, Missouri 63141

Library of Congress Cataloging in Publication Data

Brown, Donald R   1946-
  Neurosciences for allied health therapies.

  Bibliography: p.
  Includes index.
  1. Neurology. 2. Allied health personnel.
3. Physical therapists. I. Title.
QP361.B86   612.8   79-19685
ISBN 0-8016-0827-9

C/CB/B 9 8 7 6 5 4 3 2   05/B/603

My deepest gratitude and affection to

Trudy E. McDowell, who suggested the idea and gave, as always, her support

Rose Mary Asuncion, who showed great patience and helped design the tables and without whose typing skills I would still be on the preface

Frank Adamo, for his superb art work, art lessons, and an occasional push to keep going

The many students who have been through our program who have taught me so much

Preface

After teaching the neurosciences for 8 years to physical therapy students I have become convinced that a textbook is needed to present this material in an understandable and, more important, a useable manner. This text provides a practical background in neuroanatomy, neurophysiology, and neurology. Much of the extraneous detail commonly associated with the study of the nervous system has been omitted on the assumption that one can only absorb so much the first or second time around. At the same time emphasis has been placed on what a professional in the health fields must know about the neurosciences to effectively evaluate and treat neurologically impaired patients.

This text is designed for:

1. Physical and occupational therapy students as an introductory textbook
2. Physical and occupational therapists who would like a solid review of the neurosciences
3. Other health professionals such as psychologists, nurses, and speech pathologists who work with neurologically impaired patients

The text is written in a practice-while-you-read fashion. Each chapter lists at the start the major facts and concepts to be learned. This helps to organize the studying of the material. At the end of each chapter is a self-assessment quiz.

The illustrations in this text are primarily original drawings and modified drawings from current neuroanatomy texts. They have been selected and designed to provide a sense of depth and perspective to the understanding of neuroscience. Most of the drawings are accompanied by discussion to help explain their content and to identify the structures that have been labeled.

The chapters are organized in a sequence found to be the most efficient for learning. The initial emphasis is on neuroanatomy to provide a solid foundation in the structures of the nervous system, their development, and their location. Next the major aspects of neurophysiology and functional considerations of the nervous system structures are explored to provide an understanding of how the nervous system converts various stimuli into electrical impulses that can eventually result in an enormous variety of sensory, motor, visceral, or mental activities. In the last chapter, "Neurology," information from the previous chapters is applied to an understanding of neurologic case histories. The goal of this chapter is to help the reader determine from a given set of symptoms the possible site of a lesion.

At the end of this book are two appendixes. Appendix A contains the answers to the quizzes. In those questions concerning a clinical situation the answers may be debatable, as there are few consistencies in working with patients. An apparently wrong answer may actually be correct with the appropriate justification. Appendix B contains a brief biographic sketch of *some* of the well-known names in the neurosciences.

Developing a solid, useable knowledge of neurosciences requires hard work and patience. It is like trying to put a jigsaw puzzle together without seeing the picture. If confusion or frustration develops, keep reading. Many of the early difficult concepts become much clearer later in the book.

Donald R. Brown

Contents

10   Neurology, 296

Appendix

Neurosciences for
allied health therapies

CHAPTER 1

Development of nervous system

The emphasis in this chapter is on the structural organization of the nervous system. From a simple tube of cells that forms along the back of the 3-week-old embryo, billions of cells will differentiate into the various structures of the peripheral and central nervous systems. With knowledge of how the neural elements evolve from simple cells to complex units, the therapist can begin to visualize the nervous system in three dimensions. This ability, when acquired, will enhance understanding of neurologic conditions.

In this chapter special attention should be given to the following areas:

1. Structures derived from the three germinal layers of the embryo
2. Formation of the neural tube and neural crest
3. Derivatives of the alar and basal plates, neural crest, and somites
4. Components of gray matter and white matter
5. Types of neurons and neuroglial cells
6. Structures derived from the five secondary vesicles
7. Components of the basic reflex arc
8. Concept of segmental innervation with regard to a muscle, its nerve supply, and the skin overlying the muscle
9. Causes and types of spina bifida
10. Changes that occur in the nervous system during the first few years of life and with advanced age

CELL DIFFERENTIATION AND EMBRYONIC DEVELOPMENT
Germinal layers

As early as the third week after conception the embryo contains three layers of cells, which are the forerunners of the structures of the body. The embryo at this stage is shaped like an elongated disc with two cavities, one above and the other below. The cavity above the dorsum is the amnionic cavity. This cavity is filled with fluid and will eventually surround the developing embryo, providing a source of protection. The cavity below is the yolk sac, which is a source of nutrition for the young embryo (Fig. 1-1).

The three germinal layers of the embryo are the endoderm, the mesoderm, and the ectoderm.

Endoderm. The endoderm is the innermost layer and forms a lining around the *yolk sac*. This sac soon splits, with one portion migrating away from the embryo and the other portion remaining as the *gut tube* with the lining of endodermal cells (Fig. 1-3). These cells differentiate into the components of the digestive system, including the liver, spleen, and gallbladder, and the nonmuscular components of the respiratory system, such as the lungs, trachea, and bronchial tube.

Mesoderm. The mesoderm is the middle layer and also surrounds the yolk sac. These cells will differentiate into the skeletal muscles, the heart and circulatory system,

1

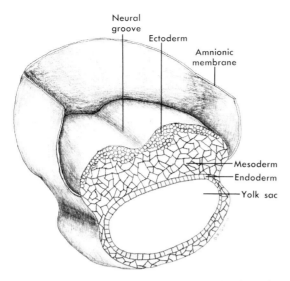

Fig. 1-1. Section through embryo in third week. Three germinal layers can be seen: ectoderm (also called neural plate), mesoderm, and endoderm. Note that ectoderm along midline is beginning to bulge, forming neural groove. Amnionic membrane has been cut. At this stage it surrounds entire dorsal half of embryo. In a few weeks it will completely surround embryo.

the skeletal system, and the genitourinary system.

Of special interest during this time are the *somites* and the *notocord*. The somites are clusters of cells located segmentally on either side of the developing neural tube (Fig. 1-2). There are usually 42 pairs of somites. They contain cells that differentiate into the skeleton (including the vertebral column), skeletal muscles, and connective tissues. The somites are involved in segmental innervation, discussed later in this chapter. The notocord is a small cylinder of cells found below the neural tube. It will differentiate into the intervertebral discs (Fig. 1-3). Also of importance are the mesodermally derived elements associated with the nervous system. These elements are the blood vessels supplying neural tissue, the microglial cells, the

meninges, the connective tissue coverings of peripheral nerves, and the capsules of some sensory receptors.

Ectoderm. The ectoderm is the outer or top layer. At 3 weeks these cells begin to show an increase in growth rate along the midline of the embryo, creating the *neural groove* (Fig. 1-1). The cells of this groove soon form the *neural tube*, which will split away from the overlying layer of remaining ectoderm. The neural tube will differentiate into the structures of the central nervous system, while the remaining ectoderm will form the skin.

Growth of embryo and fetus

The differentiation of the cells of the ectoderm into the components of the nervous system is discussed in the next section. The remainder of the embryo is considered below.

At 4 weeks the embryo is only about 4 to 7 mm in length. However, the heart is beating, most other organs are present at least in rudimentary form, and the limb buds have formed. It is apparent even at this stage that development proceeds in a cephalocaudal direction, since the upper limb buds appear first and are more advanced than the lower limb buds.

In a few weeks the embryo will triple in length, the first bones and skeletal muscles will have formed, the finger buds can be seen, and the sex can be recognized. By the end of 13 weeks the embryo (now called the fetus) is quite recognizably human. The eyes, lips, and nose are formed, and stimulation of the lips causes withdrawal movements such as rotation of the head, bending of the trunk, and extension of the shoulder. It is no wonder that exposure of the fetus to certain drugs or diseases during the first 3 months can produce severe deformities such as those associated with thalidomide or German measles.

At the end of 7 months the fetus will have grown to about 35 cm. It is capable of sucking and swallowing, and there are continuous

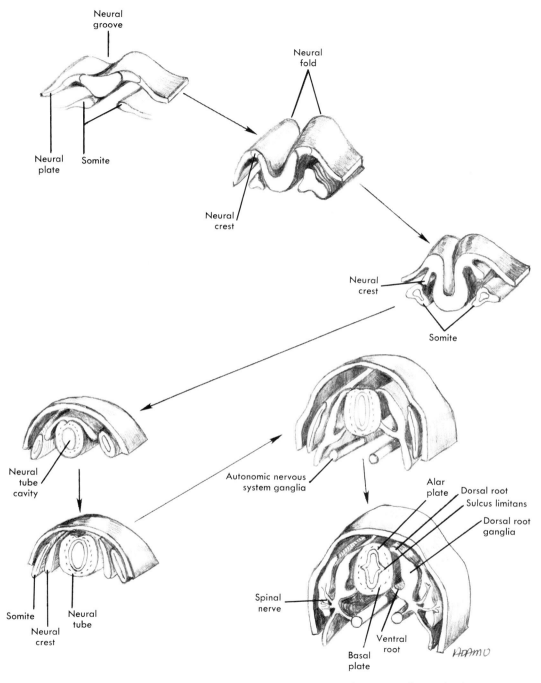

Fig. 1-2. Progressive cross-sections through embryo showing formation of neural tube. Note that neural folds come together to form neural tube. Also note position of neural crests and somites and invasion of somites by spinal nerve(s). (Adapted from House, E., and Pansky, B.: A functional approach to neuroanatomy, New York, 1967, McGraw-Hill Book Co.)

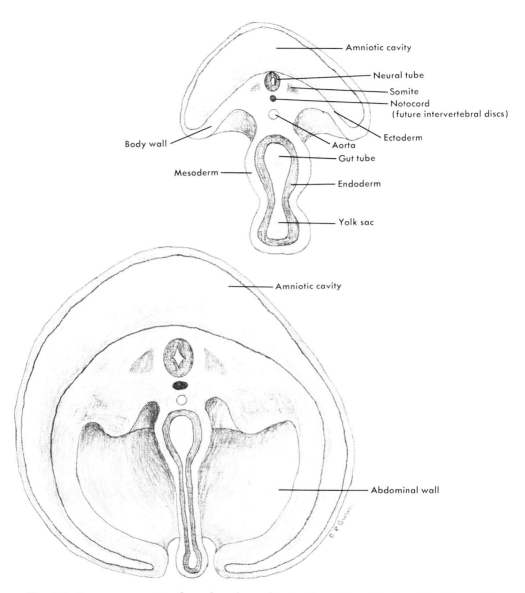

Fig. 1-3. Transverse section through embryo showing formation of body wall and gut tube. (Adapted from Passmore, R., and Robson, J., editors: A comparison of medical studies, vol. 1, Philadelphia, 1968, F. A. Davis Co.)

respiratory movements. All other systems are capable of functioning at a somewhat crude level. In fact, the fetus can survive at this time (30 to 32 weeks) outside the mother, although a great deal of special care is needed.

The fetus continues to develop during the remaining few months, although at a much slower rate than before. It is important to keep in mind that birth does not terminate the development of the organs and systems of the body. The visual and nervous systems, for example, continue to develop up to the age of 7 or 8 years.

DEVELOPMENT OF NEURAL TUBE AND NEURAL CREST
Formation of neural tube and crest

As mentioned, at 3 weeks the layer of ectoderm is growing rapidly and thickening,

forming the *neural plate* (Fig. 1-1). All structures of the nervous system will be derived from this plate. Along the midline the cells of the neural plate grow even more rapidly, creating the *neural groove* and the *neural folds* on either side (Figs. 1-1 and 1-2). During the third and fourth weeks the neural folds come together and fuse, creating a tube. This *neural tube* then splits off from the overlying ectoderm (Figs. 1-2 and 1-4). The neural tube thus lies within the embryo near its dorsal surface. At this stage it extends the full length of the embryo. All structures of the central nervous system and some of the peripheral nervous system will be derived from this tube. It is interesting to note that the cavity formed by the closure of the tube remains throughout life in the adult brain as the ventricular system.

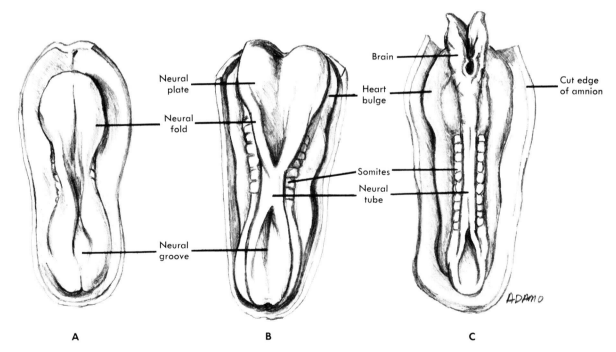

Fig. 1-4. Developing embryo as seen from above. Note in **C** that cranial (*top*) end of tube is developing more rapidly. Rest of embryo is developing beneath neural tube. **A,** 20 days old. **B,** 22 days old. **C,** 23 days old. (Adapted from Noback, C., and Demarest, R.: The human nervous system: basic principles of neurobiology, ed. 2, New York, 1975, McGraw-Hill Book Co.)

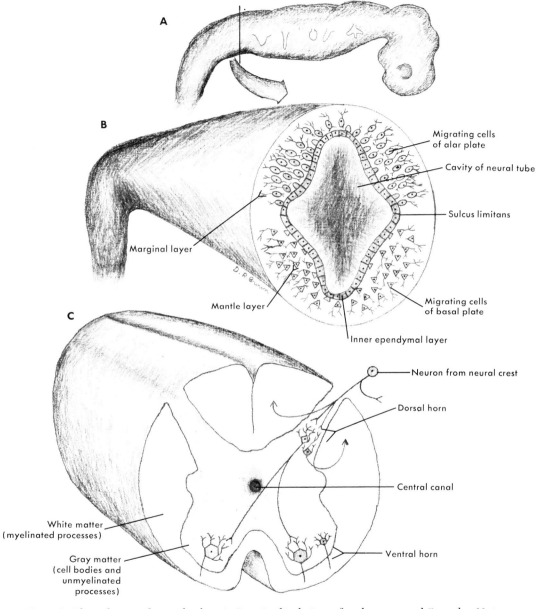

Fig. 1-5. Three layers of neural tube. **A,** Longitudinal view of embryo around 5 weeks. Note appearance of five of the cranial nerves. Compare with Fig. 1-7. **B,** Cross-section portion through upper spinal cord of neural tube as indicated by vertical line in **A.** Note migration of cells from inner ependymal layer to mantle layer. Then note how processes from these cells form marginal layer. Bend of caudal part of tube is more acute than normally found. **C,** Cross-section of full-term spinal cord showing relationship between alar plate and dorsal horn and between basal plate and ventral horn. Note that neural tube cavity is much narrower and is now called the central canal. Layer of ependymal cells still remains as lining.

As the tube is being formed the neural folds give off a loosely organized sheet of cells known as the *neural crest*, which lies over the dorsolateral portion of the tube (Fig. 1-2). This sheet of cells soon becomes organized into discrete pairs of segmented clusters that correspond to the developing somites. The cells in the neural crest develop into components of the peripheral (somatic and autonomic) nervous system.

Components of tube and crest

Both the neural tube and the neural crest contain two types of undifferentiated cells: *neuroblasts* and *glioblasts.* From these cells all of the structures of the peripheral and central nervous system develop. Neuroblasts produce all of the neurons, while glioblasts (sometimes called spongioblasts or medulloblasts) develop into a variety of cells that support, insulate, and metabolically assist the neurons. These cells are collectively known as *neuroglial cells.*

Structure of tube

Unlike the neural crest, the wall of the neural tube becomes rapidly organized after closure into three layers (Fig. 1-5). The inner layer is the *ependyma.* It contains the undifferentiated neuroblasts and glioblasts. As these cells divide many of them migrate out to the middle of the wall, forming the *mantle layer.* As the neurons develop in the mantle layer many of their cellular processes (axons and dendrites) migrate outward along with some insulating neuroglial cells, forming the *marginal layer.* Invading axons from the neural crest also contribute to the marginal layer. These insulating cells in the marginal layer surround the axons and dendrites and in some cases form a whitish substance known as myelin. Myelinated processes are often referred to as *white matter.* The mantle layer is composed of cell bodies of neurons and neuroglial cells and processes lacking myelin. The mantle layer has a grayish appearance and is often referred to as *gray matter.* In the cortices of the cerebrum and cerebellum the cells of the mantle layer migrate out to form the marginal layer, while their cellular processes remain behind. Thus the outer portion of the cerebrum and cerebellum is composed of gray matter, while in the spinal cord the outer portion is composed of white matter. The cause of this is unknown.

The neural tube is further distinguished by a longitudinal groove present on either side of the inner surface. This groove is called the *sulcus limitans* (Fig. 1-4). It divides the mantle layer into dorsal and ventral halves, known respectively as the *alar* and *basal plates.* The neurons of the alar plate subserve sensory functions such as relaying sensory data to an integrative center in the brain and contributing to the reflex activity. The cells of the basal plate are involved with motor functions. Although the alar plate, basal plate, and sulcus limitans extend throughout the neural tube, the distinctions become less clear at the cranial end because of the more rapid proliferation of cells.

Differentiation of neuroblasts and glioblasts

The mature neurons that come from neuroblasts are manifested in a wide variety of shapes and sizes. However, they can be classified into three types based on their structure (Fig. 1-6). *Bipolar neurons* have two processes extending from opposite ends of the soma (cell body). One process (dendrite) conveys sensory data from the outside world to the soma. The other process (axon) relays these data to a dendrite or soma of another neuron. Bipolar cells are primarily associated with the special senses. *Unipolar neurons* have only one process attached to the soma. This process splits a short distance from the soma into two long processes. One is a dendrite and the other an axon. Unipolar neurons (sometimes called pseudounipolar) are found in the cortices of the cerebrum and

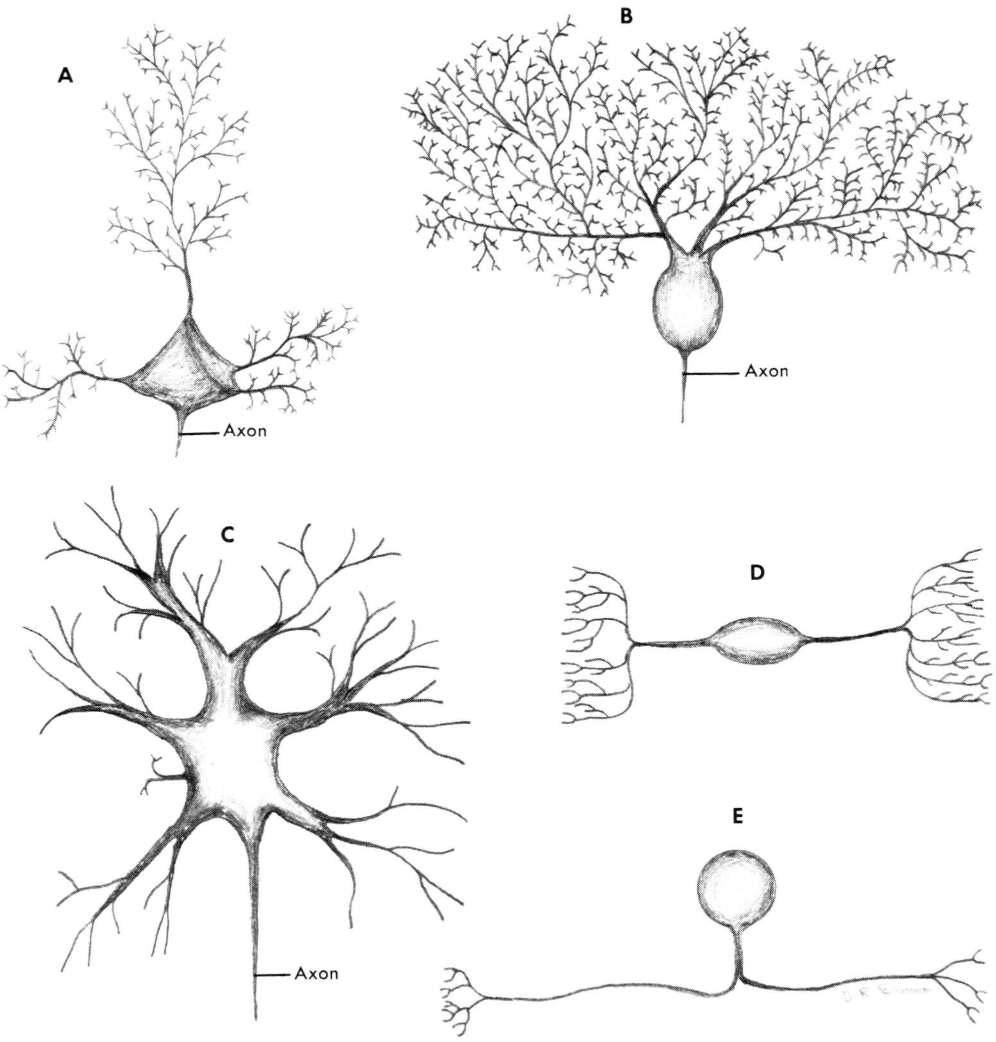

Fig. 1-6. Types of neurons. Keep in mind that true length of some axons and dendrites is not represented. For example, some unipolar neurons can extend 150 cm or more. **A,** Pyramidal neuron. **B,** Purkinje neuron. **C,** Alpha motoneuron. **D,** Bipolar neuron. **E,** Unipolar neuron.

the cerebellum and in the dorsal root ganglia. *Multipolar neurons* have a large number of dendrites and one axon attached to the cell body. Within this category are several variations:

    1. Motoneurons are found in the ventral

horn of the spinal cord and the motor nuclei of cranial nerves. They have a long axon that travels out to innervate muscle cells.

    2. Purkinje neurons are found only in the cerebellar cortex and are distinguished

by an extensive arborization (branching) of the dendrites in one plane (like vines on a trellis).

3. Pyramidal neurons are found in the cerebral cortex and are distinguished by the pyramidal shape of the soma and a long descending axon.
4. Golgi type II neurons are distinguished by many branching dendrites and one short branching axon.
5. Golgi type I neurons are multipolar with a long axon, like motoneurons and pyramidal neurons.

Neurons can be classified according to three general functions. *Sensory (afferent) neurons* convey data from the receptors of the skin, muscles, tendons, joints, special senses, and internal organs to the central nervous system (CNS). *Motor (efferent) neurons* convey commands from the CNS to muscles and glands. *Internuncials (interneurons)* connect one neuron to another and are primarily involved with reflexes and the ascending (sensory) and descending (motor) fiber tracts. In general sensory neurons tend to be bipolar or unipolar, while motoneurons and internuncials tend to be multipolar.

It is important to point out, however, that many neurons, such as those of the cerebral cortex, are involved in very complex functions that defy classification. Activities such as making judgments and creative planning require very complex neural circuits that are not purely sensory, motor, or reflex in function.

Although neurons are of primary importance, they require the support of the glioblasts. In the neural tube the supportive cells develop into three types. *Ependymal cells* remain behind in the ependymal layer and form a lining along the cavity. This lining will be part of a barrier to keep foreign substances in the ventricles from entering the brain tissue. *Astrocytes* surround much of the exposed surface of the soma. They act to support the neuron, as a link between the capil-

laries and neurons, as a metabolic helper, and as a protective barrier for neurons. So numerous are astrocytes that hardly any extracellular space exists in the CNS. The *oligondendrocytes* surround the axon and dendrites of a neuron, acting to insulate these processes from one another. For some processes the oligodendrocytes form a coating of myelin, which allows for faster conduction of the neural impulse.

In the neural crests the glioblasts develop into two types. *Satellite cells* encapsulate the neurons of the dorsal root ganglia, cranial nerve ganglia, and autonomic ganglia derived from the crest. Although their role is uncertain, it is probably similar to that of the astrocytes. *Schwann cells* are the peripheral nervous system (PNS) counterpart to the oligodendrocytes. Schwann cells migrate out with the developing axons and dendrites that form the peripheral nerves. The Schwann cells insulate these processes and in some cases form a coating of myelin.

Differentiation of tube and crest

During the first month the cranial end of the neural tube proliferates more rapidly than the caudal end and by the end of 4 weeks exhibits three bulges known as the primary vesicles. They are the *prosencephalon* (forebrain), *mesencephalon* (midbrain), and *rhombencephalon* (hindbrain). All of the structures of the brain and brainstem will be derived from these primary vesicles and will be housed within the confines of the skull. The remainder of the tube will become the *spinal cord* (Fig. 1-7).

At the end of 5 weeks the three primary vesicles have developed into the five secondary vesicles (Fig. 1-7). The prosencephalon develops into two secondary vesicles: the *telencephalon* and *diencephalon*. The telencephalon is the most rostral portion of the neural tube. It will become the two cerebral hemispheres, or cerebrum as they are often called. Unlike the other parts of the tube,

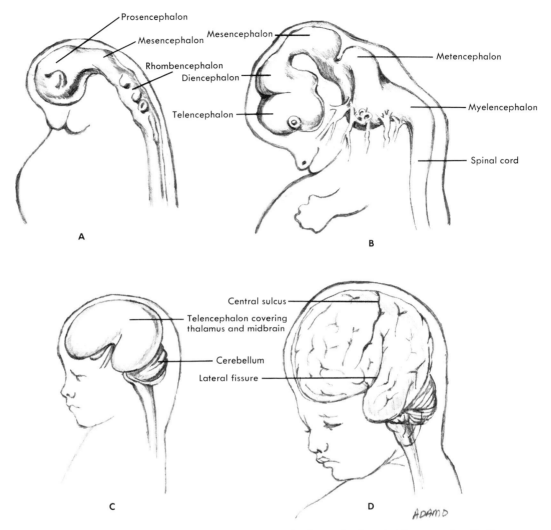

Fig. 1-7. Progressive development of cranial end of neural tube. **A,** Three weeks. **B,** Seven weeks. **C,** Four months. **D,** Newborn. (Adapted from Noback, C., and Demarest, R.: The human nervous system: basic principles of neurobiology, ed. 2, New York, 1975, McGraw-Hill Book Co.)

most of the neuroblasts of the mantle layer in the telencephalon migrate out into the marginal layer. This outer layer of cells is called the *cerebral cortex*. Other major structures are the *olfactory lobe* (rhinencephalon or smell brain), the *basal nuclei*, which have a motor function, and the *corpus callosum*,

which is a broad bundle of fibers that connects the two hemispheres. These and other structures to be mentioned are discussed in greater detail in Chapter 2. The remaining four secondary vesicles will produce structures of the brainstem and cerebellum.

The diencephalon is considered to be the

rostral or top portion of the brainstem. Important structures derived from the diencephalon are the *thalamus*, which is a sensorimotor integration center, the *hypothalamus*, which helps to regulate the internal environment of the body, and the *subthalamus*, which is a motor control center. Of interest, most of these structures are hidden from view as a result of the overlapping of the cerebral hemispheres. This overlapping is caused by a very high growth rate in the cells of the telencephalon within a confined space. This causes the hemispheres to double back over themselves and the diencephalon as they grow.

The mesencephalon does not further divide. It retains the same name as the third of the five secondary vesicles. It is located caudal to the diencephalon and is often called the *midbrain*. Important structures derived from the mesencephalon are the *substantia nigra*, *red nucleus*, and *crus cerebri*, all of which have a motor function.

The rhombencephalon divides into the final two secondary vesicles. The *metencephalon* develops into the *pons*, which is a caudal continuation of the brainstem, and the *cerebellum*, which regulates the coordination of motor activities. Like the cerebrum, most of the neuroblasts of the mantle layer of the cerebellum migrate out into the marginal layer to form the cerebellar cortex. The last of the secondary vesicles is the *myelencephalon*. It becomes the *medulla oblongata* (medulla), which is the caudal part of the brainstem.

The rest of the neural tube (spinal cord) is not inactive during this stage. Although it retains its basic shape, much is happening inside. Many of the neuroblasts in the basal plate differentiate into multipolar cells called *alpha* and *gamma motoneurons*. The axons from these cells leave the tube and form a *ventral root*. These ventral root axons join the dendrites from the neural crest cells to form spinal nerves (Figs. 1-2 and 1-9).

Many of the neuroblasts of the alar plate become internuncials. They relay incoming sensory data for reflexes or conscious awareness. The mantle layer loses its oval shape and becomes **H** shaped. There are two *ventral* (anterior) *horns* from the basal plates and two *dorsal* (posterior) *horns* from the alar plates (Fig. 1-5). The marginal layer increases in size as a result of a large number of axons and dendrites traveling up and down the cord. The cavity decreases in diameter and becomes the *central canal*. In the adult it is often obliterated, although the ependymal lining remains.

The neural crest, which will be a major contributor to the peripheral nervous system, is also undergoing extensive changes. Many of the neuroblasts remain where they are and differentiate into unipolar cells. These cells then form a discrete cluster known as *dorsal root ganglia*, which are arranged as 31 segmented pairs on either side of the spinal cord. One process from the unipolar cells (the axon) grows into the spinal cord; the axon may synapse with another neuron or enter the marginal layer, forming a fiber tract. The other process (the dendrite) forms a spinal nerve with the ventral root axons and travels out to the skin, muscle, or other tissue to end as a sensory receptor.

The other neuroblasts migrate out from the crest area and form clusters of neurons (ganglia) associated with the cranial nerves and the autonomic nervous system. Some of the cells migrate to the kidneys to form the medulla of the adrenal glands. It is important to remember that any neural structure outside the neural tube, including the dorsal and ventral roots, is considered part of the peripheral nervous system.

The glioblasts of the neural crest do one of two things. They either encapsulate the neurons in the dorsal root or other ganglia (satellite cells) or they follow the growing axons and dendrites in the dorsal root, ventral root, and spinal nerve (Schwann cells).

Refer to Fig. 1-8 for an overview of the development of the nervous system.

The clustering of the neural crest neuroblasts into dorsal root ganglia and their differentiation into unipolar neurons is an im-portant advancement for the nervous system. This is because as the peripheral and central processes of the unipolar neurons grow, they complete a vital neural circuit known as the *reflex arc*, which is considered the functional

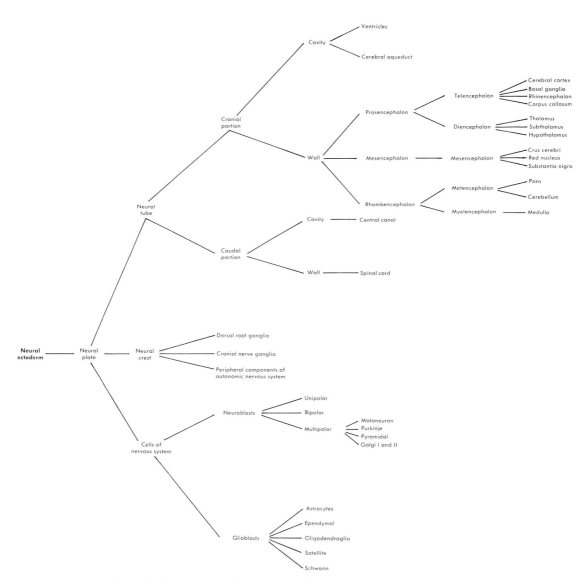

Fig. 1-8. Developmental progression of major components of nervous system.

cornerstone of the nervous system. The simplest reflex involves only two neurons, while other more complex reflexes can involve a great number of neurons. Reflexes can involve simple movements of one joint or more complex movements involving the whole body. The earliest reflex activity that can be elicited from the embryo is at around 7 or 8 weeks. It involves the withdrawal of the head when the area of the lips is stimulated. Other reflexes such as sucking and swallowing evolve later and allow the newborn to survive outside the mother. Addi-

tional information about the different reflexes is presented in later chapters.

The components of a basic reflex arc (Fig. 1-9) are as follows:
1. Cutaneous, proprioceptive (muscle, joint, tendon), or visceral receptor that has been stimulated
2. Afferent input conductor, which is usually a unipolar neuron in a dorsal root or cranial nerve ganglia
3. Internuncial connecting the afferent and efferent neurons (not all reflexes involve an internuncial)

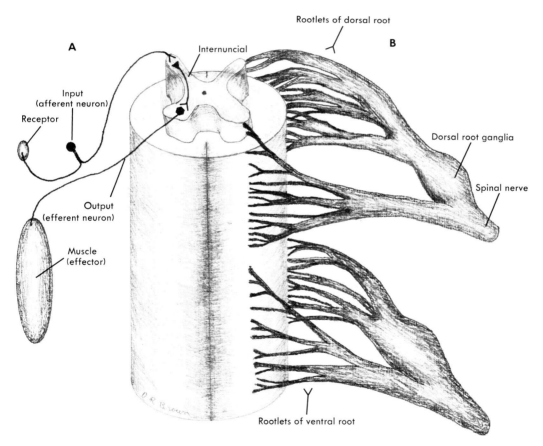

Fig. 1-9. Segmental innervation and reflex arc. View of spinal cord shows components of reflex arc, **A,** and how dorsal and ventral roots from one cord segment unite to form a spinal nerve, **B.**

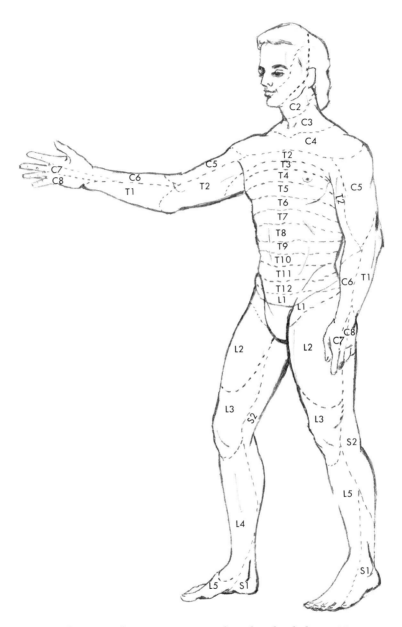

Fig. 1-10. Pattern of segmental innervation. Spinal cord is divided into 31 segments corresponding to vertebrae. Each segment is connected to pair of spinal nerves that innervates portion of skin and usually muscles underlying skin. While trunk easily demonstrates progressive flow of segmentation, extremities do not. Instead they show irregular mapping of segmental innervation. (Adapted from Noback, C., and Demarest, R.: The human nervous system: basic principles of neurobiology, ed. 2, New York, 1975, McGraw-Hill Book Co.)

4. Efferent (output) neuron, which is usually a motoneuron in the spinal cord or cranial nerve nuclei
5. Effector organ, either a muscle that will contract or a gland that will secrete

The reflex arc as described above and drawn in Fig. 1-9 shows the activity occurring within one segment of the spinal cord. It is important to realize that most reflexes, and indeed most motor activity, involve more than one cord segment. The concept of *spinal cord segmentation* is based on the fact that the mature spinal cord has 31 pairs of dorsal and ventral roots and 31 pairs of spinal nerves. Each cord segment is usually between 1.0 to 1.5 cm thick (depending on the individual's height). As a functional unit each cord segment receives sensory data from a particular area of skin and supplies specific muscles or a portion of a muscle. During development of the neural tube (Fig. 1-2) it can be seen that as the spinal nerve grows it is attracted to and invades the somites. As the somites migrate out to the periphery and into the limb buds they "tow" with them the spinal nerve. When the somites reach their respective destination the spinal nerve innervates the muscle(s) developed in that somite and generally supplies sensory receptors to the ectoderm (skin) overlying the somite. The concept of segmental innervation thus involves the following:

1. The area of skin supplied by dendrites from one dorsal root ganglia that project into one spinal cord segment (This area of skin is known as a dermatome.)
2. The muscles or portion of a muscle innervated by the motoneurons of one cord segment (one ventral root) (Fig. 1-9)

An important aspect of this concept of segmental innervation is that stimulation of the receptors in a dermatome can influence the activity of the motoneurons in the corresponding spinal cord segment. For example, scratching the volar surface of the arm (which is innervated by C_5) is often used by ther-apists to help activate the biceps brachii, which is innervated by C_5 and C_6. For the most part the innervation of a muscle and the skin overlying that muscle will be from the same cord segment. There are exceptions to this, such as the diaphragm, latissimus dorsi, gluteus medius, and hip adductors. Fig. 1-10 is a dermatomal map of the body. Note that C1 does not have a cutaneous distribution. This figure can be used with Fig. 1-11 to determine the segmental innervation between muscle and skin.

Congenital anomalies

Of all of the malformations and congenital defects found in humans about half involve the nervous system. The causes of these anomalies are varied but generally can be attributed to genetic factors, malnutrition of the mother, abnormal drug or hormone levels, or insufficient levels of oxygen. Perhaps because neural tissue is so highly specialized it is much more sensitive to adverse conditions than other tissues.

The primary signs of congenital involvement of the nervous system are (1) mental retardation such as associated with Down's syndrome and hydrocephalus (see Fig. 7-9), (2) excess or abnormal movements, (3) delayed or retarded sensorimotor development, (4) paralysis of one or more extremities, and (5) hyperirritability.

One of the more common congenital anomalies is *spina bifida* (Fig. 1-12). This wide-ranging defect can vary from simple lack of formation of the lamina and spinous process of a vertebra (spina bifida occulta) to incomplete closure of the neural plate (spina bifida with myeloschisis). In the former there may be no symptoms, while in the latter there can be severe paralysis, sensory loss, and bowel and bladder disturbances.

Maturational changes

The rate of growth of the embryo and fetus is truly amazing. At 4 weeks the embryo is about 0.4 cm long. By the fifth month the

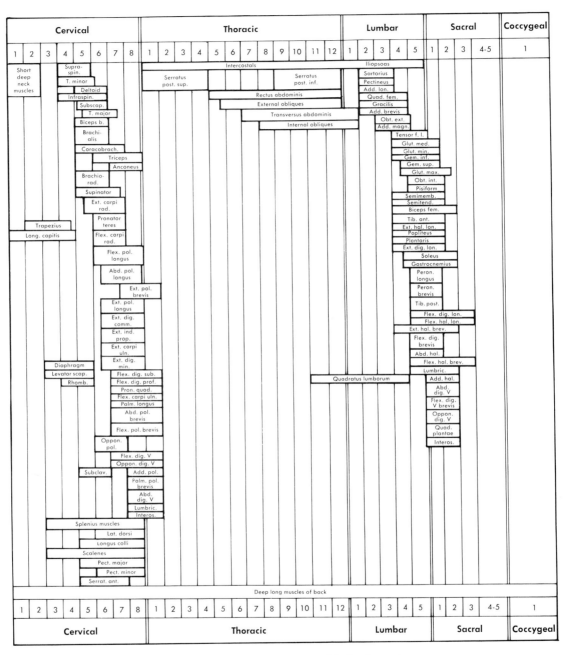

Fig. 1-11. Segmental innervation of skeletal muscles.

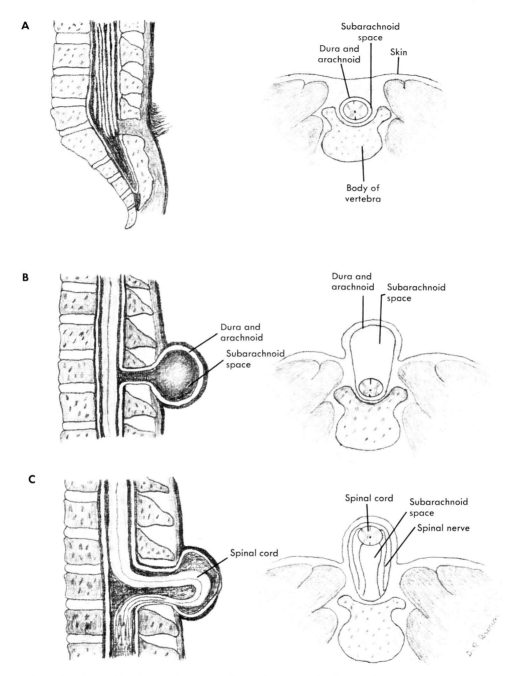

Fig. 1-12. Anomalies in development of neural tube in lumbosacral region. **A,** Spina bifida occulta. **B,** Spina bifida with meningocele. **C,** Spina bifida myelomeningocele. Note potential for extensive neural damage in **C.** (Adapted from Netter, F.: The CIBA collection of medical illustrations. I. The nervous system, Summit, N.J., 1972, CIBA Chemical Co.)

fetus has grown to about 15 cm, which represents a 375% increase in length, and it contains most of its adults structures. At birth the neonate is about 50 cm long, roughly a 233% increase.

Of all of the types of tissues in the body neural tissue has the fastest growth rate. This is substantiated in part by the fact that the fetal nervous system must produce 45,000 neurons per minute to attain the full complement of almost 25 billion neurons present at birth. It is also substantiated by the fact that the nervous system is quite functional at birth. The newborn can generally locate the mother's nipple and nurse, show protective reactions to pain, turn his or her head to avoid suffocating, and react to light. In fact, the neonate's brain (cerebrum, brainstem, and cerebellum) accounts for a full 10% of the birth weight.

The higher growth rate of the nervous system continues after birth to around 5 years of age. At this point the nervous system is about 85% developed and begins to slow down. Comparatively, the rest of the body is only 40% developed at this stage. The changes within the nervous system during these early years center around the growth and myelination of axons and dendrites and the convolution of the cerebral and cerebellar cortices. The growth and myelination of axons and dendrites allow for the completion and efficient use of additional neural circuits. The convolution of the surface of the cerebrum and cerebellum involves the development of depressions (sulci) and ridges (gyri), which increase the total surface area without requiring a corresponding increase in the size of the skull. This allows for the growth and maturation of existing neurons within a confined space. These two factors operate together to provide the infant and young child with improved responses to environmental conditions and a remarkable increase in skilled motor activities.

We are born with all the neurons we will

ever have. Myelination is generally complete at around 7 years, and growth of axons and dendrites is generally complete by the late teens. It is during this period between childhood and early adulthood that the nervous system operates at its peak of sensorimotor integration ability as cognitive and motor skills are more easily acquired during this time. During the next 50 years (between 20 and 70 years of age) there is a loss of about one tenth of the existing neurons. This averages to a daily loss of about 100,000 neurons. This loss is partly due to histologic changes within the neuron and a reduction in blood flow (up to 20%) to the brain. Accompanying the neuronal loss during this period is 10% reduction in conduction velocity of the neural impulse, a loss of brain weight, an increase in size of the ventricles, widening of the cortical sulci, and calcification of the meninges. It is important to mention that these changes occur at different rates for different individuals. Many persons in their seventies and even eighties are alert and active; others in their forties show signs of mental and physical deterioration. While part of the cause for these differences can be attributed to heredity, a person's environment, diet, and general health must also be considered. Whatever the cause, it is important as a therapist to have extra patience with older persons. They generally learn slower (especially motor skills), they react less adeptly to changes in their environment, their sensory systems (such as hearing) are less effective, their movements are slower, and they fatigue more quickly. Allowing extra time for repetition, practice, and rest periods is beneficial for all concerned.

Myelination of major systems

The process of myelination is vital to the functioning of the nervous system because it increases the velocity of the electrical impulses up to 100 times. This increase in speed of conduction not only adds a great deal of efficiency to the nervous system but also provides the following:

1. Rapid protective withdrawal of a part of the body from a noxious stimulus
2. Coordinated skilled motor activities
3. Rapid alterations in muscular contractions in response to loss of balance
4. Coordination of protective visceral reactions such as vomiting
5. Ability to think quickly and make sudden decisions

In comparison with the rest of the nervous system, myelin formation begins relatively late: not until the end of the third fetal month. By the end of the fifth month the spinal reflex arc, vestibulospinal tract, and tectospinal tracts have begun myelination and will be complete by the ninth fetal month. These areas are partly responsible for the early movements produced by the fetus and neonate. The dorsal column–medial lemniscus pathway also begins myelination in the fifth month but does not become complete until after the first postnatal year. This system is primarily used for exploratory touch and a sense of body position (kinesthesia). The spinocerebellar (coordination) and spinothalamic (crude touch, pain, and temperature) tracts become myelinated between the seventh fetal and third postnatal months. The pyramidal tracts, which are the primary pathways for voluntary skilled movements, do not begin myelination until just before birth and continue into the second year. Finally, the myelination of the cerebrum and cerebellum, which is minimal at birth, can continue into early adulthood.

TEST QUESTIONS

For each question select *all* correct choices.

1. Neuroblasts:
   a. Are found in the alar plate.
   b. Differentiate into Schwann cells.
   c. Are derived from the ependymal layer.
   d. Are found in the marginal layer of the cerebral and cerebellar cortices.

2. Schwann cell is to oligodendrocyte as:
   a. Neural crest is to neural tube.
   b. Peripheral nervous system is to central nervous system.
   c. Telencephalon is to prosencephalon.
   d. Ventral root is to dorsal root.

3. Cutting spinal nerves from C5 and C6:
   a. Produces anesthesia over the anterior surface of the arm.
   b. Destroys the motoneurons in the C5 and C6 ventral horns.
   c. Causes weakness of forearm flexion.
   d. Interferes with the C5 and C6 reflex arc.

4. The basal nuclei:
   a. Contain neurons and neuroglia.
   b. Are derived from the diencephalon.
   c. Are derived from the ependyma.
   d. Are part of the cerebral cortex.

5. If a cerebrum developed completely flat and contained as many cells as a normal cerebrum, the skull would be about:
   a. The same.
   b. Twice as large.
   c. Four times as large.
   d. Eight times as large.

6. The brain of a 70-year-old person compared with that of a 15-year-old person:
   a. Is lighter in weight.
   b. Has wider convolutions.
   c. Has lost 10% of the original neurons.
   d. Operates at about 45% efficiency.

7. The myelencephalon:
   a. Develops into the pons and cerebellum.
   b. Has an outer cortex of gray matter.
   c. Is continuous with the spinal cord.
   d. Contains the crus cerebri.

8. Although the nervous system is derived from the ectoderm, some components are of mesodermal origin. These include:
   a. Myelin.
   b. Notocord.
   c. Capsules of some receptors.
   d. Ependymal lining of the ventricles.

9. In the formation of the neural tube:
   a. Mesodermal somites organize as segmented pairs on either side.
   b. A cavity is formed that will later become part of the ventricular system.
   c. Neural crest splits off from the wall of the tube.

    d. Neural folds fuse along the midline of the dorsum of the embryo.

10. Which part(s) of the following statement is(are) false? A receptor in a muscle relays data to a *unipolar neuron* derived from the *alar plate*, which then relays the data to a *multipolar neuron* of the *ventral horn*. The *dendrite* of this multipolar neuron travels via the *ventral root* and *spinal nerve* to cause a *muscle to contract*, thus completing a reflex arc.
    a. Unipolar neuron.
    b. Alar plate.
    c. Multipolar neuron.
    d. Ventral horn.
    e. Dendrite.
    f. Ventral root.
    g. Spinal nerve.
    h. Muscle to contract.

11. In spina bifida with myelomeningocele:
    a. Sac begins to form within 2 weeks after birth.
    b. Part of the spinal cord is displaced into the sac.
    c. There is incomplete closure of the neural tube in the lumbar region.
    d. Motor and sensory loss in the lower extremities is expected.

12. Which of the following cells compose the gray matter of the spinal cord?
    a. Multipolar motoneurons.
    b. Pyramidal cells.
    c. Schwann cells.
    d. Astrocytes.

13. The diencephalon:
    a. Produces the thalamus.
    b. Is on top of the midbrain.
    c. Is hidden by the developing cerebral hemispheres.
    d. Forms the boundaries for one of the ventricles.

14. When the patellar tendon is briskly tapped, a sudden extension of the leg is seen. This movement:
    a. Involves the reflex arc.
    b. Is diminished if the L_3 dorsal root is severed.
    c. Is abolished if the femoral nerve is severed.
    d. Requires reflex inhibition of the hamstrings.

SUGGESTED READINGS

House, E., and Pansky, B.: A functional approach to neuroanatomy, ed. 2, New York, 1967, McGraw-Hill Book Co.

Minckler, J.: Introduction to neuroscience, St. Louis, 1972, The C. V. Mosby Co.

Noback, C.: The human nervous system, New York, 1975, McGraw-Hill Book Co.

Passmore, R., and Robson, J., editors: A companion to medical studies, vol. 1, Philadelphia, 1968, F. A. Davis Co.

Schade, J., and Ford, H.: Basic neurology, ed. 2, New York, 1973, Elsevier Scientific Publishing Co.

CHAPTER 2

Gross anatomy of central nervous system

In this chapter the external and internal components that develop from the five secondary vesicles and from the spinal cord are examined. Each of the major structures and points of reference is presented in terms of its location and its relationship to neighboring structures. The cerebrum, brainstem, cerebellum, and spinal cord are examined from the outside in different views and from the inside using cross-sections taken in the horizontal, frontal, or sagittal plane. The therapist must be able to identify the location of any structure or point of reference in three dimensions. In other words, he or she must not only be able to *identify* a structure from a given diagram but to *determine* where one structure is located in relation to another structure (e.g., medial, caudal, dorsal). This precise knowledge of neuroanatomy assists the therapist in the following:

1. Discussing a neurologic patient with the referring physician
2. Determining a patient's symptoms from a given location of the lesion
3. Evaluating a neurologic patient
4. Designing an effective treatment program

This chapter presents only the more fundamental components of the central nervous system. Other components such as the fiber tracts and cranial nerves are discussed in detail in later chapters. A brief statement of the function of many of the structures is mentioned. Further functional and pathologic considerations are presented later.

Before proceeding, a quick review of important terms is essential.

General vocabulary

CNS Central nervous system, composed of cerebrum, brainstem, cerebellum, and spinal cord.

PNS Peripheral nervous system, composed of dorsal and ventral roots, dorsal root ganglia, spinal and peripheral nerves, cranial nerves and their ganglia, and autonomic nerves and their ganglia.

autonomic nervous system Regulates internal environment of the body.

afferent Carries information toward the cell body via dendrites.

efferent Carries information away from the cell body via axons.

somatic Pertaining to the body wall: muscles, skin, bones, and joints.

visceral Pertaining to internal organs.

neuron Single cell body and all of its processes (axons and dendrites).

nerve Collection of neuronal processes.

nucleus Collection of neuron cell bodies within the CNS.

ganglia Collection of neuron cell bodies outside of the CNS.

gray matter composed of cell bodies and unmyelinated processes.

white matter Composed mostly of myelinated processes.

sulcus Depression or groove.

fissure Deep sulcus.

fossa Wide depression.

gyrus Ridge.

fasciculus Bundle of fibers.

Cross-sectional planes

sagittal plane Divides an object into right and left parts. A midsagittal section divides the parts equally. In a midsagittal slice of a person, the cut is made from head to pelvis, producing two parts, each with an arm and leg and half of a face.

frontal (coronal) plane Divides an object into front and back parts. In a midfrontal section of a person, the slice is made from the head down through the feet. Result is one part with a face, abdomen, kneecaps, and toes and another part with the back of the skull, buttocks, and heels.

horizontal (transverse) plane Divides object into top and bottom parts. In a transverse section through the waist of a person one part would have the head, trunk, and arms and the other part the pelvis and legs.

Directional references

anterior-ventral Front or belly portion.
posterior-dorsal Back portion.
medial Near midline of structure.
caudal-inferior Lower or tail part. Spinal cord is caudal to the medulla.
rostral-superior Uppermost part. Thalamus is the rostral portion of the brainstem.

CENTRAL NERVOUS SYSTEM

The *central nervous system* (CNS) is composed of the brain and spinal cord. The *brain* is made up of the cerebrum, brainstem, and cerebellum, or those structures derived from the five secondary vesicles. The *spinal cord* is derived from the remaining portion of the neural tube. The following discussion of the anatomy of the central nervous system begins with the spinal cord and ascends progressively to the cerebrum.

Spinal cord

The spinal cord is a long cylindrical extension of the medulla. It begins anatomically at the foramen magnum and continues in the vertebral canal to about the level of vertebra L-2 (this may vary, from T-12 to L-3). The

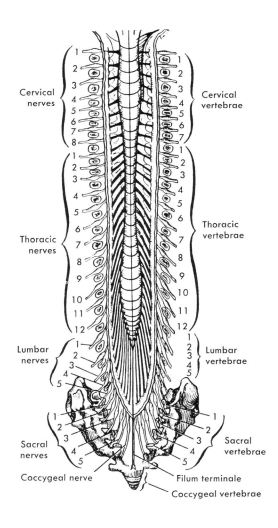

Fig. 2-1. Dorsal view of spinal cord. Note hourglass shape resulting from cervical and lumbar enlargements. Also note that spinal nerves angle down (especially lumbar and sacral nerves) to exit through corresponding vertebral foramen as a result of differential growth factors between spinal cord and vertebral column. Conus medullaris is not labeled but easily located. (From A functional approach to neuroanatomy by House, E. L., and Pansky, B. Copyright 1967 by McGraw-Hill Book Co. Used with permission of McGraw-Hill Book Co.)

spinal cord is roughly the width of an ordinary pencil, although it is somewhat flattened in an anteroposterior direction.

When viewed in full length the spinal cord has somewhat of an hourglass shape because of enlargements in the cervical and lumbar portions. These *enlargements* are most prominent between C-5 and C-8 and between L-1 and S-2 (Fig. 2-1). They are caused by an increase in the number of neurons in the gray matter supplying the muscles of the upper and lower extremities. Also the cord is thicker in the cervical portion compared with the lumbar portion because of the greater number of motor and sensory fibers.

Below S-1 the spinal cord tapers and terminates as a rounded tip called the *conus medullaris.* The conus is anchored to the sacrum by a connective tissue thread called the *filum terminale.*

Below the conus in the remaining fourth of the vertebral canal is an area known as the *cauda equina.* This area is filled with cerebrospinal fluid (CSF) and the lumbar and sacral spinal nerves, which must exit via their appropriate intervertebral foramina. During the first few months of development the spinal cord and vertebral canal are fairly equal in length, and the 31 pairs of spinal nerves simply exit straight out of the vertebral canal. During the last two trimesters, however, the body and vertebral canal grow in greater proportions than the spinal cord. The result is that beginning around spinal nerve T_1 the nerves must angle downward to exit via their corresponding foramina (Fig. 2-1). Below the conus the spinal nerves give the appearance of a horse's tail; thus the name cauda equina. This area is used in a diagnostic procedure known as a lumbar puncture (spinal tap). A needle can be inserted between vertebrae L-4 and L-5 to withdraw fluid that surrounds the spinal cord for examination. There is no danger of puncturing the spinal cord at this level, and the cauda equina nerves are rarely injured.

The surface of the cord has several important markings (Fig. 2-2). The *anterior me-*

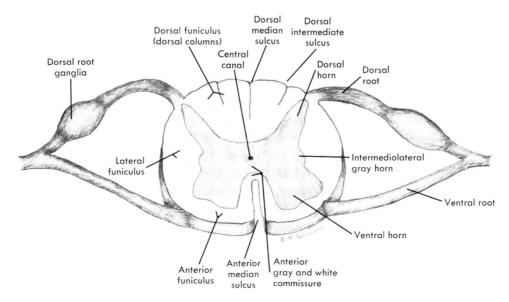

Fig. 2-2. Transverse section through upper thoracic portion of spinal cord with dorsal and ventral roots attached.

dian fissure and *dorsal median sulcus* divide the cord into right and left halves. A fissure is nothing more than a deep sulcus. In each dorsal half there is a *dorsal intermediate sulcus*. It is found only in the cervical and upper thoracic part of the cord, since it separates a sensory fiber tract from the upper extremity with a comparable tract in the lower extremity.

No external lines delineate the 31 spinal cord segments, as indicated in Fig. 2-1. The segments do exist, however, as functional motor, sensory, and reflex units that are easily identified when the cord is viewed with its dorsal and ventral roots attached (Fig. 1-7). However, what dictates that a group of neurons at C7 should extend the wrist, while the neurons caudal to these (C8) flex the fingers, can only be explained as one of the wonders of development.

The internal structure of the spinal cord is divided into the white matter and the gray matter. Anatomically the white matter of the cord is divided into three pairs of funiculi (Fig. 2-2). The *ventral funiculi* are located between the anterior median fissure and the ventral root. The *lateral funiculi* are located between the dorsal and ventral roots. The *posterior funiculi* are located between the dorsal root and the dorsal median sulcus. Different motor and sensory fiber tracts travel in the different funiculi and can create a variety of clinical symptoms, depending on which one is lesioned or diseased.

The **H**-shaped gray matter is functionally and anatomically divided into three pairs of horns and a central area (Fig. 2-2). The *ventral horns* project toward the ventral roots and are motor in function. The *dorsal horns* project toward the dorsal roots and have a sensory and reflex function. The ventral horns are larger and more rounded than the dorsal horns. The *intermediolateral horns* project into the lateral funiculus from the lateral edge of the **H**. This horn is associated with the autonomic nervous system and is

only located between cord segments T1 to L2.

The central part of the cord contains two commissures and the central canal (Fig. 2-2). The *gray and white commissures* primarily contain axons that are crossing to the opposite side for a reflex or to form a fiber tract. The difference between the two is presence of myelin in the white commissure. The *central canal* is the remains of the neural tube cavity. Although usually not open in the adult, it still retains its ependymal lining.

Rexed differentiated the gray matter into groups of cells called *laminae*. The cells in each lamina have a similar function (Fig. 2-3). Laminae I to V make up the dorsal horn. Lamina VI and part of VII form the intermediate area. Laminae VIII, IX, and the rest of VII form the ventral root. Finally, lamina X surrounds the central canal. In older terminology the gray matter was divided into nuclei, each with a separate function. The major nuclei are shown in Fig. 2-3.

Myelencephalon

Fig. 2-4 is a midsagittal (equal right and left halves) view of the brainstem in relation to the cerebellum and cerebrum. This is a point of reference for the rest of this chapter.

The *medulla oblongata* is only about 2.5 cm (1 in) long and only slightly larger in diameter than an ordinary pencil. The study of the external features of the medulla and the rest of the brainstem can be best visualized in Fig. 2-5, which shows the brainstem in anterior, lateral, and dorsal views. Frequent reference to this drawing will assist the reader with the next few sections of this text.

The external anatomy of the medulla is presented from anterior to posterior. Although Fig. 2-5 shows the cranial nerves, only brief comments are made about them in this chapter. The pons and midbrain are also discussed in a similar manner.

The anterior median fissure that dented the spinal cord is continuous with the entire

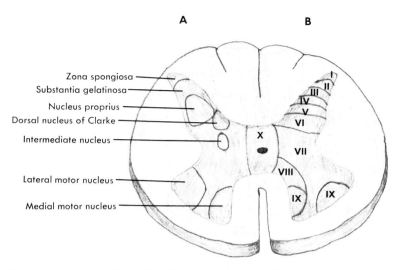

Fig. 2-3. Transverse section of spinal cord showing division of gray matter into nuclei, **A**, and laminae, **B.** (Adapted from Willis, W. D., Jr., and Grossman, R. G.: Medical neurobiology: neuroanatomical neurophysiological principles basic to clinical neuroscience, ed. 2, St. Louis, 1977, The C. V. Mosby Co.)

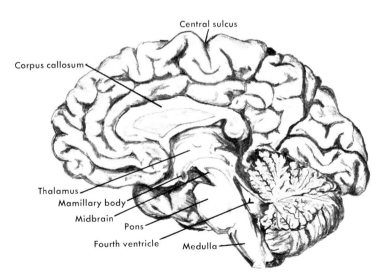

Fig. 2-4. Midsagittal section of brainstem, cerebellum, and cerebrum. (Adapted from Truex, R., and Carpenter, M.: Human neuroanatomy. ed. 6, Baltimore, 1969, The Williams & Wilkins Co.)

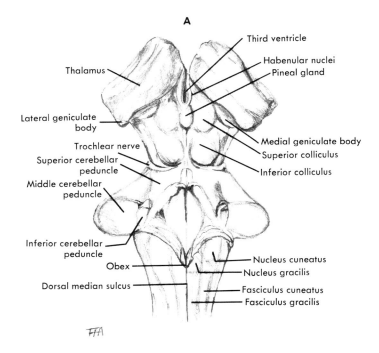

A

Third ventricle

Thalamus

Habenular nuclei

Pineal gland

Lateral geniculate body

Medial geniculate body

Trochlear nerve

Superior colliculus

Superior cerebellar peduncle

Inferior colliculus

Middle cerebellar peduncle

Inferior cerebellar peduncle

Nucleus cuneatus

Obex

Nucleus gracilis

Dorsal median sulcus

Fasciculus cuneatus

Fasciculus gracilis

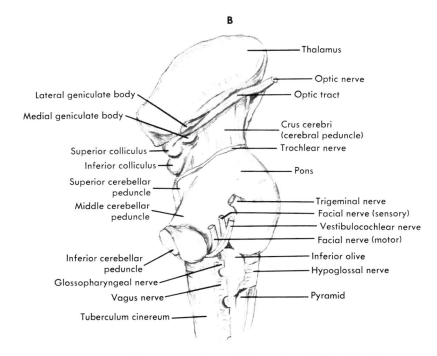

B

Thalamus

Optic nerve

Optic tract

Lateral geniculate body

Medial geniculate body

Crus cerebri (cerebral peduncle)

Superior colliculus

Trochlear nerve

Inferior colliculus

Pons

Superior cerebellar peduncle

Middle cerebellar peduncle

Trigeminal nerve

Facial nerve (sensory)

Vestibulocochlear nerve

Facial nerve (motor)

Inferior cerebellar peduncle

Inferior olive

Glossopharyngeal nerve

Hypoglossal nerve

Vagus nerve

Pyramid

Tuberculum cinereum

Fig. 2-5. External features of brainstem. Telencephalon and cerebellum have been removed to give better perspective. Cranial nerves, shown here, are discussed in Chapter 5. **A,** Dorsal view. **B,** Lateral view. **C,** Anterior view. (Adapted from Everett, N.: Functional neuroanatomy, ed. 6, Philadelphia, 1971, Lea & Febiger.)

C

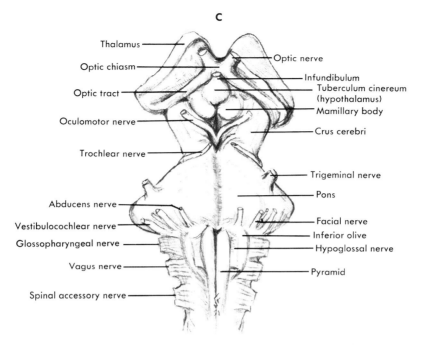

Thalamus

Optic chiasm

Optic tract

Oculomotor nerve

Trochlear nerve

Abducens nerve

Vestibulocochlear nerve

Glossopharyngeal nerve

Vagus nerve

Spinal accessory nerve

Optic nerve

Infundibulum

Tuberculum cinereum
(hypothalamus)

Mamillary body

Crus cerebri

Trigeminal nerve

Pons

Facial nerve

Inferior olive

Hypoglossal nerve

Pyramid

Fig. 2-5, cont'd. For legend see opposite page.

length of the medulla. On either side of it are the *pyramids,* which contain descending motor fibers from the cerebral cortex going to the ventral horn cells of the spinal cord. On either side of the pyramids in the upper half of the medulla are the *inferior olives.* These are important relay nuclei for proprioceptive data going from the spinal cord to the cerebellum. Lateral to the olives and pyramids are the *trigeminal eminences* (tuberculum cinereum). These provide a sensory pathway for the trigeminal nerve. In the dorsolateral corner of the upper medulla the tuberculum cinereum becomes covered with the fibers of the *inferior cerebellar peduncle* (restiform body). This is one of the three pairs of stalks that connect the cerebellum to the brainstem. Together they contain the afferent and efferent fibers going to and from the cerebellum. The *fasciculus cuneatus* and

fasciculus gracilis can be seen in the lower medulla. These are the continuation of the fibers in the dorsal funiculus of the spinal cord. The dorsal medial and dorsal intermediate sulci that separate these sensory tracts in the cord also do so in the medulla. The fasciculi terminate in bulges (tuberculi) known respectively as the *nucleus cuneatus* and the *nucleus gracilis.* These act as synaptic relay stations for ascending sensory data going from the cord to the thalamus. These nuclei mark the caudal boundary of the *fourth ventricle.* The very tip of this V-shaped boundary is the *obex.*

The internal structures of the medulla (Figs. 2-6 and 2-7) as well as the pons and midbrain are discussed in reference to horizontal sections taken through these structures. The external structures can also be seen in a horizontal section.

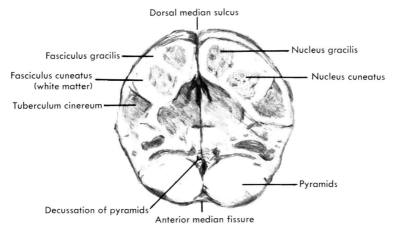

Fig. 2-6. Horizontal section through caudal medulla at level of pyramidal decussation. (After Watson, C.: Basic human neuroanatomy: an introductory atlas, ed. 2, Boston, 1977, Little, Brown & Co.)

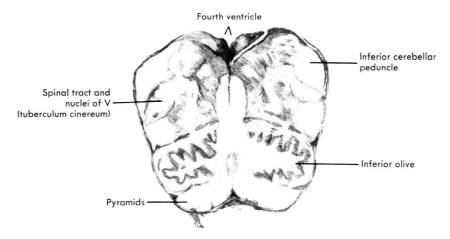

Fig. 2-7. Horizontal section through upper medulla at level of midolive. Tuberculum cinereum is now located internally. (After Watson, C.: Basic neuroanatomy: an introductory atlas, ed. 2, Boston, 1977, Little, Brown & Co.)

The internal structures of the medulla not seen from the outside include the following:

1. Ascending sensory and descending motor tracts
2. Nuclei for the reticular formation including vital cardiac and respiratory centers
3. Nuclei for cranial nerves VIII, IX, X, XI, and XII

None of these structures is labeled in Fig. 2-6 or 2-7 as they are presented in detail in later chapters.

The last structure associated with the myelencephalon is the *fourth ventricle*. It is

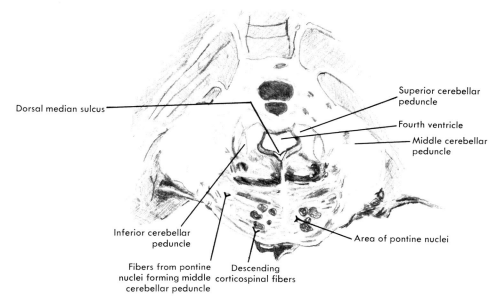

Dorsal median sulcus

Superior cerebellar peduncle

Fourth ventricle

Middle cerebellar peduncle

Inferior cerebellar peduncle

Area of pontine nuclei

Fibers from pontine nuclei forming middle cerebellar peduncle

Descending corticospinal fibers

Fig. 2-8. Transverse section through midpons and cerebellum. (After Watson, C.: Basic human neuroanatomy: an introductory atlas, ed. 2, Boston, 1977, Little, Brown & Co.)

a shallow cavity located between the back of the pons and upper half of the medulla and the cerebellum (Figs. 2-4 and 2-9). This ventricle is continuous with the central canal below and the cerebral aqueduct above. Associated with the fourth ventricle are the foramina (openings) of Magendie and Luschka. These tiny holes allow the cerebrospinal fluid (CSF) to circulate in the proper manner.

Metencephalon

Above the medulla is the *pons*. Although most of the external features of the medulla are not continuous in the pons, the pons is very easy to identify because of its large ventral swelling. This swelling is actually a large bundle of fibers known as the *middle cerebellar peduncle* (branchium pontis). Whereas the inferior cerebellar peduncle travels upward from the dorsolateral corner of the upper medulla to enter the cerebellum, the middle peduncle curves dorsally with a slight downward angle around the rest of the pons to connect to the cerebellum (Figs. 2-8 and

2-9). The back of the pons, which helps to make up the floor of the fourth ventricle, has several eminences for various nuclei. The most prominent feature, however, is the continuation of the *dorsal median sulcus*.

The internal structures of the pons are divided into an anterior or *basilar portion* and a dorsal or *tegmental portion*. The anterior part of the pons contains three important structures. First is the descending motor tract going from the cerebral cortex to the medulla and spinal cord. Most of the fibers in this tract, often called the *corticospinal fibers*, form the pyramids in the medulla. Other fibers, often called *corticobulbar fibers*, go to motor nuclei for the cranial nerves. Second are the *pontine nuclei*, which are scattered throughout the central basilar portion. They act as a relay for information going from the cerebrum to the cerebellum. Third are the fibers from these nuclei going to the cerebellum, creating the ventral swelling known as the *middle cerebellar peduncle* (Fig. 2-8).

The tegmentum of the pons is similar to

the area of the medulla dorsal to the pyramids. Ascending and descending sensory and motor tracts, part of the reticular nuclei, and nuclei for cranial nerves V, VI, VII, and VIII are all located in the tegmentum.

The final structure associated with the pons is the *fourth ventricle*. It separates the pons from the cerebellum. As mentioned, the fourth ventricle has openings to allow the flow of CSF out of the ventricular system. Blockage of these tiny openings can result in severe brain damage and death if not treated.

The *cerebellum* is sometimes called the small brain because of its similarity to the cerebrum. Both have an outer gray cortex, convolutions (called folia), and inner nuclei. The cerebellum monitors the length and ten-sion of every skeletal muscle. With this data it is able to make corrections in cortically induced movements, resulting in smooth coordinated patterns of movement that require little or no conscious effort. The cerebellum also functions in equilibrium reactions.

The cerebellum is connected to the brainstem by three pairs of stalks known as the *cerebellar peduncles*. The inferior and middle pairs have already been mentioned, and the superior pair is discussed with the midbrain. These peduncles are made up of axons traveling between the cerebellum and the rest of the CNS (Fig. 2-9).

The cerebellum is divided into two hemispheres connected to a central area known as the vermis. Each hemisphere is divided on a functional basis into three lobes: the anterior,

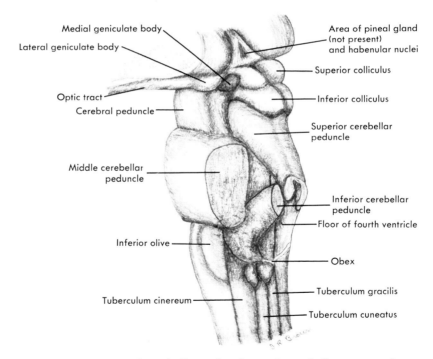

Fig. 2-9. Dorsolateral view of cerebellar peduncles with cerebellum removed. Note that peduncles help make up part of boundaries for fourth ventricle. Also note that tuberculum cinereum moves internally at level of midolive being covered by fibers of inferior cerebellar peduncle. (Adapted from Netter, F.: The CIBA collection of medical illustrations. I. The nervous system, Summit, N.J., 1972, CIBA Chemical Co.)

posterior, and flocculonodular lobes. In the midsagittal view the *anterior lobe* sits on top of the cerebellum and is separated from the posterior lobe by the *primary fissure*. The *posterior lobe* occupies the rest of what you see except for the innermost lump, which is adjacent to the fourth ventricle. This is the *flocculonodular lobe*. It is separated from the posterior lobe by the *posterolateral fissure*. Compare Figs. 2-10 and 2-11 for understand-

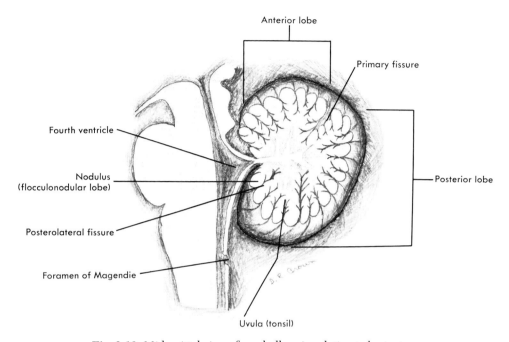

Fig. 2-10. Midsagittal view of cerebellum in relation to brainstem.

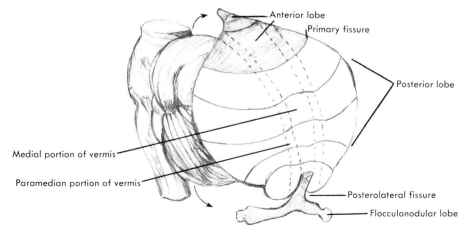

Fig. 2-11. Three lobes of cerebellum, shown as though cerebellum were opened like a clam.

ing of how the flocculonodular lobe curls under.

When looking at the dorsum of the cerebellum in Fig. 2-12 it is easy to see the two hemispheres and a middle area that connects them. This area is known as the *vermis*. The vermis has many separate components, which are discussed later. The flocculonodular lobe, anterior lobe, and vermis basically function in posture and in regulating muscle tone. The posterior lobe functions in the coordination of phasic movements.

A slice through either hemisphere will reveal the internal structure (Fig. 2-13), which includes the cortex, white matter, and nuclei.

The outer layer of *cortex* with its many folds (folia) is easily seen. The cortex receives all of its afferent input via the peduncles. After a fascinating process of computerlike data programming, signals are sent to the *deep nuclei* via the myelinated fibers that make up the *white matter*. The deep nuclei then project programs of movement to the motor centers in the cerebrum and brainstem. There are four deep nuclei in each hemisphere, which are identified later.

The cerebellum is susceptible to injury and many diseases. A wide variety of symptoms can be seen clinically, depending on the area and extent of damage.

Mesencephalon

The *midbrain* has a special intrigue because it is often the site of experimental lesions in lower animals. Unlike the pons or medulla, damage to the midbrain does not directly interfere with the vital centers. What usually results is a strong increase in extensor tone such that the animal can support itself.

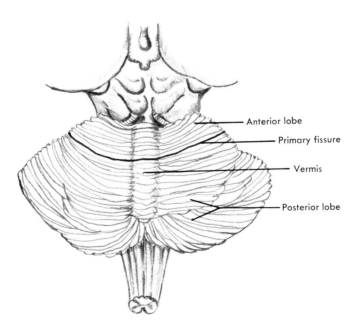

Anterior lobe

Primary fissure

Vermis

Posterior lobe

Fig. 2-12. Superior dorsal view of cerebellum in relation to brainstem. In this view both hemispheres and vermis can be seen, while flocculonodular lobe is hidden from view. (Adapted from House, E., and Pansky, B.: A functional approach to neuroanatomy, New York, 1967, McGraw-Hill Book Co.)

This condition of *decerebration* (loss of the regulating influences of the cerebrum dien-cephalon, and upper midbrain) can occur in humans after severe head trauma. In such patients exercises to maintain range of motion are very difficult because of the excess extensor tone, and contractures can quickly develop. The prognosis for such patients is poor.

The midbrain is the area between the pons and diencephalon and has its own unique shape. When viewed from the front (Fig. 2-5) two short thick stalks can be seen. These are the *cerebral peduncles*. The space between them is the *interpeduncular fossa*. From the dorsal view four prominences can be seen. These are *superior* and *inferior colliculi*. They are involved with visual and auditory reflexes respectively. The colliculi are also known as the *quadrigeminal plate* or *tectum* of the midbrain. In the dorsolateral corner of the lower midbrain the final pair of cerebellar peduncles, the *superior cerebellar peduncle* (brachium conjunctivum), can be seen (Fig. 2-9). This peduncle angles downward to reach the cerebellum.

The internal structures of the midbrain are divided into two main areas. The tectum is the area dorsal to the cerebral aqueduct, and the cerebral peduncle is the large anterolateral area in front of the cerebral aqueduct (Fig. 2-14). The tectum is composed of the four colliculi and is easily identified in a sagittal or horizontal section of the midbrain. The *cerebral aqueduct* (aqueduct of Sylvius) connects the third and fourth ventricles. Because of its narrow diameter (about the thickness of the wire in a paper clip) it is prone to obstruction and thus blockage of the flow of CSF.

The cerebral peduncle is further divided into two areas. The outer portion is known as the *crus cerebri*. It contains the descending motor fibers of the corticobulbar and corticospinal systems. Included in the former are the corticopontine fibers traveling to the pontine nuclei. Behind the crus cerebri is a thin band of cells known as the *substantia nigra*. It is always easy to identify in brain sections because of its dark grayish color. The substantia nigra has a motor function and demonstrates an interesting involvement in Parkinson's disease. The substantia nigra

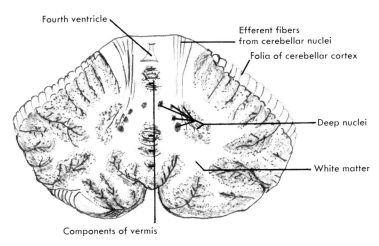

Fourth ventricle

Efferent fibers from cerebellar nuclei

Folia of cerebellar cortex

Deep nuclei

White matter

Components of vermis

Fig. 2-13. Horizontal section through middle third of cerebellum. Note extensive indentations (convolutions) of cortex. These are often called folia of cortex. (Adapted from Everett, N.: Functional neuroanatomy, ed. 6, Philadelphia, 1971, Lea & Febiger.)

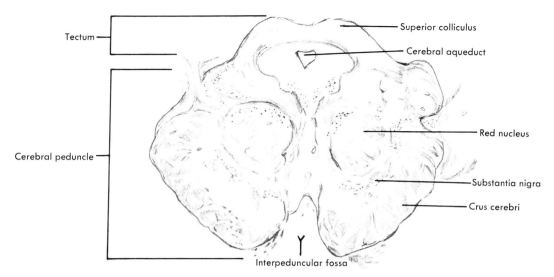

Tectum

Cerebral peduncle

Superior colliculus

Cerebral aqueduct

Red nucleus

Substantia nigra

Crus cerebri

Interpeduncular fossa

Fig. 2-14. Horizontal section through midbrain at level of superior colliculi. Note that crus cerebri fibers cover most of outer surface of cerebral peduncle. (After Watson, C.: Basic human neuroanatomy: an introductory atlas, ed. 2, Boston, 1977, Little, Brown & Co.)

separates the anterior and posterior portions of the cerebral peduncle. The posterior portion is known as the tegmentum and is continuous with that of the pons. Located in the tegmentum are cranial nerve nuclei III and IV, the nuclei for the reticular formation, motor and sensory fiber tracts, and the superior cerebellar peduncle. Also located in the tegmentum of the upper half of the midbrain near the midline is the *red nucleus*. It has a distinct round shape and in fresh dissection has a pinkish color. The red nucleus is involved with the regulation of muscle tone.

Several points should be emphasized. First, the segmentation of the brainstem into medulla, pons, and midbrain is partly based on their different appearances and partly on their different functions. However, structures such as the reticular formation and the fiber tracts are continuous throughout the brainstem. For example, the corticospinal tract is known as the crus cerebri in the midbrain and the pyramids in the

medulla. These fibers descend, however, uninterrupted. Second, the internal structures shown in the illustrations are not flat and one dimensional and seen only at the level of the cross-sections. For example, the red nucleus is not only round, it is tubular and extends from the upper half of the midbrain to the junction of the midbrain and diencephalon. Third, only a portion of the structures of the brainstem and cerebellum have been discussed thus far; more structures will be mentioned later.

Diencephalon

The diencephalon is considered the most rostral portion of the brainstem. Its many derivatives give it a highly irregular shape, thus making it hard to visualize. Most of the diencephalon is covered by tissues of the telencephalon, adding to the problem of visual identification. The structures of the diencephalon have varied functions including sensory perception, motor integration, and regulation of visceral activity. Because of its

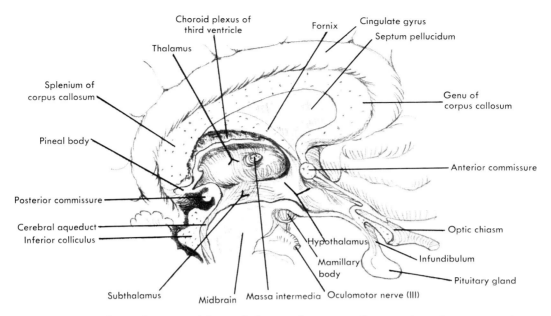

Fig. 2-15. Midsagittal section of diencephalon in relation to midbrain and cerebrum. Area of tuber cinereum (not labeled) between hypothalamus and infundibulum is somewhat exaggerated in length. Note how fornix curves through hypothalamus. (Adapted from House, E., and Pansky, B.: A functional approach to neuroanatomy, New York, 1967, McGraw-Hill Book Co.)

lack of accessibility, surgery on the diencephalon is difficult and sometimes impossible.

To comprehend the dimensions of the diencephalon, see Figs. 2-5 and 2-15. The boundaries from front to back are the infundibulum and the posterior part of the pituitary gland to the pineal gland respectively. The boundaries from top to bottom are the fornix and the area just above the level of the superior colliculus respectively. The side boundaries are the lateral edges of the thalamus.

The diencephalon has four divisions. The *epithalamus* is the area dorsal to the thalamus and is composed of the habenular nuclei, the pineal gland, and the posterior commissure. The *habenular nuclei* have an obscure visceral function and are located on either side

of the posterior portion of the third ventricle, just in front of the pineal gland. The *pineal gland* has a function in gonad development and is an easily located bulge above the superior colliculus (Figs. 2-5 and 2-15). The pineal gland is a midline structure (it is bisected in a midsagittal section). Since it may calcify after puberty, it can provide a valuable landmark for x-ray examination of the brain. The *posterior commissure* is a bundle of fibers that connect the two superior colliculi and thus functions in optic reflexes. It is located between the pineal gland above and the superior colliculi below (Fig. 2-15).

The *subthalamus* is a collection of nuclei located between the midbrain and thalamus, dorsolateral to the hypothalamus and medial to the internal capsule. Included in this area are the *subthalamic nuclei, the zona incerta,* and *Forel's field* (H field). These nuclei function in the motor system.

The third division of the diencephalon is the *hypothalamus*. The main components of the hypothalamus are the hypothalamic nuclei, the fornix, and the stria terminalis. The *hypothalamic nuclei* are located anteromedial to the subthalamus and anteroinferior to the thalamus. These nuclei form the walls and floor of the anterior portion of the third ventricle. Anteriorly the hypothalamic nuclei taper, forming an area known as the *tuber cinereum*, which then narrows into a stalk of fibers known as the *infundibulum*. This stalk relays data from the hypothalamus to the posterior portion of the pituitary gland, the *neurohypophysis*. Directly in front of the infundibulum is the optic chiasm. In the *optic chiasm* the fibers from the nasal portion of each retina decussate. The *optic tracts* that leave the chiasm thus contain visual data from each eye. The optic tract actually forms

part of the lateral boundary around the tuber cinereum and midbrain (Fig. 2-5). Directly in back of the infundibulum are two hypothalamic nuclei known as the *mamillary bodies*. These nuclei lie in the interpeduncular fossa and are easily identifiable (along with the infundibulum and optic chiasm) in the inferior view of the cerebrum (Fig. 2-24).

The functions of the hypothalamus are quite varied and include involvement in the following:
1. Regulation of the autonomic nervous system
2. Release of some hormones from the pituitary gland
3. Regulation of temperature
4. Food and water intake
5. Sleep-wake cycle
6. Expression of emotions

Although all of these functions are important, expression of emotions is most interesting because it is involved with a diffusely organized system known as the *limbic system*.

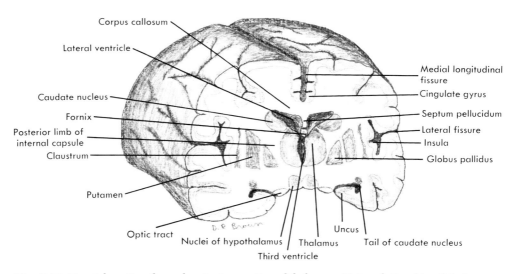

Fig. 2-16. Frontal section through anterior portion of thalamus. Note relationship of thalamus and caudate nucleus and internal capsule. Compare this with transverse section shown in Fig. 2-20. (Adapted from Netter, F.: The CIBA collection of medical illustrations. I. The nervous system, Summit, N.J., 1972, CIBA Chemical Co.)

The limbic system is composed of cortical and subcortical diencephalic structures that primarily function in the expression of overall personality. Two important fiber tracts of the limbic system are the *fornix* and the *stria terminalis*. The fornix connects a part of the cerebral cortex (the hippocampus) with the mamillary bodies. The stria terminalis connects a cerebral nucleus (the amygdala) with the anterior hypothalamic nuclei. The courses of these two tracts are similar and are discussed in the next section. Additional information on the limbic system is presented in Chapter 9.

The last of the four divisions of the diencephalon is the *thalamus*. The thalamus is composed of two egg-shaped clusters of nuclei, each about the size of a robin's egg. It is considered the top of the brainstem, and al-

though it is easy to identify, its anatomic relationship to surrounding structures is somewhat confusing. To best understand this relationship, first see Fig. 2-5 to visualize its shape; then see Figs. 2-15 to 2-17, which show how the thalamus is related to the caudate nucleus, internal capsule, and fornix. Both the fornix and caudate nucleus curve along the anterior, superior, and dorsal surfaces of the thalamus. Note that the internal capsule makes up the lateral borders of the thalamus. Figs. 2-18 to 2-20 further clarify this relationship. Note that the anterior (head) portion of the caudate nucleus covers the front of the thalamus, while the tail of the caudate nucleus and fornix cover part of the back of the thalamus. Keep in mind that the components of the epithalamus, subthalamus, and hypothalamus surround the

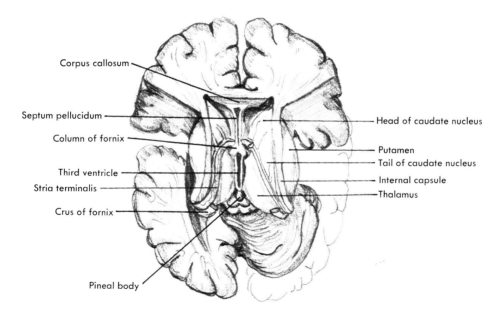

Fig. 2-17. Superior view of thalamus in relationship with basal nuclei and internal capsule. Note that caudate head tapers backward along top of thalamus. Internal capsule fibers have been cut to show caudate medial and putamen lateral. Portion of fornix between crus and third ventricle has also been cut to show how it travels along top of thalamus medial to caudate. (Adapted from Truex, R., and Carpenter, M.: Human neuroanatomy, ed. 6, Baltimore, 1969, The Williams & Wilkins Co.)

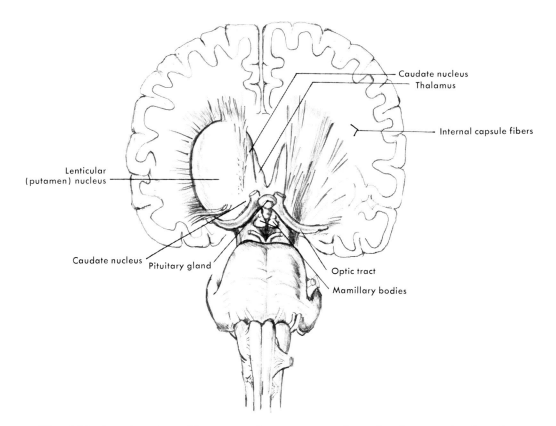

Fig. 2-18. Anterior view of brainstem demonstrating relationship between basal nuclei, thalamus, and internal capsule. Note that lenticular nucleus has been removed on left side to demonstrate fibers of internal capsule. (Adapted from Noback, C., and Demarest, R.: The human nervous system: basic principles of neurobiology, ed. 2, New York, 1975, McGraw-Hill Book Co.)

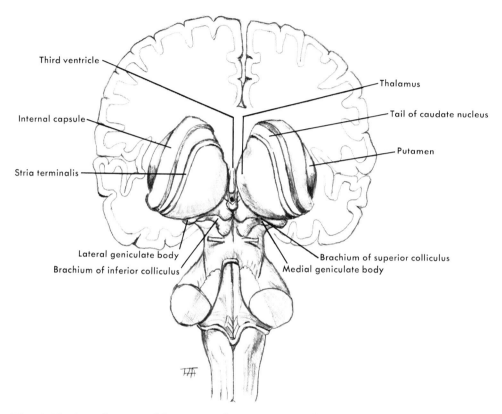

Fig. 2-19. Dorsal view of brainstem demonstrating relationship between basal nuclei, thalamus, and internal capsule. Note that fornix has been omitted. (Adapted from Noback, C., and Demarest, R.: The human nervous system: basic principles of neurobiology, ed. 2, New York, 1975, McGraw-Hill Book Co.)

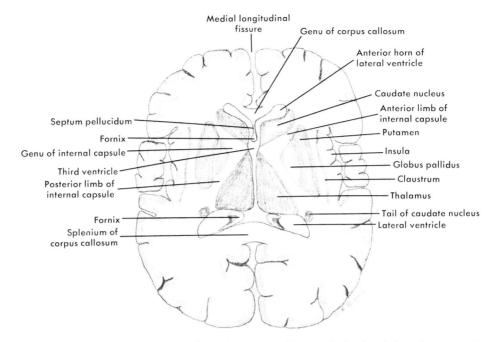

Fig. 2-20. Horizontal section through cerebrum at middiencephalon level, just above anterior commissure. Note relationship between caudate and thalamus, internal capsule, and lenticular nucleus. (Adapted from Netter, F.: The CIBA collection of medical illustrations. I. The nervous system, Summit, N.J., 1972, CIBA Chemical Co.)

lower ventral and dorsal portions and the undersurface of the thalamus. Also note that each half of the thalamus forms the walls of much of the third ventricle.

Each half of the thalamus is divided anatomically into three major nuclear groups by a thin band of fibers called the *internal medullary lamina* (Fig. 2-21). These three groups and their subnuclear divisions are as follows:

1. Anterior nuclei: anteromedial, anterodorsal, and anteroventral
2. Medial nuclei: dorsomedial and centromedial
3. Lateral nuclei: ventral anterior, ventral lateral, ventral dorsal (ventral posterolateral and ventral posteromedial), lat-

eral dorsal, lateral posterior, reticular, and posterior nuclei (pulvinar, medial geniculate body, and lateral geniculate body) (Note in Fig. 2-21 that the reticular nuclei are separated from the other lateral nuclei by the external medullary lamina.)

Functionally two additional nuclear groups can be identified:

1. Midline nuclei, located between the third ventricle and the medial nuclei
2. Intralaminar nuclei, formed by neurons irregularly clustered in relation to the internal medullary lamina

The functions of the thalamus are quite varied and include the following:

1. Integration and relay of motor data from

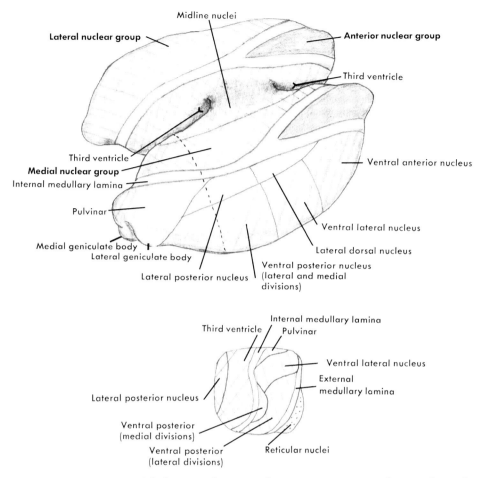

Fig. 2-21. Superior view of thalamus with surrounding structures removed. Note that a thin band of white matter separates each half into three nuclear divisions. Massa intermedia has not been labeled. *Dotted line* indicates area where cross-section was taken. (Adapted from Netter, F.: The CIBA collection of medical illustrations, I. The nervous system, Summit, N.J., 1972, CIBA Chemical Co.)

the cerebellum and basal nuclei to the motor cortex

2. Relay of all sensations except olfaction to specific cortical areas
3. Perception of touch, temperature, and pain
4. Level of cortical arousal and patterns of sleep

5. Moods such as pleasant or disagreeable

Further information on the functional aspects of the thalamus is presented in Chapter 8.

Unlike the rest of the brainstem, each of the internal structures of the diencephalon has its own makeup. There are, however, two midline structures that deserve mention. The *massa intermedia* is a stalk of fibers and nuclei that connect the two halves of the thalamus. It is located in the middle third of the thalamus (Figs. 2-15 and 2-21). The *third ventricle* is completely surrounded by the diencephalic structures and is an important link between the lateral ventricles and the cerebral aqueduct.

Telencephalon

Examining the surface of the cerebrum is fascinating. Most striking is its smallness. The entire adult cerebrum is only around 12 cm from side to side, 13.5 cm front to back, and 7.5 cm from top to bottom. The surface is dented with numerous convolutions that seem to wander aimlessly. The color varies from pale pink in fresh brains to dull gray in fixed brains. No other organ has such overall complexity, efficiency, and variety of functioning as the human cerebrum.

The external anatomy of the cerebrum is quite extensive, and all of the sulci and gyri have names. However, this section focuses only on the major markings and structures that are discussed in later chapters or are useful points of reference.

The most identifiable marking of the cerebrum is the deep *medial longitudinal fissure* seen in the superior view (Fig. 2-22). It separates the two hemispheres and extends both anteriorly and posteriorly to the undersurface of the cerebrum. At the bottom of the medial longitudinal fissure is a broad bundle of fibers known as the *corpus callosum*. These fibers connect the two hemispheres and actually lie over many internal structures of the cerebrum and diencephalon (Figs. 2-22 and 2-23, *B*). Also visible in the superior view is the *central sulcus of Rolando*. It is located about halfway back from the front of the cerebrum. The central sulcus dips into the medial longitudinal fissure and also runs along the lat-

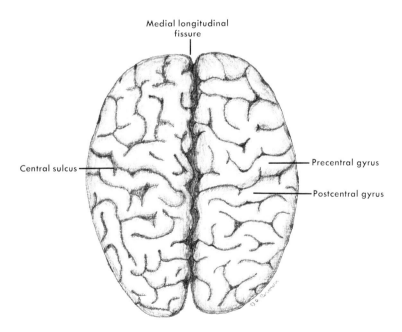

Fig. 2-22. Superior view of adult brain. (Adapted from Shade, J., and Ford, D.: Basic neurology, New York, 1973, Elsevier Scientific Publishing Co.)

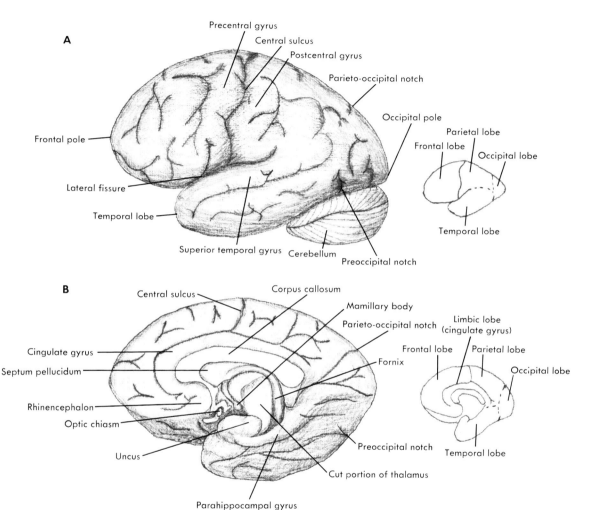

Fig. 2-23. A, Lateral view of brain. Major reference points are labeled, as are lobes of cerebrum. **B,** Medial inferior view of one cerebral hemisphere with brainstem and cerebellum removed. (Adapted from Netter, F.: The CIBA collection of medical illustrations, I. The nervous system, Summit, N.J., 1972, CIBA Chemical Co.)

eral surface to the *lateral fissure of Sylvius*. The central sulcus is important for two reasons. First, it separates the *precentral gyrus* in front from the *postcentral gyrus* in back. The former is the primary source of skilled voluntary movements, while the latter is the primary perception center for many of the body's sensations. These gyri are often lesioned in cerebrovascular accident (CVA). Second, the central sulcus separates the *frontal lobe* in front from the *parietal lobe* in back.

In the lateral view the *lateral fissure of Sylvius* is readily visible as it travels in a dorso superior direction to about two thirds of the way back. The lateral fissure separates the frontal and parietal lobes above from the *temporal lobe* below. It also continues along the inferior surface of the hemisphere as it separates the temporal and frontal lobes (Fig. 2-24). As mentioned, the central sulcus can also be seen in the lateral view. It is often difficult to locate on a real brain because of

the inconsistent shape of the surrounding gyri. It generally travels from the superior border of the hemisphere to the lateral fissure with a slight anterior angle (Fig. 2-23). At the junction of the central sulcus and lateral fissure (in the frontal lobe) is the area for motor speech. Below the lateral fissure is the *superior temporal gyrus*, which contains the area for receiving and understanding the spoken word. These areas are often involved in the communication deficit known as aphasia seen in patients with CVA of the dominant hemisphere.

The *preoccipital notch* and *parieto-occipital notch* are found in the posterior third of the lateral view. A line drawn between these two points separates the *occipital lobe* dorsally from the parietal and temporal lobes anteriorly. A second line drawn from the end of the lateral fissure to this first line separates the parietal lobe above from the temporal lobe below (Fig. 2-23). Each hemisphere therefore contains four lobes

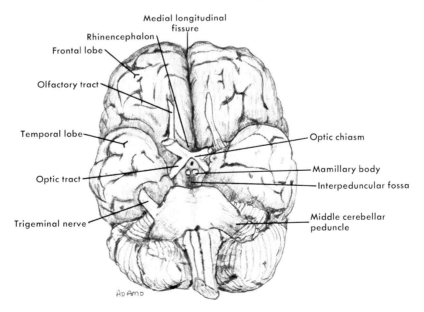

Fig. 2-24. Inferior view of cerebrum with brainstem and cerebellum attached. (After Watson, C.: Basic human neuroanatomy: an introductory atlas, ed. 2, Boston, 1977, Little, Brown & Co.)

with separate but interrelated functions. The frontal lobe functions in intelligence, personality, motor activities, and speech. The parietal lobe functions in somatic sensory perception and interpretation of this data. The temporal lobe functions in memory, hearing, and comprehending the spoken word. The occipital lobe functions in vision and the integration of visual data such as in reading. These four lobes constantly communicate with each other within a hemisphere and between both hemispheres. Many activities such as learning, thinking, and creating involve wide areas of all four lobes and even of both hemispheres.

It is important to mention that each hemisphere, although similar in appearance, has different functions. In 95% of persons the left hemisphere controls movement predominantly on the right side, perceives sensations predominantly from the right side, is more analytical, and is responsible for speech. The right hemisphere is predominantly responsible for motor and sensory activities on the left side, is adept at spatial relations, and is more artistic-creative. More information about hemispheric specialization is presented in Chapter 9.

If the lateral fissure is pulled apart a fifth lobe, *the insula* (island of Riel), can be seen (Fig. 2-16). The insula has an uncertain visceral function and is relatively unimportant.

Two additional structures are seen on the inferior surface (Fig. 2-24). The *olfactory tract* lies just lateral to the medial longitudinal fissure. It conveys data from chemoreceptors in the nose to the *rhinencephalon*. The rhinencephalon (Figs. 2-23, *B* and 2-24) is composed of the olfactory tract and the areas to which these fibers project. The primary cortical areas are the *anterior perforated substance*, which is just lateral to the optic chiasm, the rostral portion of the *uncus* (the medial tip of the temporal pole), and the *septal area*, which is the medial surface of the frontal lobe below the rostrum of the corpus callosum (see Fig. 4-11). These components of the rhinencephalon work with components of the limbic system to provide an emotional reaction to the world of odors that we constantly experience.

Examining one hemisphere from a midsagittal view (Fig. 2-23, *B*) reveals additional external components and provides a beginning to the study of the organization of the internal structures. The remaining external structures involve cortical components of the limbic system. These include the *cingulate gyrus*, which lies over and follows the course of the corpus callosum, and the *uncus* and *parahippocampal gyrus*, which are found in the medial surface of the temporal lobe. These three structures form a ring of primitive cortex on the medial surface of each hemisphere. Together they are sometimes referred to as the limbic lobe. Note that the four lobes, the central sulcus, and the pre- and postcentral gyri seen in the lateral view can also be found in the medial view.

The internal structures of the cerebrum include nuclei, bundles of fibers, cavities (ventricles), and gyri. In the midsagittal view several internal structures can be seen. The first is the *corpus callosum*. It is composed of a broad flat bundle of fibers that allow one hemisphere to communicate with the other hemisphere. The corpus callosum has several divisions. The *body* lies on top of the thalamus. It curves anteriorly to bend around the front of the caudate nucleus (forming the *genu* of the corpus callosum). It then curves back under the caudate nucleus to end at the anterior commissure. This forms the *rostrum* of the corpus callosum. Dorsally the body curves downward around the back of the thalamus, forming the *splenium* of the corpus callosum (Fig. 2-15).

Below the corpus callosum is the *fornix*. It is derived from both the diencephalon and telencephalon, and its course is somewhat complex. The fornix, as mentioned, is an important fiber tract for the limbic system. The

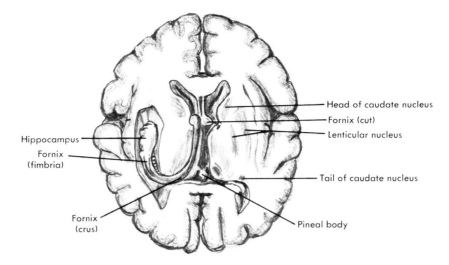

Fig. 2-25. Superior view of fornix showing direction of its fiber. Note that lenticular nucleus and surrounding white matter on left side have been "scooped out" to show hippocampus in temporal lobe.

axons that make up this tract originate in the medial portion of the temporal lobe in a cortical area known as the *hippocampus* (Fig. 2-16). The hippocampus is primarily composed of the *dentate* and *hippocampal gyri*, which form the floor of the inferior horn of the lateral ventricle; the *parahippocampal gyrus*, which forms part of the medial border of the temporal lobe; and the *uncus*, which is the anterior bend of the parahippocampal gyrus (Figs. 2-23, *B*, and 9-13). Neurons in this area project axons dorsally as a thin band of white matter known as the *fimbria of the hippocampus*. This band travels in a dorsomedial direction to the region of the pulvinar. Here it courses upward around the back of the thalamus, forming the *crus* of the fornix. The crus then moves anteriorly on the superomedial border of each half of the thalamus, forming the *body* of the fornix (Fig. 2-25). Directly on top of the fornix is the corpus callosum (Fig. 2-16). Each body of the fornix curves downward as the *columns* of the fornix following the curve of

the thalamus. The columns of the fornix are directly behind the anterior commissure. At this point the two columns split somewhat and travel inferiorly to terminate in the *mamillary bodies* (Fig. 2-15). Neurons in the mamillary bodies then relay cortically originating limbic data to the thalamus via the *mamillothalamic fasciculus* (Figs. 2-31 and 2-33). The limbic system is both fascinating and complex; more information is presented in Chapter 9.

Between the corpus callosum and fornix are two thin sheets of neural tissue known as the *septum pellucidum*. The septum is a midline structure that acts to separate the two lateral ventricles (Figs. 2-15 and 2-16). The septum also contains the *septal nuclei*, which are a component of the limbic system (Fig. 2-51).

The final internal structure seen in the midsagittal view is the *anterior commissure*. This is a thin bundle of fibers that connects the two temporal lobes (commissure fibers always connect one side of the nervous system with a corresponding structure on the other side). It has somewhat of a handlebar shape with a slight downward angle. It is always in front of the columns of the fornix and under the caudate and lenticular nuclei (Figs. 2-41 and 2-42).

The remaining internal structures include the lateral ventricles, the basal nuclei, and the internal capsule (Figs. 2-14, 2-15, and 2-26 to 2-50). It might be beneficial to read the narrative for each of these three components first and then look at the diagrams for visual reinforcement. The serial cross-sections are especially helpful in adding depth to these structures.

The *lateral ventricles* are the rostral extent of the original neural tube cavity. There is one lateral ventricle in each hemisphere. The irregular shape of the lateral ventricle can be broken down into a central portion, the body, and three horns. The *body* is found between the corpus callosum and the thalamus. It projects forward into the frontal lobe as the *anterior horn* and is surrounded by the corpus callosum, head of the caudate nucleus, and septum pellucidum. Dorsal to the thalamus the body enlarges and is known as the *atrium*. It is surrounded by the tail of the caudate nucleus, crus of the fornix, posterior limb of the internal capsule, and splenium of the corpus callosum. The atrium projects dorsally into the occipital lobe, forming the *posterior horn*, and inferoanteriorly into the temporal lobe, forming the *inferior horn*. It is along the floor of the inferior horn that axons from the hippocampal components form the thin band of myelinated fibers known as the *alveus* of the hippocampus. The alveus consolidates into the fimbria of the hippocampus, which then forms the crus of the fornix. At the junction of the body and anterior horn (at the anterior limit of the thalamus) is the *interventricular foramen*. These channels connect the two lateral ventricles with the third ventricle and allow for the flow of cerebrospinal fluid into the latter.

Blockage of this flow can cause severe complications. For a better perspective of the lateral ventricles see Fig. 7-7.

By definition the *basal nuclei* include all of the subcortical clusters of neurons within the telencephalon, that is, the claustrum, amygdala, caudate nucleus, and lenticular nucleus. The *claustrum* is a thin plate of cells between the insula and the putamen portion of the lenticular nucleus. Its functions and connections remain obscure. The *amygdala* (amygdaloid nucleus) is found in the medial portion of the temporal lobe just in front of the inferior horn of the lateral ventricle. It is somewhat almond shaped and usually blends in with the cortex along the medial surface of the temporal lobe (uncus). Its anterior limit is at the level of the optic chiasm. The amygdala functions with the limbic system.

The *caudate and lenticular nuclei* are often grouped together functionally and referred to as the basal nuclei or basal ganglia. The subthalamus, substantia nigra, and red nucleus are sometimes included as basal ganglia because they have strong functional connections with the caudate and lenticular nuclei. These five structures operate within the extrapyramidal motor system, which basically involves stereotyped patterns of movement and postural adjustments (see Chapter 9).

The caudate and lenticular nuclei are collectively known as the *corpus striatum.* The lenticular nucleus is subdivided into the *putamen* and *globus pallidus.* The caudate nucleus and putamen are similar histologically, and since they are newer components on the developmental ladder, they are sometimes called the *neostriatum.* The neostriatum is the main receiving station for afferent data. The globus pallidus, which is older developmentally, is sometimes called the *paleostriatum.* It is the primary source of efferent projections to other motor areas from the basal ganglia.

The corpus striatum is unusual anatomically. Anteriorly the caudate nucleus and putamen are fused and extend along with the anterior horn of the lateral ventricle to the genu of the corpus callosum. The two nuclei become divided by the fibers of the *internal capsule*, with the caudate nucleus medial and the lenticular nucleus lateral. This occurs as the caudate nucleus tapers dorsally into a long *tail* that travels along the lateral edge of the top of the thalamus. The tail then curves downward along the back of the thalamus and then travels anteriorly to terminate in the amygdala (Fig. 2-26). In doing so it hooks around the back of the internal capsule and helps form the roof of the inferior horn of the lateral ventricle. In spite of the apparent anatomic connection, there does not seem to be any functional relationship between the caudate nucleus and amygdala.

The *putamen* is shaped somewhat like a seashell. It extends from the genu of the corpus to the geniculate bodies of the thalamus and from just below the corpus callosum to the level of the subthalamus. Its lateral border is a thin band of white matter known as the external capsule, and its medial border is the internal capsule and globus pallidus.

The globus pallidus is so named because of its natural pallor in the fresh preparation. It is much smaller than the putamen and is completely hidden by it in the lateral view (Fig. 2-26). The globus pallidus extends from the anterior commissure to the midthalamus and from the level of the massa intermedia to the level of the subthalamus. Its medial border is the internal capsule; its lateral border is the putamen. It is anatomically divided into globus pallidus I and II.

The white matter of the telencephalon falls into four categories:

1. Commissure fibers connecting both hemispheres, such as the corpus callosum and anterior commissure
2. Association fibers that connect different areas within one hemisphere, such as the external and extreme capsules on either side of the claustrum
3. Radiation fibers, which ascend from the thalamus to the cerebral cortex
4. Projection fibers, which descend from the cerebral cortex to lower structures.

The *internal capsule*, which makes up a large portion of the white matter of the cere-

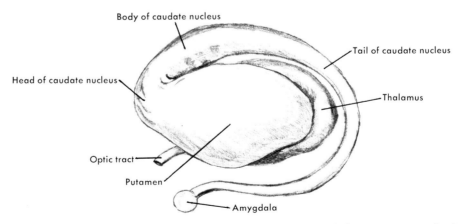

Fig. 2-26. Lateral view of basal nuclei and thalamus. In this view thalamus appears in back of putamen. Globus pallidus is smaller than putamen and thus is not visible. Try to picture internal capsule between caudate body and tail and putamen. (Adapted from Truex, R., and Carpenter, M.: Human neuroanatomy, ed. 6, Baltimore, 1969, The Williams & Wilkins Co.)

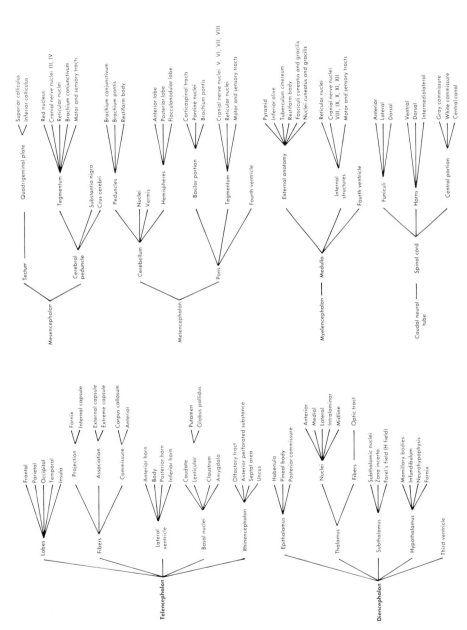

Fig. 2-27. Derivation of major components of nervous system.

brum, is composed of both radiation and projection fibers going to and coming from all areas of the cerebral cortex. In the lateral view (Fig. 2-18) the fibers can be seen fanning out to all parts of the cerebral cortex. At the level of the upper border of the thalamus (Fig. 2-20) these fibers consolidate into a V-shaped handle that has three parts. The *anterior limb* separates the lenticular and caudate nuclei. The middle area of the bend is the *genu*. The *posterior limb* separates the lenticular nucleus from the thalamus. At the junction of the subthalamus and midbrain the internal capsule contains only projection fibers and forms the crus cerebri portion of the cerebral peduncle. Many of these fibers pass through the pons to form the pyramids of the medulla and then to the spinal cord as the corticospinal tracts.

An overview of the major components derived from the five secondary vescicles and the spinal cord is given in Fig. 2-27.

CROSS-SECTIONAL ATLAS OF HUMAN BRAIN

In this final section are presented 25 drawings of the human brain adapted from the *Atlas of the Human Brain in Section* by Roberts and Hanaway. This atlas presents a series of photographs of three human brains sliced in the frontal, horizontal, and sagittal planes. Each slice was stained to provide a greater contrast between gray and white matter. In preparing the drawings some degree of realism was lost from the photographs. However, an extra degree of simplification was added to aid the viewer in identifying the more important structures. To aid in understanding where each slice was made in relation to the whole brain, two small line drawings of the brain in a midsagittal, frontal, or horizontal view are provided with each brain section drawing. The solid line drawn through these smaller drawings indicates where the slice was made.

Two points need to be mentioned concerning these drawings. First, the right and left sides in some of the drawings may not appear entirely symmetric. This assymmetry may be due to a natural assymmetry of the brain or an imperfect slice made through the brain. Second, other drawings or photographs made through a brain may show different relationships between some structures. This may be because many brains show individual variations in the size and shape of some components or because the slice was made at an angle other than the 90° or 0° used in preparing these three brains. It is hoped that any variations encountered in future reading of neuroanatomy texts will arouse curiosity to question why and thus enhance the reader's knowledge of neuroanatomy.

The illustration on the opposite page will assist in identifying some of the structures of the smaller line drawings.

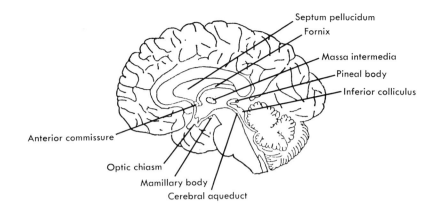

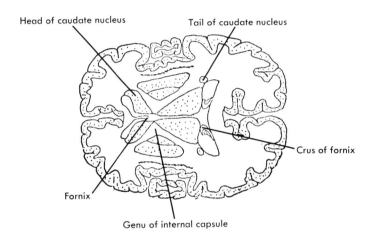

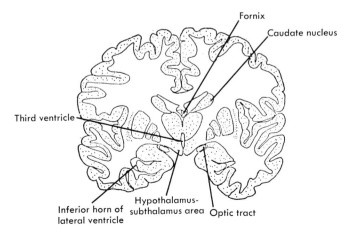

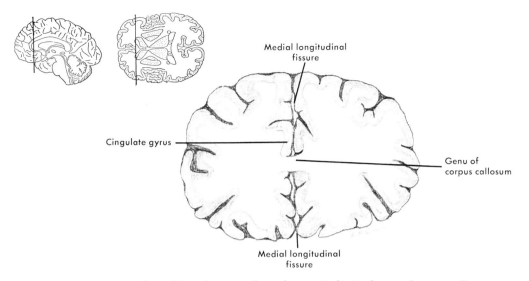

Fig. 2-28. Anterior surface of frontal section through anterior limit of genu of corpus callosum.

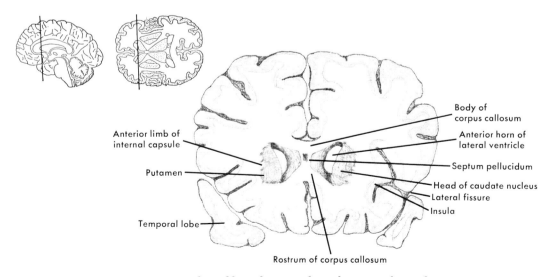

Fig. 2-29. Posterior surface of frontal section through anterior limit of putamen.

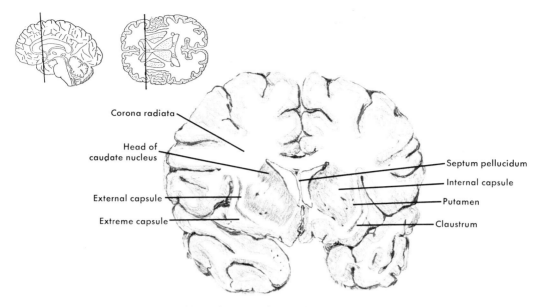

Fig. 2-30. Anterior surface of frontal section through head of caudate nucleus and putamen.

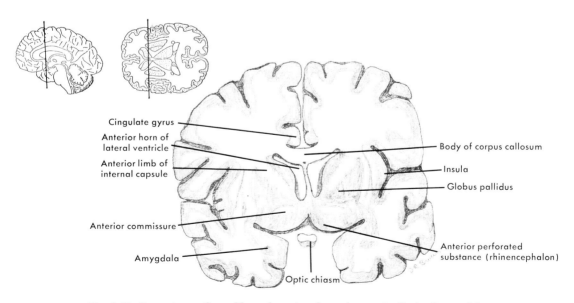

Fig. 2-31. Posterior surface of frontal section through anterior limit of amygdala.

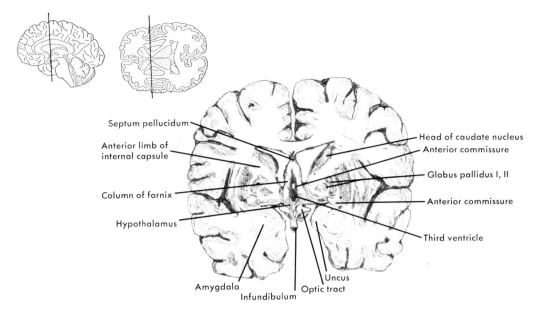

Septum pellucidum

Anterior limb of
internal capsule

Column of fornix

Hypothalamus

Head of caudate nucleus

Anterior commissure

Globus pallidus I, II

Anterior commissure

Third ventricle

Amygdala

Infundibulum

Optic tract

Uncus

Fig. 2-32. Posterior surface of frontal section through anterior limit of hypothalamus.

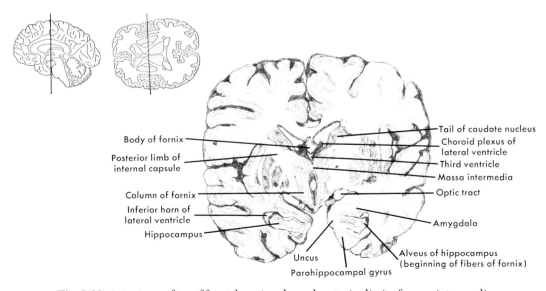

Body of fornix

Posterior limb of
internal capsule

Column of fornix

Inferior horn of
lateral ventricle

Hippocampus

Tail of caudate nucleus

Choroid plexus of
lateral ventricle

Third ventricle

Massa intermedia

Optic tract

Amygdala

Alveus of hippocampus
(beginning of fibers of fornix)

Uncus

Parahippocampal gyrus

Fig. 2-33. Anterior surface of frontal section through anterior limit of massa intermedia.

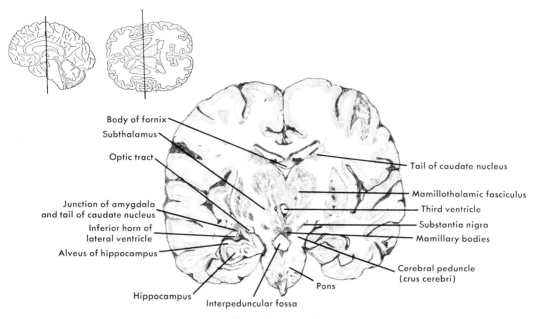

Body of fornix

Subthalamus

Optic tract

Junction of amygdala
and tail of caudate nucleus

Inferior horn of
lateral ventricle

Alveus of hippocampus

Hippocampus

Interpeduncular fossa

Tail of caudate nucleus

Mamillothalamic fasciculus

Third ventricle

Substantia nigra

Mamillary bodies

Cerebral peduncle
(crus cerebri)

Pons

Fig. 2-34. Anterior surface of frontal section through mamillary bodies.

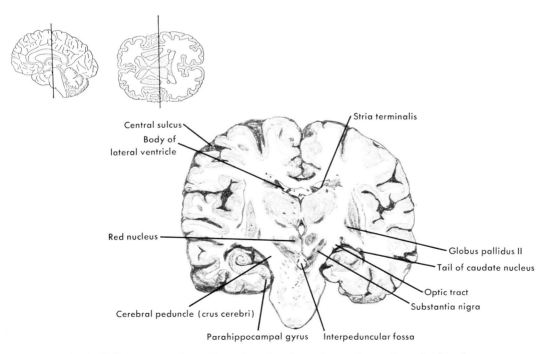

Central sulcus

Body of
lateral ventricle

Stria terminalis

Red nucleus

Globus pallidus II

Tail of caudate nucleus

Optic tract

Substantia nigra

Cerebral peduncle (crus cerebri)

Parahippocampal gyrus

Interpeduncular fossa

Fig. 2-35. Posterior surface of frontal section through anterior portion of red nucleus.

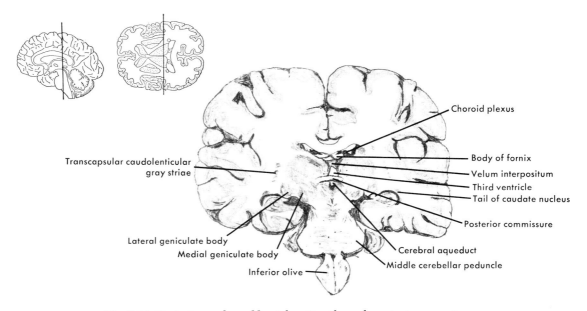

Fig. 2-36. Posterior surface of frontal section through posterior commissure.

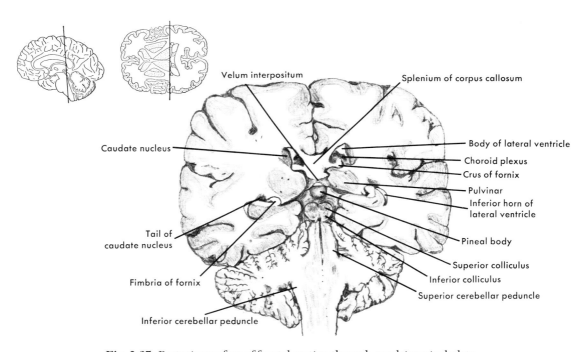

Fig. 2-37. Posterior surface of frontal section through quadrigeminal plate.

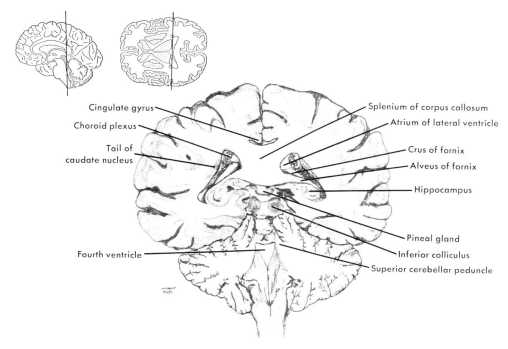

Fig. 2-38. Posterior surface of frontal section through posterior limit of tail of caudate nucleus.

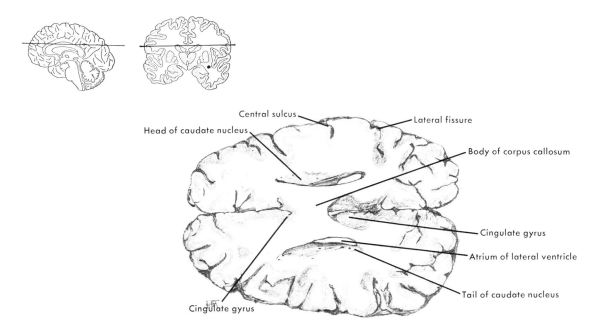

Fig. 2-39. Superior surface of horizontal section through superior limit of caudate nucleus.

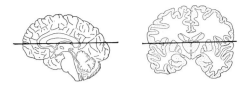

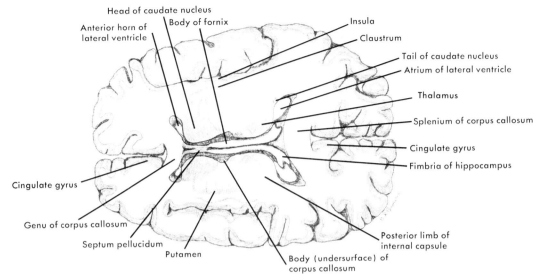

Fig. 2-40. Inferior surface of horizontal section through superior limit of putamen.

Head of caudate nucleus
Body of fornix
Anterior horn of lateral ventricle
Insula
Claustrum
Tail of caudate nucleus
Atrium of lateral ventricle
Thalamus
Splenium of corpus callosum
Cingulate gyrus
Fimbria of hippocampus
Cingulate gyrus
Genu of corpus callosum
Septum pellucidum
Putamen
Body (undersurface) of corpus callosum
Posterior limb of internal capsule

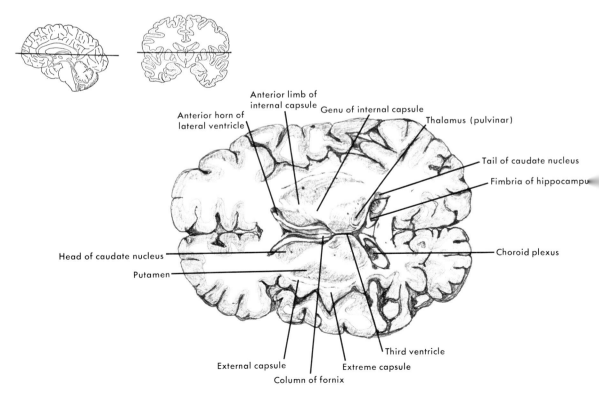

Anterior limb of internal capsule
Genu of internal capsule
Thalamus (pulvinar)
Anterior horn of lateral ventricle
Tail of caudate nucleus
Fimbria of hippocampus
Head of caudate nucleus
Putamen
Choroid plexus
External capsule
Column of fornix
Extreme capsule
Third ventricle

Fig. 2-41. Superior surface of horizontal section through upper border of third ventricle.

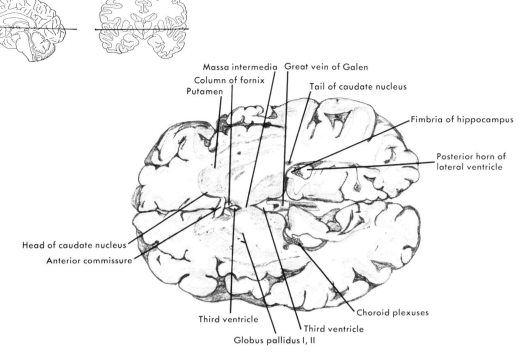

Fig. 2-42. Inferior surface of horizontal section through anterior commissure.

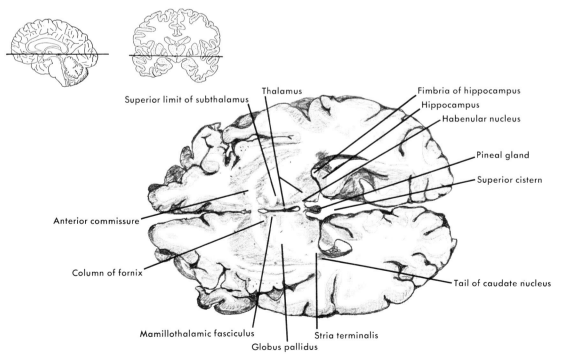

Fig. 2-43. Inferior surface of horizontal section through superior limit of subthalamus.

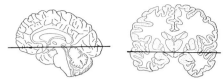

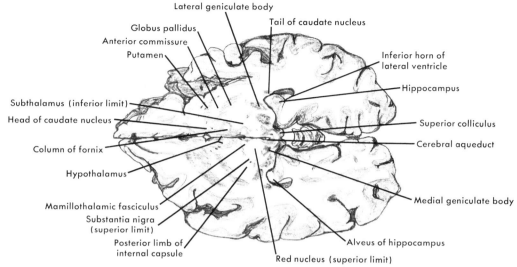

Lateral geniculate body

Tail of caudate nucleus

Globus pallidus

Anterior commissure

Putamen

Inferior horn of
lateral ventricle

Hippocampus

Subthalamus (inferior limit)

Head of caudate nucleus

Superior colliculus

Column of fornix

Cerebral aqueduct

Hypothalamus

Mamillothalamic fasciculus

Medial geniculate body

Substantia nigra
(superior limit)

Posterior limb of
internal capsule

Alveus of hippocampus

Red nucleus (superior limit)

Fig. 2-44. Inferior surface of horizontal section through superior colliculus.

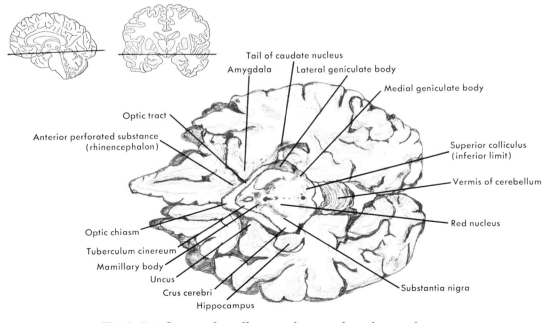

Tail of caudate nucleus

Amygdala

Lateral geniculate body

Medial geniculate body

Optic tract

Anterior perforated substance
(rhinencephalon)

Superior colliculus
(inferior limit)

Vermis of cerebellum

Red nucleus

Optic chiasm

Tuberculum cinereum

Mamillary body

Uncus

Crus cerebri

Hippocampus

Substantia nigra

Fig. 2-45. Inferior surface of horizontal section through optic chiasm.

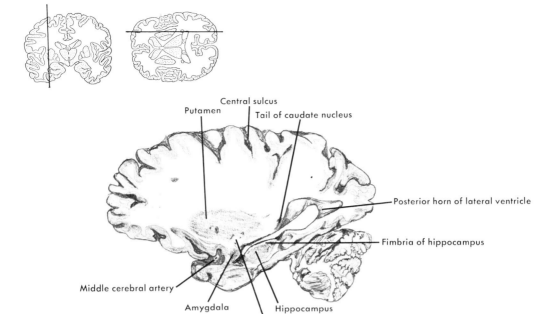

Fig. 2-46. Medial surface of sagittal section through lateral portion of putamen.

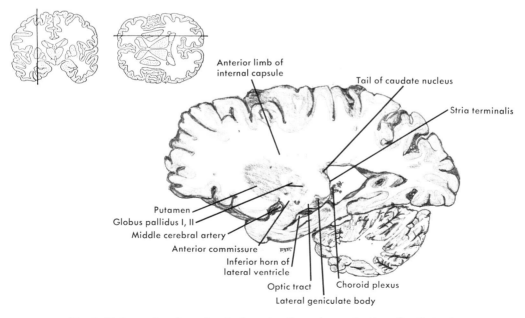

Fig. 2-47. Lateral surface of sagittal section through termination of optic tract.

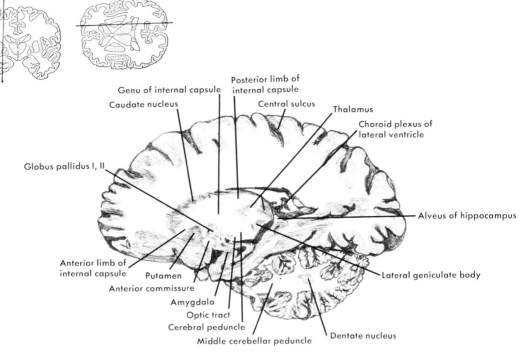

Fig. 2-48. Lateral surface of sagittal section through lateral portion of pulvinar.

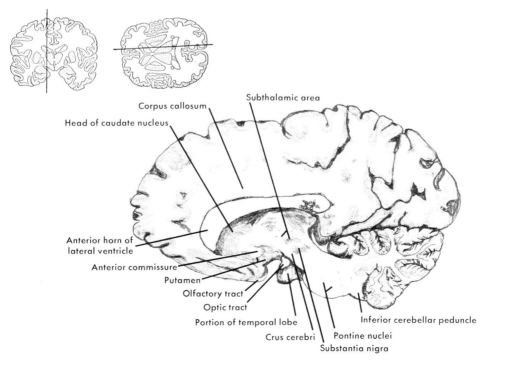

Fig. 2-49. Medial surface of sagittal section through lateral limit of red nucleus.

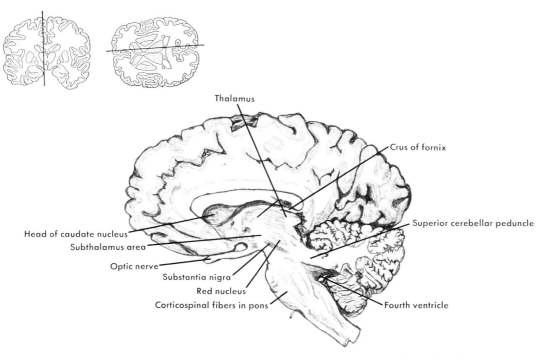

Fig. 2-50. Medial surface of sagittal section through superior cerebellar peduncle.

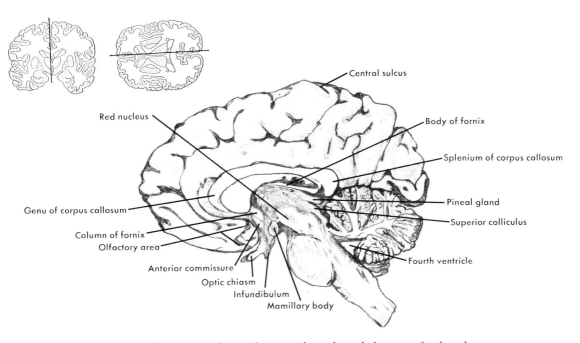

Fig. 2-51. Medial surface of sagittal section through medial extent of red nucleus.

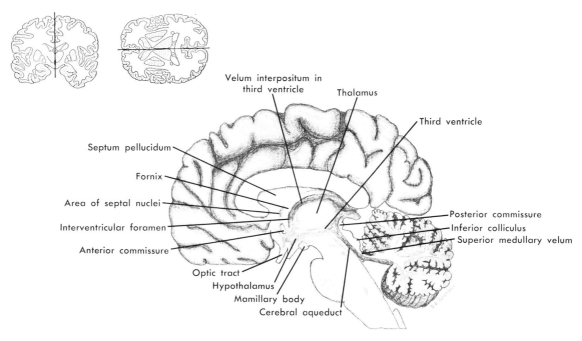

Fig. 2-52. Medial surface of sagittal section through cerebral aqueduct.

TEST QUESTIONS

For each question select *all* correct choices.

1. A tumor of the filum terminale:
   a. Interferes with the functioning of the cauda equina nerves.
   b. Possibly compresses the sacral and coccygeal cord segments.
   c. Produces bowel and bladder symptoms.
   d. Initially interferes with ambulation.

2. The gray matter of the spinal cord:
   a. Is composed of three pairs of horns of equal size.
   b. Is larger at C6 than at T8.
   c. Will be bisected if a line is drawn from the dorsal median sulcus and anterior median fissure.
   d. Terminates at spinal cord segment L2.

3. The fibers of the fasciculi cuneatus and gracilis:
   a. Contain sensory data from the arms and legs.
   b. Terminate in the nuclei cuneatus and gracilis.

   c. Lie on either side of the anterior median fissure.
   d. Continue directly into the pons.

4. A tumor growing on and compressing the lateral surface of the medulla will initially interfere with:
   a. Descending motor pathway of the pyramids.
   b. Information reaching the cerebellum from the inferior olive and the restiform body.
   c. Data relayed from the nucleus gracilis.
   d. Circulation of cerebrospinal fluid in the fourth ventricle.

5. The fourth ventricle is surrounded by:
   a. Six cerebellar penuncles.
   b. Inferior olives.
   c. Fasciculi cuneatus and gracilis.
   d. Quadrigeminal plate.

6. In a midsagittal view of the cerebellum:
   a. Deep nuclei are visible.

b. Slice will go through the vermis.

c. Slice will be at right angles to the primary fissure.

d. Flocculonodular lobe is hidden.

7. In a midfrontal slice through the brainstem can be seen:

a. Pyramids.

b. Red nucleus.

c. Superior colliculi.

d. Middle cerebellar penduncle.

8. The superior colliculi:

a. Are involved with auditory reflexes.

b. Are part of the tectum.

c. Can be viewed from the front of the brainstem.

d. Are dorsal to the cerebral aqueduct.

9. Deviation of the pineal gland to the left side as seen on an antero posterior x-ray film of the skull could indicate:

a. Absorption of the cerebral cortex on the left, creating a vacuum.

b. Tumor growing in the right hemisphere.

c. Concussion on the left side of the head.

d. Hematoma between the skull and brain on the right side.

10. In a series of frontal sections of the diencephalon the order of appearance of the following structures from anterior to posterior is:

a. Subthalamus.

b. Infundibulum.

c. Hypothalamus.

d. Pineal gland.

11. In a series of lateral to medial sagittal sections of the diencephalon the proper order of appearance of the following structures from lateral to medial is:

a. Subthalamus.

b. Mamillary bodies.

c. Massa intermedia.

d. Lateral geniculate body.

12. A part of the thalamus is sometimes surgically destroyed to relieve the tremor of Parkinson's disease. If the approach were from the lateral side (just above the ear), what structure(s) would be affected in reaching the thalamus?

a. Internal capsule.

b. Caudate nucleus.

c. Corpus callosum.

d. Putamen.

13. An inferior view of the cerebrum without the brainstem or cerebellum attached reveals:

a. Mamillary bodies.

b. Temporal lobe.

c. Corpus callosum.

d. Central sulcus.

14. The longitudinal cerebral fissure:

a. Reveals the insula, if separated.

b. Has convoluted gray matter in its walls.

c. Separates the right and left hemispheres.

d. Is a useful midline structure to help detect space-occupying lesions.

15. In a frontal section through the anterior commissure can be seen:

a. Thalamus.

b. Globus pallidus.

c. Amygdala.

d. Red nucleus.

16. The structure(s) forming the boundaries of the lateral ventricles is(are):

a. Caudate nucleus.

b. Thalamus.

c. Globus pallidus.

d. Internal capsule.

17. A horizontal section through the massa intermedia reveals:

a. Mamillary bodies.

b. Amygdala.

c. Anterior commissure.

d. Caudate body.

18. A sagittal section through the lateral edge of the thalamus (pulvinar) reveals:

a. Amygdala.

b. Claustrum.

c. Anterior commissure.

d. Globus pallidus.

19. The fornix fibers:

a. Travel over the top of the thalamus.

b. Are connected to the septum pellucidum.

c. Connect the cerebral cortex to the diencephalon.

d. Form part of the lateral boundary of the internal capsule.

20. A sagittal section through the genu of the internal capsule reveals:

a. Caudate nucleus anteriorly.

b. Thalamus posteriorly.

c. Third ventricle.

d. Optic tract.

SUGGESTED READINGS

Barr, M.: The human nervous system, New York, 1972, Harper & Row, Publishers.

Everett, N.: Functional neuroanatomy, ed. 6, Philadelphia, 1971, Lea & Febiger.

House, E., and Pansky, B.: A functional approach to neuroanatomy, New York, 1967, McGraw-Hill Book Co.

Netter, F.: The CIBA collection of medical illustrations. I. The nervous system, Summit, N.J., 1972, CIBA Chemical Co.

Noback, C., and Demarest, R.: The human nervous system: basic principles of neurobiology, ed. 2, New York, 1975, McGraw-Hill Book Co.

Roberts, M., and Hanaway, J.: Atlas of the human brain in section, Philadelphia, 1971, Lea & Febiger.

Shade, J., and Ford, D.: Basic neurology, New York, 1973, Elsevier Scientific Publishing Co.

Truex, R., and Carpenter, M.: Human neuroanatomy, ed. 6, Baltimore, 1969, The Williams & Wilkins Co.

Watson, C.: Basic human neuroanatomy: an introductory atlas, ed. 2, Boston, 1977, Little, Brown & Co.

Willis, W. D., Jr., and Grossman, R. G.: Medical neurobiology: neuroanatomical neurophysiological principles basic to clinical neuroscience, ed. 2, St. Louis, 1977, The C. V. Mosby Co.

CHAPTER 3

Neuron: origin, structure, surroundings, and electrical activity

In this chapter attention is focused on the basic unit of the nervous system, the neuron. The structural components of a neuron, its insulating coat of myelin, its responses to injury, the production of the action potential, and the propagation of the action potential are discussed. The neuron is a fascinating cell. It is present in an enormous variety of shapes and sizes, generates an electrical current, and initiates all of our movements, thoughts, and emotions. In this chapter special attention should be given to the following areas:

1. Functional classification of neurons
2. Structure of a neuron and its intracellular components
3. Formation and function of myelin
4. Reaction of a neuron to injury both within the cell body and to the axon
5. Recovery process of a crushed or severed nerve
6. Production of the resting membrane potential in terms of concentration gradients, voltage gradients, and membrane permeability
7. Events responsible for the production and propagation of an action potential
8. Effects of hyperpolarization, prolonged stimulation, and subthreshold stimulation on the production of an action potential
9. Early symptoms produced from low levels of sodium, potassium, and calcium in the blood

EMBRYOLOGY OF NEURONS
Derivatives of neuroblasts from neural tube and neural crest

As discussed in Chapter 1, the neurons and neuroglial cells originate from cells in the inner ependymal layer of the neural tube and from cells in the neural crest. The neuroblasts that migrate from the ependymal layer mature into a variety of multipolar neurons, including (1) the alpha and gamma motoneurons of the ventral horn, (2) preganglionic neurons of the autonomic nervous system (intermediolateral horn) and some cranial nerve nuclei, (3) the pyramidal cells of the cerebral cortex, (4) the Purkinje cells of the cerebellar cortex, and (5) the Golgi I and II neurons found throughout the CNS. A few neuroblasts form unipolar neurons that become a midbrain nucleus associated with the fifth cranial nerve. The axons of the motoneurons and the peripheral processes of these unipolar neurons migrate from the CNS via the ventral roots and cranial nerves. The cell bodies and remaining processes stay within the CNS with the other multipolar neurons.

The neuroblasts of the neural crest mature into unipolar neurons that form dorsal root and cranial nerve ganglia, bipolar neurons of the special senses, and postganglionic (multipolar) neurons of the autonomic nervous system. These cell bodies remain outside the CNS. Only the central process of the unipolar and bipolar neurons enter the CNS.

67

Classifications of neurons

In Chapters 1 and 2 a neuron was classified on the basis of its structure or into the general functional categories of afferent, efferent, or internuncial. In this chapter a more detailed functional classification is considered, based on the major neural systems of the body: the somatic, visceral, and proprioceptive systems (Fig. 3-1). The somatic system is concerned with relaying sensory data from the outside environment and activating skeletal muscles. The visceral system is concerned with the sensory and motor activity of the internal organs (smooth muscle), the secretions from glands, and the muscles of the face, jaw, throat, and heart. The proprioceptive system is concerned with the input from muscle, tendon, joint, and inner-ear receptors for the regulation of tone, equilibrium, and posture. The specific classification of neurons with these three categories is as follows:

1. General somatic afferents that convey sensory data from the skin
2. General somatic efferents that activate the skeletal muscles of the body including the extrinsic eye muscles and the tongue
3. Special somatic afferents that convey data from the eye (vision) and ear (hearing)
4. General visceral afferents that convey data from the internal organs
5. General visceral efferents that activate the smooth muscles of the internal organs, the cardiac muscle of the heart, and all glands
6. Special visceral efferents that activate the muscles of facial expression and mastication and of the larynx and pharynx
7. Special visceral afferents that convey the sensations of taste and smell
8. General proprioceptive afferents that

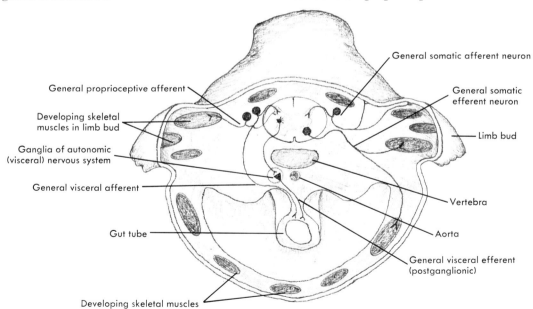

Fig. 3-1. Schema of some types of neurons found in developing fetus (transverse section). (Adapted from Noback, C., and Demarest, R.: The human nervous system: basic principles of neurobiology, ed. 2, New York, 1975, McGraw-Hill Book Co.)

convey data from the muscles, tendons, and joints

9. Special proprioceptive afferents that convey data from the labyrinthine mechanism

This classification is not only useful in explaining the embryology of different types of neurons but is also useful in explaining the functional characteristics of the spinal and cranial nerves.

MORPHOLOGY OF NEURONS
Intracellular components

In many ways a neuron is similar to other body cells. There is an outer membrane around the cytoplasm, and within the cell are the following components (Fig. 3-2):

1. Nucleus and nucleolus, containing DNA and RNA, which function in hereditary characteristics, protein synthesis, and repair

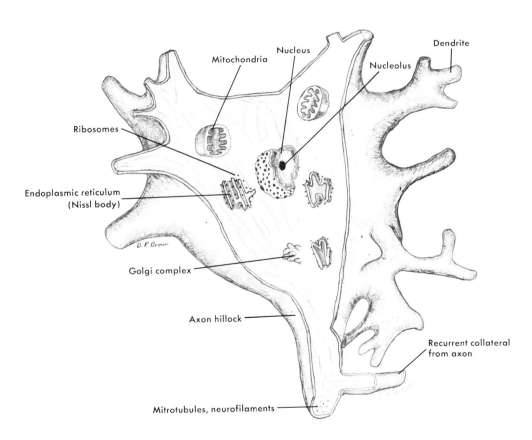

Fig. 3-2. Three-dimensional view of neuron cell body with wedge slice taken out. Major internal structures are shown. Note central location of nucleus. (Adapted from Willis, W. D., Jr., and Grossman, R. G.: Medical neurobiology: neuroanatomical neurophysiological principles basic to clinical neuroscience, ed. 2, St. Louis, 1977, The C. V. Mosby Co.

of the cell after injury. The nucleus normally is centrally located in the cells.

2. Mitochondria, the site of metabolism of acetylcoenzyme A (the end product of glucose, fat, or amino acid breakdown) into the energy-rich molecules of adenosine triphosphate (ATP). ATP is essential to the production of the action potential as well as other cellular functions.

3. Golgi complex, a storehouse for RNA-produced proteins. The Golgi complex is normally a storage place for hormones such as the insulin-producing cells of the pancreas. In neurons, however, the Golgi complex may be useful in regeneration after injury when proteins are needed for new cytoplasm.

4. Endoplasmic reticulum, a tubular system that travels throughout the cell and is continuous with the nuclear membrane, Golgi complex, and plasma membrane. This system is responsible for the transport of synthesized products such as proteins and enzymes throughout the cell.

5. Nissl bodies, found in all neurons and most glandular cells. Nissl bodies are composed of portions of endoplasmic reticulum dotted with tiny granules on the outer surface. These granules are called ribosomes and are composed of RNA. Nissl bodies are the production sites for the proteins and enzymes needed for neurotransmitters and cellular regeneration.

6. Microtubules and neurofilaments, fine filamentous protein structures found within the soma, dendrites, and axon. They function in mechanical support of the slender axons and dendrites and are probably associated with the axoplasmic flow from the soma to the synaptic endings.

Structural design

Neurons are distinguished from other cells by their ability to generate electricity (which is discussed later) and by their cell processes. A typical neuron consists of a *cell body* (soma, perikaryon) and a varying number of threadlike processes. The cell body, which contains all of the above organelles, is usually oval or pyramidal (Fig. 3-3) and has a diameter that can vary from 4 to 135 mμm for different neurons. The soma is the "life support" center for the entire cell, and any injury to the soma will result in death of the neuron.

Usually a neuron has two or more short stalks attached to the soma that extend a short distance and branch into many short fine threads. These are called *dendrites* and function in conveying the converging action potentials toward the cell body. Most dendrites have special sites known as "spines" at which these synapses occur. Dendrites have the same organelles as the soma, and they are unmyelinated.

The cell body usually gives off one long process known as the *axon*. The axon conveys the action potential from the soma to another neuron, muscle cell, or gland. Axons are usually much longer than dendrites. In fact, some of the motoneuron axons extend from the lumbar part of the spinal cord to the muscles of the foot. The junction of the axons and soma is known as the *axon hillock*. The axon hillock is the origin for the efferent action potentials. Axons usually do not branch until they reach their effector, although some types such as those from motoneurons have a branch a short distance from the cell body known as a *recurrent collateral*. The terminal portions of these branches are known as *synaptic knobs* or *boutons* (Fig. 3-3). Here is contained the neurotransmitter substance used in synaptic activity. In order for the bouton to participate in synaptic activity, a constant flow of supplies (enzymes and chemicals) must come from the soma. This transport is partly accomplished by a system of microtubules distributed through the axoplasm. These microtubules are probably a continuation of the endoplasmic reticulum. Thus substances produced by the Nissl bodies are transported to the synaptic knob,

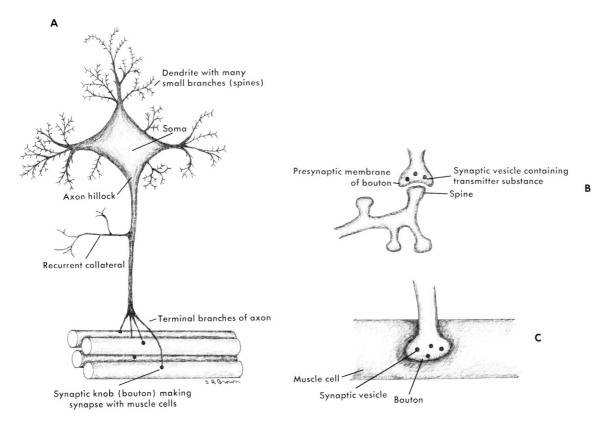

Fig. 3-3. A, Structure of neuron. **B,** Closeup of synapse between axon and spine of dendrite. Junction between the two is synaptic cleft. **C,** Closeup of synapse between bouton and muscle cell. Junction between the two is neuromuscular junction.

where they are formed into neurotransmitters and neurosecretory compounds. The only other structures within the axon are mitochondria, which are located periodically along the axon and in boutons, and neurofilaments, which are very fine threads that provide support for the axon to maintain its shape. Axons may or may not be myelinated. The axon hillock, however, is never myelinated.

It is important to clarify one deviation of axons and dendrites. The unipolar neuron has only one process attached to the soma. This process splits a short distance from the soma into a central process and a peripheral process. The peripheral process conveys data from receptors to the soma and can be called a dendrite even though it is quite long and often myelinated. The central process carries the data away from the soma into the CNS and can be called an axon.

The relationship between the cell body and its processes is interesting. First, the total volume of axons and dendrites may exceed the volume of the soma more than 1000 times. Second, if the cell body of a motoneuron were enlarged to the size of a baseball, the axon would be a mile long and the branching of the dendrites would fill a large amphitheater!

It is important to mention that the cytoplasm of the cell body and all of the processes is surrounded by a highly organized dynamic three-layered structure known as the *plasma (cell) membrane*. The plasma membrane plays a major role in the electrical activity of a neuron. It has pores for the diffusion of particles in and out of the cell and an active transport system to regulate this flow.

SURROUNDING ENVIRONMENT OF A NEURON
Neuroglial cells

The neuroglial cells have an intimate relationship with the neurons. They outnumber the neurons from five to ten times and account for about half of the total volume of the CNS. The neurons "float" in a sea of neuroglial cells that function to support the delicate soma and its processes, insulate the electrical fields between different neurons, and provide homeostasis in the extracellular fluid. After a brief review of the embryology of neuroglial cells, attention is focused on the formation of the insulating sheath known as myelin. The homeostatic functions of neuroglial cells are discussed in greater detail in Chapter 7.

The glioblasts that migrate from the ependymal layer are the oligodendrocytes and astrocytes. Oligodendrocytes are found around the cell body to act as support and around the nerve fibers to form an insulating sheath of myelin. The astrocytes are found around the cell body and function in structural support of the neuron, in isolating synaptic areas so that the activity at one synapse does not interfere with activity at another synapse, and as a barrier system to help regulate the extracellular fluid around the neuron, and they have a possible role in transporting electrolytes, nutrients, and waste products between the capillaries and neurons (Fig. 3-4).

One of the glioblasts remains in the ependymal layer and develops into the ependymal cell, which forms a lining of the neural tube

activity and eventually of the ventricular system.

One other glial cell found throughout the CNS is known as a *microglial cell*. These cells are scavengers and act to phagocytize and remove damaged (destroyed) neurons. Microglia are of mesodermal origin and thus are often not considered neuroglial cells.

The glioblasts of the neural crests develop into satellite cells and Schwann cells. Satellite cells become supportive cells for the peripheral ganglia. Thus some of them remain in the crest with the unipolar neurons and form the dorsal root ganglia, and some migrate out with the autonomic neurons to form other ganglia. The Schwann cells migrate out with the unipolar cell processes and the axons from motoneurons of the neural tube. Schwann cells form an insulating sheath of myelin around many of these processes.

One of the most fascinating relationships among cells can be found between oligodendrocytes in the CNS and the long processes of the neurons and the Schwann cells in the PNS. These two neuroglial cells act to insulate the long axons and dendrites of the nervous system. In some cases these neuroglial cells simply engulf an axon or dendrite on three sides like a hot dog roll around a hot dog. These neuron processes are called *unmyelinated processes*. In the remaining cases these neuroglial cells wrap their cell membranes around the neuron processes in jelly-roll fashion (Fig. 3-5). These processes are called *myelinated fibers*. It is important to realize that all long axons and dendrites have neuroglial cells along the long axis at approximately 1 mm intervals. In unmyelinated processes the fiber is exposed on one side to the extracellular fluid along its entire length. In myelinated fibers only the junctions between the neuroglial cells are exposed to the extracellular fluid. These junctions are known as the *nodes of Ranvier*. The nodes of Ranvier are important in terms of speed of

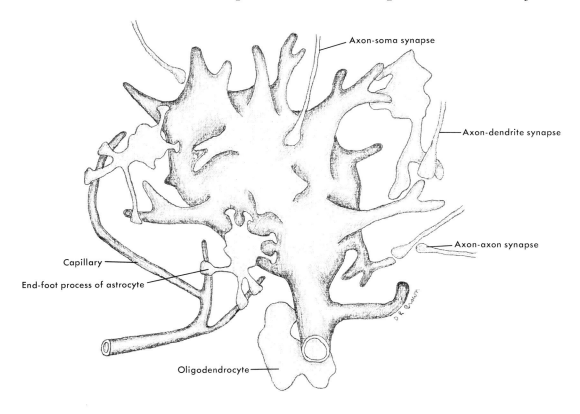

Fig. 3-4. Relationship between neuron, neuroglial cell, synapses, and capillaries.

conduction of the electrical impulses, as discussed later in this chapter. For now, suffice it to say that in myelinated fibers the neuroglial cell wraps itself around the axon or dendrite like one might wrap an empty toothpaste tube around a pencil. These concentric layers of neuroglial cell membrane are known as *myelin.*

Connective tissue coverings

In the CNS the neurons and neuroglial cells are so closely related that only 15% of the total volume is available for the extracellular fluid. In the peripheral nervous system the neuron processes not only are closely related to the Schwann cells but are surrounded by three layers of connective tissue covering. These coverings function to protect the axons and dendrites and help form the spinal, cranial, and peripheral nerves. As can be seen in Fig. 3-5, all axons and dendrites are related in some manner to neuroglial cells. In the peripheral nerves the Schwann cell is sometimes referred to as a *neurilemma cell.* The neurilemma cell that surrounds one or more neuron processes is surrounded by a layer of connective tissue called the *endoneurium.* The neuron processes with their neurilemma and endoneurium coverings are grouped in clusters surrounded by more connective tissue called the *perineurium.* These clusters are further grouped together into nerve trunks and are surrounded by even more connective tissue called the

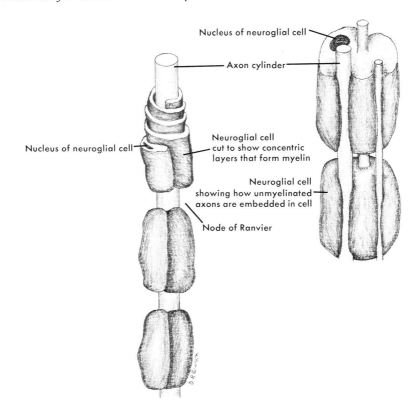

Fig. 3-5. Relationship of axon cylinder to its neuroglial cell. Note that each axon (or unipolar dendrite), whether myelinated or not, is at least partly surrounded by neuroglial cells. Also note that in myelinated fibers, only area exposed to extracellular fluid is at node of Ranvier.

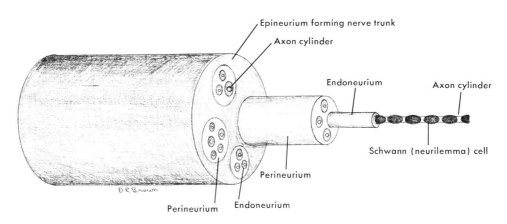

Fig. 3-6. Connective tissue coverings of neuron processes in peripheral nerves. Only small portion of axons and unipolar dendrites are represented in this drawing.

epineurium (Fig. 3-6). Thus nerves are not simple axons or dendrites or even myelin. They are instead a complex arrangement that also includes coatings of connective tissue.

Responses of neurons to injury

Any damage to the cell body such as from traumatic, ischemic, or toxic causes will result in the death of that neuron and eventual phagocytosis by the microglia. If the insult is to the axon away from the soma, the neuron may be able to undergo some degree of repair. In general, the farther the insult is from the soma, the better the chances are for recovery. When an axon is injured the following changes occur (Fig. 3-7):

1. Retrograde chromatolysis:
   a. Cell body swells with water to three times normal size.
   b. Nucleus moves to an eccentric position.
   c. Nissl bodies disperse (undergo chromatolysis) and are unable to take up Nissl stain.
   d. There is an increase in RNA and protein synthesis.
2. Primary degeneration:
   a. Area proximal to the injury shows breakdown of the neuroglial cell (including the myelin sheath, if present) and the axon cylinder for several internodes from the injury.
3. Secondary (wallerian) degeneration:
   a. Fragmentation of the distal axon cylinder within the first week.
   b. Myelin sheath breaks up into fragments, which are eventually phagocytized.
   c. Neuroglial cells remain intact within the endoneurial tube.

After the neuron or its axons has been injured there are immediate and prolonged effects on the postsynaptic cell. First, there is a change in function. A glandular cell may cease secretion or become overactive, a muscle cell will cease to function, and a postsynaptic neuron may alter its production of action potentials, depending on the influence of the injured axon. These effects can persist until reinnervation can occur. Second, muscle and glandular cells can have denervation hypersensitivity during this period of denervation. This phenomenon is explained further in Chapter 5. Finally, there are trophic effects. Although still somewhat unclear, it seems that a neurotrophic substance is synthesized in the soma and transported via axoplasmic flow to the boutons. This substance is not involved with synaptic activity but is apparently secreted continuously by the bouton into the synaptic cleft. This neurotrophic substance seems to have an influence on the vitality of the postsynaptic cell. This influence is most pronounced in skeletal muscle. When these cells lose their neural innervation they not only cease to contract but also undergo severe atrophy and will eventually die. Other trophic effects seen with loss of innervation are dry, flaky, inelastic skin and brittle nails.

As a further point of interest, it is generally believed that this neurotrophic substance can influence the contractile properties of a muscle cell. Muscles that contract slowly but fatigue less are known as tonic or red muscle cells. Those that contract quickly but fatigue readily are called phasic or white muscle cells. The motoneurons to red muscles are smaller than those to white muscles. Experiments have been performed in which red motoneuron axons have been severed and placed near white muscle cells, and vice versa. After reinnervation occurred the former red muscles contracted faster and fatigued quicker, while the former white muscles contracted slower and fatigued less.

The recovery process of an injured neuron is painstakingly slow. Within 1 week postinjury the cell body has synthesized and transported enough new cytoplasm to the distal portion of the remaining axon that several sprouts appear (Fig. 3-7). The main obstacle

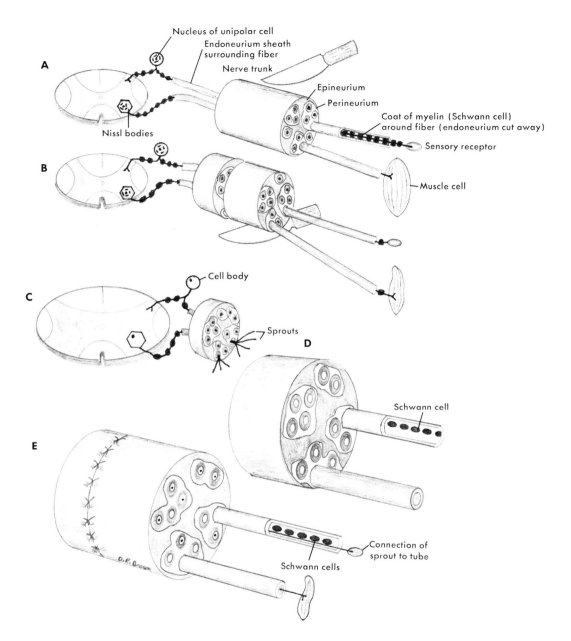

Fig. 3-7. Axon degeneration and regeneration. **A,** Normal neurons as part of nerve trunk. **B,** Sectioning of nerve trunk. Immediate effects are loss of sensation and muscle paralysis. **C,** Proximal effects of denervation. Within a few days, cell body swells, Nissl bodies disperse, and nucleus moves to eccentric position. Within a few weeks, proximal portion of cut nerve is growing sprouts. These sprouts are attempting to locate their appropriate end-organ. **D,** Secondary (distal) effects of denervation. Axons disintegrate within endoneural tube. Myelin sheath breaks up, but Schwann cell remains intact in tube. **E,** Effect of suturing two portions of nerve trunk. Many sprouts can locate "tube" to grow into, to guide them to a muscle or receptor. If sprout is able to find its correct tube (motor or sensory), then appropriate connection can be made and Schwann cells can remyelinate.

facing these sprouts is that one of them must find its former pathway within the distal nerve trunk. For example, if the median nerve were severed at the elbow the proximal portion would have viable axons, each with many sprouts. The distal portion in the forearm would contain only the connective coverings and the Schwann cells that normally surround the axons. If the two ends of the nerve are not sutured together, the regenerating sprouts will have great difficulty finding a "tunnel" in the distal stump to grow into. At least part of the cause of this difficulty is that the neuroglial cells form a scar in the gap between the two ends, preventing the sprouts from crossing the gap. The sprouts cease to grow and often form a small lump known as neuroma. Neuromas can be very painful because they have many afferent pain fibers. Neuromas are very common after amputation of a limb because the distal portion of the nerve is removed.

If the two ends of the stump can be sutured together, the amount of scarring is minimal and the sprouts can easily cross the gap. Although it is easy to suture the two ends of a cut nerve, it is very difficult to reunite both ends so they match. (Think of cutting a telephone cable with 10,000 people making calls and then trying to rejoin both ends so that each caller is talking to the correct person.) With a sutured nerve many of the sprouts do find a tunnel down which they can grow. The rate of growth is slow and averages 3 to 4 mm per day. Although several sprouts may enter a tunnel, as soon as one makes an appropriate contact with an end organ the others degenerate. It is important to mention that if a motor sprout grows down a former sensory tunnel, or vice versa, no functional connection will be made and that sprout will degenerate. However, when a sprout finds an appropriate end organ (muscle cell or receptor) the neuroglial cells that form the inner lining of the tunnel begin to surround this new axon either on three sides (unmyelinated fiber) or

in concentric layers forming a myelin sheath. Regenerated and remyelinated fibers have internodal distances, diameters, and conduction velocities about 80% of normal.

The degree of recovery will vary from excellent to poor, depending on severity of the injury, proximity of the injury to the soma, and whether the distal and proximal ends are sutured together. A nerve that has been crushed but not severed will show good return because the regenerating sprouts are still within their original tunnels. A crushed nerve can occur from leaning on crutches with the axilla or falling asleep in a chair with an arm over the back. Severed nerves, even if reunited, usually have some loss of function as a result of axons growing down inappropriate tunnels. Also influencing the degree of recovery is the extent of collateral innervation from uninjured axons or dendrites (Fig. 3-8). Although the reason is uncertain, it is probable that an injured process releases a chemical that stimulates undamaged axons or dendrites to produce collateral sprouts to aid in reinnervation.

The above discussion of the recovery process centers around peripheral nerve regeneration. The same degeneration and regeneration processes occur in the CNS, but the extent of CNS regeneration is much less. This unfortunate fact is clearly seen in patients with spinal cord injury. Even if the patient is immobilized so that the separated cord segments approximate each other, the degree of recovery is minimal. This is partially the result of inability of the sprouts to penetrate the glial scar, inability of many cell bodies to sustain the metabolic activity for regeneration, and poor ability of the oligodendrocytes to form a new myelin sheath.

As with incomplete peripheral nerve lesions, incomplete CNS lesions can induce collateral sprouting from undamaged axons. In both cases the degree of return of motor and reflex activities and the reduction of sensory loss are quite variable.

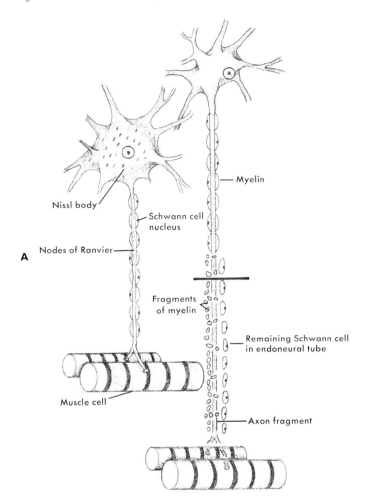

A

Nissl body

Schwann cell
nucleus

Nodes of Ranvier

Myelin

Fragments
of myelin

Remaining Schwann cell
in endoneural tube

Muscle cell

Axon fragment

Fig. 3-8. Collateral sprouting in axonal regeneration. **A,** Early effects of severence include myelin and axonal fragmentation both distal to cut and proximal to next node of Ranvier, shifting of nucleus, and chromatolysis (dispersion of Nissl bodies).

One final point about axon regeneration concerns time. The recovery process is slow and can extend for 12 to 18 months. During this time the muscles involved must be maintained in a viable state to keep them "ready" for regenerating axons. A muscle that has atrophied or a joint that has become contracted may prevent return of useful function after reinnervation. It is imperative that the therapist be acutely sensitive to any subtle changes that indicate recovery has occurred. Once reinnervation has occurred, the patient can be guided through the appropriate sequence of functional development.

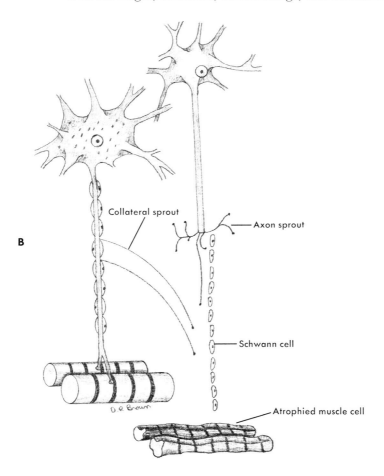

Collateral sprout

Axon sprout

Schwann cell

B

Atrophied muscle cell

D. R. Brown

Fig. 3-8, cont'd B, Collateral sprouts from undamaged axons can grow into "tunnel" to aid in renervation. Note atrophy associated with denervation.

ELECTRICAL ACTIVITY OF NERVOUS SYSTEM

An understanding of the electric current generated by neurons is very important to the therapist. Every movement, all cerebral functions, and most visceral activity involves the production and propagation of an electri-cal impulse. This impulse is known as an *action potential (AP)*, and it can be produced only by neurons and muscle cells. While most neurologic conditions do not interfere with these electrical impulses, some peripheral nerve lesions, metabolic disorders, and severe burn injuries can significantly alter the production or conduction of the action potential.

Basic principles of electricity

Before discussing how an action potential is produced, several basic principles concerning the nature of charged particles (ions) need to be reviewed. Ions with similar charges repel each other, while opposite charges attract one another. When opposite charges are separated an electrical force exists that attempts to pull them together. This force varies directly with the number of ions and indirectly with the distance. This force represents a potential source of work energy because if the oppositely charged ions were allowed to come together a movement of particles over a distance (work) will have been performed. The same principle applies when one tries to hold the north and south poles of two magnets apart but very close together. The force used to hold them apart can be thought of as a potential source of work energy. This energy becomes available when one of the magnets is released and allowed to move toward the other magnet. In the same manner, when oppositely charged ions are separated by something such as a nonporous membrane, the result is a potential for work to be done if the membrane were removed. This potential is measured in units called volts. *Voltage,* then, is defined as the amount of work that can be done by an electric charge when moving from one point in a system to another. Voltage is often used to measure the potential for work that can exist when oppositely charged ions are separated. This potential is often referred to as the *potential difference* between two oppositely charged areas. A common example of potential difference is the sockets of any wall outlet. These sockets each have a potential work energy (potential difference) of 120 V. When something is plugged into the outlet and turned on there is a flow of charged particles. This flow of ions is known

as *current,* and it is measured in units called amperes.

The movement of ions (current) is always hindered to some degree, depending on the medium through which they are traveling. This hindrance is known as *resistance.* Substances such as copper have low resistance and thus are good conductors. Rubber has a high resistance and is a poor conductor. In the body the neuron (and muscle cell) is bathed in an extracellular fluid of water and salts (electrolytes) and contains an inner fluid (cytoplasm) that also contains water and electrolytes. Both of these fluids are very good conductors. Separating these two fluids is the cell membrane, which is a poor conductor because it is partly composed of lipids. This membrane is the key to production of the action potential. It not only separates ions but also acts as an insulator around which a current can flow.

Factors influencing diffusion of ions

The production of an AP depends on the flow of ions through the membrane of the neuron or muscle cell. This diffusion of charged ions is influenced by three factors: (1) membrane permeability, (2) concentration gradients, and (3) voltage gradients. Suppose two compartments are filled with water and separated by a nonpermeable membrane. If a salt such as sodium chloride (NaCl) is added to compartment 1, a *diffusion potential* exists on the basis of a *concentration gradient* (Fig. 3-9). Keep in mind that substances such as salts, which dissolve in water, break up into their separate ions. If the membrane between these two compartments is removed there will be a flow of ions (current) into the region of less concentration. Suppose, however, that instead of removing the membrane small pores are made allowing the membrane to be selectively permeable to Cl^- ions and not Na^+ ions (Fig. 3-10). In this example the Cl^- ions will diffuse into compartment 2 as a result of a concentration gradient. However,

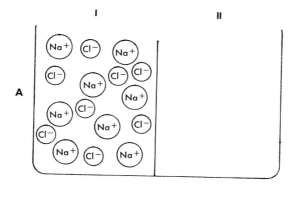

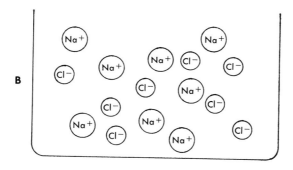

Fig. 3-9. Diffusion of ions from regions of high concentration to regions of low concentration. **A,** Salt (NaCl) is placed in compartment I and water is added to both I and II, allowing salt to dissolve into its separate ions. **B,** Nonpermeable membrane that separated I and II is removed, allowing ions to move as result of concentration gradients (forces). This movement represents work that has been done.

as Cl⁻ leaves compartment 1 a positive *voltage gradient* will be formed in compartment 1 as a result of the loss of the negative ions. The Cl⁻ ions will leave compartment 1 until the electrical attraction builds up to prevent any further outflow. Cl⁻ ions will then be in equilibrium; that is, the concentration gradient drawing them out of compartment 1 will be balanced by the electrical gradient holding them in. The end result of this situation is the establishment of a voltage gradient between 1 and 2 that can be measured by a voltmeter (Fig. 3-10). Compart-ment 1 can be said to have a positive charge in respect to 2. If the membrane were to become permeable to Na⁺ as well as Cl⁻, there would be an initial flow of Na⁺ ions into 2 as a result of both a concentration and a voltage gradient.

To summarize the above:
1. The permeability of a membrane that separates two compartments of ions will influence the diffusion of those ions across the membrane (between the two compartments).
2. A selectively permeable membrane can

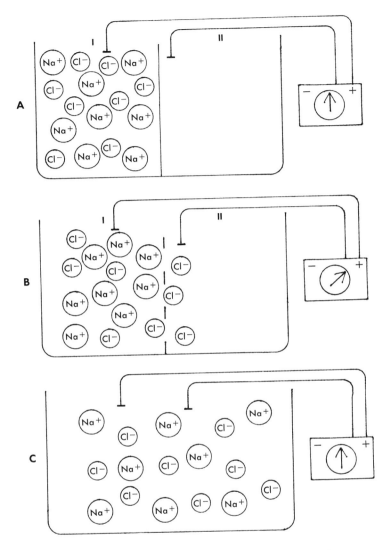

Fig. 3-10. Factors influencing diffusion of ions. **A,** NaCl has been added to compartment I. I and II are filled with water and separated by nonpermeable membrane. Voltmeter will record zero voltage gradient (potential difference) between I and II. **B,** Membrane is now selectively permeable to small ions. Some Cl⁻ ions diffuse into II as result of concentration gradient, leaving I more positive. This creates voltage gradient that will balance flow of Cl⁻ out of I. Voltmeter will show that I is more positive with respect to II. **C,** All ions are free to move. Na⁺ ions will move as result of both concentration and voltage forces. Voltmeter will show zero voltage gradient after ions have reached equilibrium.

allow a voltage gradient to develop between the two compartments as a result of the diffusion of one ion but not the other.

3. This voltage gradient represents a potential for work to be done if the membrane were to become completely permeable.

Development of resting membrane potential

As was mentioned, the neurons and muscle cells of the body are not only filled with a solution of charged ions (cytoplasm) but also lie in a sea of extracellular fluid that also contains charged ions. It is important to remember that many substances such as salts dissociate into their individual positive ions (cations) and negative ions (anions) when placed in an aqueous solution. Ions move freely when in solution and thus are the sole carriers of electric charge.

If a thin recording electrode is placed inside an excitable cell and another recording electrode is placed in the extracellular fluid around the cell, a voltage gradient of around 70 mV can be recorded (Fig. 3-11). This voltage difference (potential difference) is known as the *resting membrane potential (RMP)*. Although the value may seem small, keep in mind that the cell membrane is only one millionth of a centimeter thick. This would be equivalent to a field of 100,000 V if the membrane were 1 cm thick.

The RMP is described as the inside of the cell being negative compared with the outside as a result of loss of positive ions from the cell. To aid in understanding how a negative RMP can develop, the following hypothetical model is presented. Keep in mind, however,

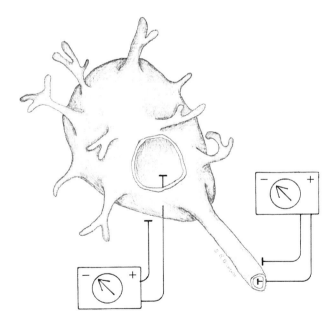

Fig. 3-11. Procedure for recording resting membrane potential. When one electrode from voltmeter is placed outside cell membrane and other electrode is placed inside cell, negative voltage gradient can be recorded, with inside of cell more negative than outside. Same voltage gradient can be recorded from any area of neuron.

that the precise origin of the RMP is as yet uncertain.

1. A neuron (or muscle cell) containing positively charged potassium ions (K^+) and negatively charged chloride ions (Cl^-) and protein molecules (A^-) is placed in a fluid containing positively charged sodium ions (Na^+) and negatively charged chloride ions (Cl^-). The membrane around the cell is nonpermeable to any ion.

2. A voltmeter recording will show 0 mV potential difference between the inside and outside of the cell (Fig. 3-12), indicating an isoelectric state. Also the number of anions and cations will be balanced both inside and outside the cell.

3. The membrane is then made very permeable to K^+ and Cl^-, slightly permeable to Na^+, and impermeable to A^-.

4. K^+ will leave the cell as a result of a high concentration gradient, leaving

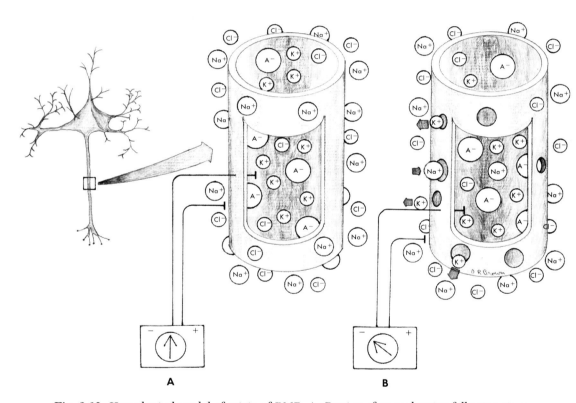

A   **B**

Fig. 3-12. Hypothetical model of origin of RMP. **A,** Portion of axon showing following situation: (1) inside of axon contains large amounts of potassium ions and negatively charged protein molecules (A^-) and some chloride ions; (2) extracellular fluid contains large amounts of sodium and chloride ions; (3) axon membrane is nonpermeable; (4) there is no voltage gradient between inside and outside area; and (5) cations and anions are balanced inside and outside. **B,** Similar hypothetical situation except that axon membrane is now very permeable to K^+ and Cl^-, slightly permeable to Na^+, and impermeable to A^-. More K^+ will leave cell that Na^+ entering, leaving interior negative with respect to outside.

the inside of the cell negatively charged.

5. Some Cl^- will leave the cell as a result of the negative electrical gradient. However, concentration gradients prevent the Cl^- ions from balancing the loss of K^+

6. The A^- ions attempt to leave the cell because of electrical and concentration gradients but are blocked by the cell membrane.

7. The Na^+ ions attempt to enter the cell because of electrical and concentration gradients, but only a few ions are allowed in by the membrane.

8. A situation thus develops in which the cell has lost more positive ions than it has gained from the outside and is unable to eject sufficient negative ions to balance this loss. Thus a negative potential difference can be recorded between the inside and outside of the cell (Fig. 3-12).

While this model is not based on scientific fact, it does provide a possible explanation as to how neurons and muscles have a negative voltage gradient with their surrounding extracellular fluid. In reality the neuron and muscle cell have high concentrations of K^+ and A^-, and the extracellular fluid has high concentrations of Na^+ and Cl^-; the exact numbers can be seen in Fig. 3-13. Also the membrane is freely permeable to Cl^-, 30 times more permeable to K^+ than Na^+, and impermeable to A^- (K^+ and Cl^- are smaller ions).

The model is, however, incomplete. Since ions move freely in an aqueous solution, disequilibrium in charge cannot exist inside or outside the cell. In other words, the number of cations and anions must be balanced in the fluid within the cell and outside the cell. The problem of excess anions within the cell and excess cations outside the cell is solved by the cell membrane. The cell membrane, which is largely composed of lipid, is a poor conduc-

tor. It acts as a capacitor between two conducting mediums. The excess negative ions in the cell lie along the inner surface of the membrane, while the excess positive ions outside the cell lie along the outer surface of the membrane. This ion–cell membrane–ion complex forms an electrically neutral unit that in essence "neutralizes" the intracellular and extracellular fluids (Fig. 3-13).

The cell membrane also has a role in maintaining the RMP. Within the membrane is an active transport system known as the *sodium-potassium pump*. This "pump" is an enzyme system that is powered by molecules of *adenosine triphosphate (ATP)*, which are derived from the metabolic breakdown of ingested food (Fig. 3-14). If the sodium-potassium pump were blocked, Na^+ ions would slowly enter the cell as a result of powerful electrical and chemical forces. As the Na^+ ions entered the cell there would be less negative voltage difference. As the potential difference declined K^+ would leave the cell. Eventually the concentration gradients would decline, and the RMP would drop to 0 mV. The sodium-potassium pump acts to eject any additional Na^+ that may diffuse into the cell. As Na^+ ions are ejected from the cell, K^+ ions that may have escaped from the cell are brought back in by the pump. Thus the concentration and voltage gradients are maintained. As will be seen, these gradients are essential to the production of the action potential.

As shown in this section, the neuron and muscle cell never really rest. There is a continual movement of ions across the cell membrane, which is regulated by the sodium-potassium pump. The driving force behind this ionic flow is that all ions attempt to reach an equilibrium between the electrical and concentration forces that act on them. Many of the ions involved with excitable cells are not in equilibrium. For example, if the cell membrane were freely permeable to K^+, K^+ ions would leave the cell to balance the con-

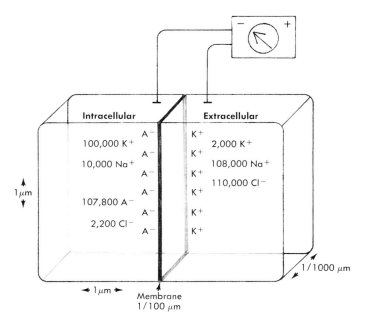

Fig. 3-13. Distributions of ions in intracellular and extracellular fluid. Number of ions inside and outside cell equals 220,000 for area of $1 \times 1 \times 1/1000$ μm. Cell membrane is occupied by six anions (A^-) and six cations (K^+) in area of $1 \times 1/1000$ μm. This allows both compartments to be electrically neutral by themselves, but when one is compared with other, voltage gradient can be recorded with voltmeter. (Adapted from Schmidt, R. editor: Fundamentals of neurophysiology, New York, 1975, Springer-Verlag.)

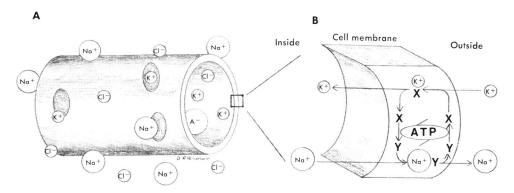

Fig. 3-14. Hypothetical model of sodium-potassium pump. **A,** Cut section of axon. **B,** Enlarged section of portion of cell membrane. Sodium-potassium pump is property of excitable cell membranes that enable cell to maintain its RMP. This is accomplished by retrieving K^+ ions lost from cell and removing any Na^+ ions that have leaked in. Pump is powered by ATP and involves two ion carrier molecules (*X* and *Y*), which aid in diffusion of Na^+ and K^+ across cell membrane.

centration gradient. The concentration force would soon be balanced by an increasing negative voltage, keeping K^+ in the cell. The voltage gradient that achieves this balance of forces is known as the *equilibrium potential.* The equilibrium potential for any ion can be determined by using the Nernst equation. The equilibrium potential for K^+ is -90 mV, for Na^+ it is $+65$ mV, and for Cl^- it is -70 mV. The meaning of these equilibrium potentials will become clearer after discussion of the action potential.

Development and propagation of action potential

The formation of the electrical impulses produced by neurons and muscle cells is not only fascinating but also relatively easy to understand. Action potentials are produced in the body by the sensory receptors, synaptic activity between two or more neurons, and at the junction between the motor axon and a muscle cell. The precise nature of these mechanisns is discussed later. For now, consider how an action potential is produced by electrical stimulation of a single axon. The sequence of events is similar to that in the body. Fig. 3-15 represents the setup used to produce and record the AP. The sequence of events is as follows:

1. The "resting" axon, like any part of the neuron, has a voltage gradient of -70 mV. The anions lie along the inside of the membrane, and the cations lie along the outside of the membrane.

2. The cathode (negative pole) is the stimulating electrode.

3. As stimulus intensity is slowly increased, positive ions from adjacent areas along the membrane flow toward the cathode. This is known as *local current flow*. The amount of local current flow varies directly with stimulus intensity and inversely with the distance from the stimulus.

4. The local current flow from adjacent areas causes that portion of the axon to have less of a voltage gradient; that is, the loss of positive ions from the outside of the membrane makes that area less positive (or more negative) and reduces the voltage difference between inside and outside. This loss of polarity across the membrane is known as *depolarization* and is reflected in a decline of the RMP.

5. If the stimulus strength is strong enough to cause local current flow to depolarize the membrane to -50 mV, a sudden alteration of membrane permeability occurs. The *conductance* of ions across a membrane is often written as g. When the membrane is depolarized to -50 mV, *the critical firing level (CFL)* is reached. At this point gNa^+ is increased more than 100 times. The reason for this is unknown.

6. The sudden increase in gNa^+ causes a large flow of Na^+ ions into the cell as the Na^+ ions try to reach their equilibrium potential.

7. This marked increase in gNa^+ will reach a peak in a very short time, usually within 1 msec. During this time the membrane will *reverse polarity.* Enough Na^+ ions will enter the cell to cause the inside to become positive with respect to the outside. A voltmeter recording taken at this time will have a value of around $+30$ mV.

8. After the peak of gNa^+ the membrane enters a recovery phase. This period usually lasts several milliseconds. During this time the gNa^+ decreases rapidly and the gK^+ increases. This effect removes positive ions in the cell and allows the membrane to repolarize from $+30$ mV back toward the RMP of -70 mV. During the increase in gK^+, K^+ ions attempt to reach their equilibrium potential.

9. When the gK^+ returns to normal the

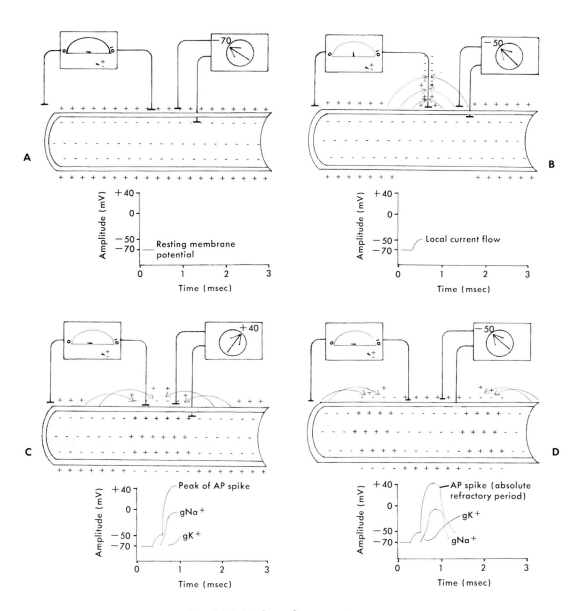

Fig. 3-15. For legend see opposite page.

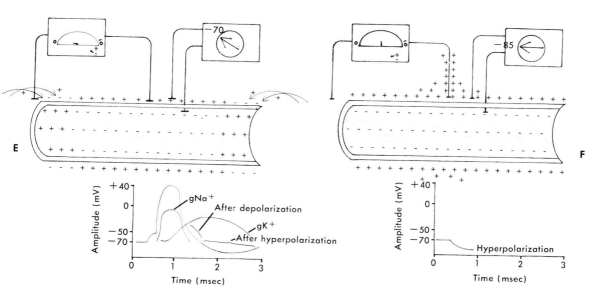

Fig. 3-15. Procedure for recording production of action potential. **A,** Recording resting membrane potential. With stimulator off, voltmeter will record −70 mV difference across axon membrane. Axon has been cut in half lengthwise to show movement of ions. For simplification, inside of axon shows only negative (−) ions and outside shows only positive (+) ions. Realize that following event will occur around entire circumference of axon. **B,** Local current flow. As stimulator is turned on, using negative (−) electrode as active electrode, positive ions along outside of membrane will move toward negative field. This results in local depolarization of membrane. Amount of local current flow will decrease with distance from cathode. **C,** Reversal of polarity. If stimulus strength is strong enough to cause membrane to depolarize to critical firing level there will be sudden increase in gNa+. This will cause inside to become positive compared with outside, even if stimulator is shut off. Void left by inrushing positive ions can now act as cathode to attract positive ions from adjacent areas. **D,** Repolarization of membrane. Within 1 msec increased permeability to Na+ falls back to normal and permeability to K+ is increased. This acts to remove excess positive charge inside and enables inside of axon to become negative. Until this point, axon is said to be in absolute refractory period. Note that two adjacent areas have reversed polarity, causing their respective adjacent areas to have local current flow. **E,** Afterpotentials. After repolarization to −50 mV, membrane enters brief phase known as afterdepolarization. Early part of this phase is relative refractory period; latter part is supernormal period. Succeeding this phase is period of subnormal excitability known as afterhyperpolarization. Differences in excitability during afterpotentials may be related to increase in gK+ and cell membrane. **F,** Hyperpolarization of membrane. If active electrode is positive, positive ions will flow to adjacent areas. This additional positive charge on outside of membrane increases potential difference.

sodium-potassium pump can restore the resting voltage and concentration gradients.

The above steps summarize the events that lead to the production of an action potential. These stages are diagrammed in Fig. 3-15. There are several points of interest. First, once the critical firing level (threshold) has been reached, there is no need for any further stimulation because the membrane will undergo the changes in conductance necessary for the reversal of polarity. Second, the time course for the AP will vary depending on the type of cell (neuron or muscle) and the type of animal. Finally, the values used for the RMP, CFL, and peak of the AP will also vary depending on the cell and animal studied.

During the time span of the AP the ability of the membrane to be excited by another stimulus and produce a second AP will vary. From the initiation of the AP at the CFL to about 1 msec after the peak of the AP, the membrane cannot be reexcited no matter how large a stimulus is applied. This period is known as the *absolute refractory period* and is related to the decline in gNa^+. In other words, once the membrane has produced an AP and reversed its polarity, it must recharge itself before a second AP can be generated. During this recharging Na^+ ions are kept from entering the cell, while K^+ ions leave the cell.

When the repolarization of the membrane exceeds -50 mV and the decline of gNa^+ is over, it is possible to reexcite the same area of the membrane if a stronger than threshold stimulus is used. This period is known as the *relative refractory period* and may last for several milliseconds. It is related to the sodium system, which has not fully recovered from the inactivation that followed the first AP, thus requiring a stronger stimulus to be activated, and to the increase in gK^+.

The relative refractory period is actually the first part of a phase of membrane excita-bility known as the *afterpotential period*. Immediately following the relative refractory period is the first of these afterpotentials, the *afterdepolarization*. This is a brief period in which the membrane is more excitable than in the resting state. During this excitable phase the membrane is slightly more positive than in the resting state. Possibly the sodium system, having been recently triggered and recovered, is in a temporary heightened state of readiness.

Following the so-called supernormal period the membrane may enter a period of subnormal excitability. During this period, which is called the *afterhyperpolarization*, the membrane is more negative than the RMP. This subnormal period may be related to excess K^+ ions leaving the cell during the increased gK^+. It is important to note that the increase in gK^+ starts later and lasts longer than the increase in gNa^+ (Fig. 3-15, *E*). The time course of these afterpotentials will vary depending on the type of neuron.

The conduction of the AP along the cell membrane is *self-propagating*. Once one action potential has been initiated, whether from synaptic activity, a receptor, neuromuscular transmission, or artificial means, the peak of the AP becomes the stimulus to initiate another AP in the adjacent area. In other words, the reversal in polarity to $+30$ mV is strong enough to cause positive ions from adjacent areas to flow into the void left by the influx of Na^+ ions into the cell. This local current flow then depolarizes the adjacent areas to -50 mV (CFL), and an entire AP develops that then becomes the stimulus for the next area of the membrane.

The AP is conducted along the axon in a wavelike fashion, analogous to the ripple or wave seen to travel along a length of rope when one end of the rope is quickly flipped up and down. Fig. 3-16 shows the propagation of the AP in a myelinated and an unmyelinated axon. Notice that in unmyelin-

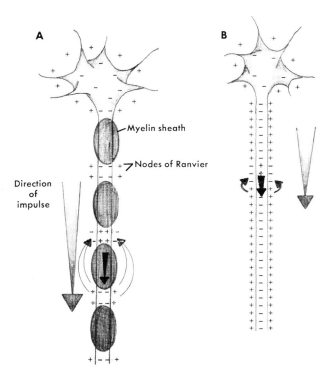

Fig. 3-16. Propagation of action potential in myelinated (**A**) and unmyelinated (**B**) axons. Note that AP can "jump" from node to node in myelinated axons, while in unmyelinated axons each segment must be involved in impulse. *Arrows* inside and around axon show movement of Na$^+$, K$^+$, Cl$^-$, and A$^-$ ions involved in electrical events. Of interest, AP is not electrical current such as passes through electric wire. Rather, it is sequence of ion exchanges that alters electric field established by momentary imbalance of ions. (Adapted from Noback, C., and Demarest, R.: The human nervous system, ed. 2, New York, 1975, McGraw-Hill Book Co.)

ated fibers every area of the fiber is involved in conducting the AP. In myelinated fibers only the nodes of Ranvier are involved in the AP propagation. The AP thus "jumps" from one node to the next. This type of conduction is known as saltatory conduction and is very effective in speeding up the velocity of conduction. The distance between each node will vary from 0.5 to 1.5 mm. The larger the diameter of a fiber the more myelin there will be. The more myelin a fiber has the farther apart the nodes of Ranvier will be, and the velocity of conduction will be faster.

The number of ions that diffuse across the membrane to produce an AP is relatively small. In fact, in a $1 \times 1/1000$ μm area of the membrane (1 μm equals 0.001 mm or 0.00004 in), eight Na$^+$ ions enter the cell and six K$^+$ ions leave the cell. The excess Na$^+$ ions in the cell account for the membrane potential of +30 mV. Considering that the number of ions in the cell and the extracellular fluid in 1 μm^2 totals 440,000, the shift of 14 ions for the AP is indeed small. Normally the sodium-potassium pump maintains the proper balance of Na$^+$ and K$^+$. Even if this system is blocked an axon can be stimulated thousands of times before the cell is rendered inexcitable.

The role of the neuroglial cells in the electrical activity of neurons is not definite. Neuroglia do have an RMP, but their membranes do not develop action potentials if depolarized. This may result because neuroglial membranes are only permeable to small ions such as K^+ and Cl^-. A possible role of neuroglial cells may be to act as a drain for excess K^+ in the extracellular fluid. In the event of prolonged neural activity the K^+ ions ejected from the cell to repolarize the membrane could be absorbed by the surrounding neuroglial cells, thus the proper extracellular concentration of K^+.

Several characteristics of an AP distinguish it from other electrical events associated with the nervous system. These characteristics are as follows:

1. Action potentials require a minimal length of stimulus intensity and duration. A single subthreshold stimulus will not produce an AP, nor will a suprathreshold stimulus of very brief duration. In other words, the membrane must be depolarized to a specific level for a specific time before an AP results.

2. An AP is either produced or not produced. This is the "all or none" response. Once the CFL has been reached the entire AP results.

3. For a given excitable cell the amplitude and time course of the AP are constant.

4. The absolute refractory period limits the maximum refractory of stimulation.

5. An AP is self-propagating and does not decrease in amplitude with distance.

6. An AP produced in the body is unidirectional, since the absolute refractory period prevents the reexcitement of the previously excited area. Artificial stimulation will produce two action potentials from the point of stimulation. One will travel peripherally, the other centrally.

7. Action potentials cannot be summated.

If in the original setup of Fig. 3-15 the anode were made the stimulating electrode, the area of the membrane under and around the anode would show an increase in voltage gradient. This is known as *hyperpolarization* and is caused by a buildup of positive ions on the outside of the membrane. These ions will be repelled by the positive electrode (anode) as the stimulus is turned on. A hyperpolarized membrane is less excitable and will not respond to a threshold stimulus (Fig. 3-15, *F*). This concept of hyperpolarization is an important component of synaptic activity (see Chapter 6).

Another important component of synaptic activity involves *subthreshold stimulation*. Normally a subthreshold stimulus will not produce an AP. However, if two subthreshold stimuli are applied very close together in time (within a few milliseconds) an AP can be produced. This can occur because the first subthreshold stimulus will cause some degree of local current flow and the second stimulus can cause local current flow that can be added to the first. The effect is that the CFL can be reached.

One final point about producing an AP concerns the phenomenon of *accommodation*, in which the membrane becomes very difficult to excite. Accommodation occurs from a prolonged subthreshold depolarization or a slowly rising stimulus. In prolonged subthreshold depolarization the sodium system becomes weakly active, as it does for any depolarization. It then follows the gNa^+ curve and becomes inactive and will remain so as long as the stimulus lasts. A second superimposed stimulus may not be able to cause an AP even if above threshold in strength. With a slowly rising stimulus the peak in current strength occurs as the sodium system is becoming inactive. Normal stimulation occurs in a square wave pattern, in which the peak of strength occurs immediately after the stimulus is turned on. A slowly rising stimulus has the appearance of a sawtooth blade.

A few clinical points are of interest. The first concerns calcium (Ca^{++}) ions. These ions are responsible for the stability of the cell membrane. They are located along the outside of the membrane and act to help keep Na^+ ions from leaking inside. When the level of extracellular Ca^{++} is low, enough Na^+ ions can leak in to depolarize the membrane and cause an AP. Early symptoms of low Ca^{++} levels include muscle cramping (tetany).

Second, some patients have altered levels of Na^+ and K^+ in their blood and thus in the extracellular fluid. For example, a severely burned patient will lose a lot of body fluid containing Na^+, K^+, and Cl^-. These electrolytes (and others lost) must be replaced and carefully monitored. If the extracellular concentration of K^+ is too low the membrane will become hyperpolarized because the K^+ ions in the cell will flow out (from concentration forces). If the extracellular concentration of Na^+ is low the amplitude of the AP will be smaller or the AP will be absent. Remember, it is the sudden influx of Na^+ ions that produces the AP spike. In both instances synaptic activity will be reduced. Early symptoms often include general fatigue or paralysis.

If there are excess extracellular K^+ ions the membrane will depolarize as there will be a flow of K^+ ions into the cell (from electrical forces). This will reduce the RMP and may reduce the amplitude of the AP or prevent it from developing. Remember, the influx of Na^+ is based on concentration *and* voltage forces. This can also produce general fatigue or paralysis. Excess extracellular Na^+ does not seem to be a complication clinically.

Finally, local anesthetics such as procaine hydrochloride (Novocain) block the flow of ions across the membrane, thus preventing the development of an AP. An injection of such a compound to a portion of a nerve trunk can block pain impulses from distal receptors.

TEST QUESTIONS

1. A cross-section of the median nerve reveals the following type(s) of fibers:
   a. General visceral afferents.
   b. General proprioceptors.
   c. General somatic efferents.
   d. General somatic afferents.
2. Nissl bodies:
   a. Are composed entirely of endoplasmic reticulum.
   b. Are involved with axoplasmic flow.
   c. Assist in the production of new cytoplasm after injury.
   d. Are the storage place for neurotransmitters.
3. Boutons:
   a. Are found in the terminal branches of the axon.
   b. Are actually the "spines" of a dendrite.
   c. Are intracellular components of the soma.
   d. Are structures strictly associated with neurons.
4. The myelin sheath formed by the Schwann cell:
   a. Is composed of Schwann cell membrane.
   b. Leaves no part of the fiber exposed to the extracellular fluid.
   c. Is wrapped around the fiber in concentric layers.
   d. Is thicker in larger fibers.
5. The connective tissue coverings of peripheral nerves:
   a. Include the perineurium, endoneurium, and neurilemma.
   b. Provide tunnels for regenerating fibers.
   c. Allow for the suturing of severed nerve trunks.
   d. Regulate the ionic balance of the extracellular fluid.
6. Wallerian degeneration involves:
   a. Dispersion of the Nissl bodies and chromatolysis.
   b. Fragmentation of the myelin sheath.
   c. Loss of the axon or dendrite cylinder.
   d. Loss of the Schwann cells.
7. The nodes of Ranvier are an area of nerve fibers where:
   a. Capillaries penetrate the fiber.
   b. There is no myelin sheath.
   c. Fiber is in contact with the extracellular fluid.

d. There is no endoneurium.

8. The separation of charges by a high-resistance membrane will prevent the production of:
   a. Voltage.
   b. Current.
   c. Potential difference.
   d. Equilibrium.

9. The critical first event in natural impulse production is:
   a. Decreased Na^+ permeability.
   b. Activation of the sodium pump.
   c. Membrane depolarization from local current flow.
   d. A sudden increase in gNa^+.

10. The membrane will become depolarized if:
    a. Concentration of Na^+ outside is lowered.
    b. Concentration of K^+ outside is lowered.
    c. Anodal current is applied.
    d. K^+ gates are opened.

11. The amplitude of an AP can be increased by:
    a. Increasing Na^+ outside.
    b. Using an anodal current.
    c. Using a stronger current.
    d. Using two threshold stimuli successively.

12. The size of the resting membrane potential depends on:
    a. Concentration of ions.
    b. Ability of K^+, Na^+, and Cl^- to diffuse across the membrane.
    c. Equilibrium potential for each ion.
    d. Type of cell studied.

13. The sodium pump:
    a. Keeps Na^+ in.
    b. Keeps K^+ out.
    c. Maintains the RMP.
    d. Is present in all cells.

14. A nerve fiber will be less excitable if:
    a. Concentration of Ca^{++} is increased on the outside.
    b. Procaine hydrochloride or other local anesthetic is administered.
    c. Concentration of K^+ is decreased outside.
    d. Prolonged subthreshold stimulus is applied.

15. In saltatory conduction an AP at one node:
    a. Causes local flow of positive ions from the adjacent node.
    b. Depolarizes the adjacent node to threshold.
    c. Produces another AP of equal height at the adjacent node.
    d. Flows in a wave pattern to the next node.

16. Which of the following is false about an action potential?
    a. It can be summated by using two subthreshold stimuli.
    b. It normally travels in one direction along a fiber.
    c. It is propagated without a decrease in amplitude.
    d. It is self-generating once CFL is reached.

17. The increase in permeability to K^+ during the AP:
    a. Occurs simultaneously with the increase in gNa^+.
    b. Helps repolarize the membrane.
    c. Allows K^+ ions to leave the cell.
    d. Is the cause of the absolute refractory period.

18. An increase in extracellular concentration of K^+:
    a. Will occur if the cell is stimulated and the sodium-potassium pump is blocked.
    b. Will hyperpolarize the membrane.
    c. Can be absorbed by certain neuroglial cells.
    d. Can cause muscle weakness.

SUGGESTED READINGS

Bishop, B.: Neurophysiology study guide, ed. 2, Flushing, N.Y., 1973, Medical Examination Publishing Co.

Ezyaguirre, C., and Fidone, S.: Physiology of the nervous system, Chicago, 1975, Year Book Medical Publishers, Inc.

Noback, C., and Demarest, R.: The human nervous system, ed. 2, New York, 1975, McGraw-Hill Book Co.

Schmidt, R, editor: Fundamentals of neurophysiology, New York, 1975, Springer-Verlag.

Willis, W. D., Jr., and Grossman, R. G.: Medical neurobiology: neuroanatomical and neurophysiological principles basic to clinical neuroscience, ed. 2, St. Louis, 1977, The C. V. Mosby Co.

CHAPTER 4

Receptors and nature of their activity

The central nervous system receives a continuous barrage of impulses from structures known as sensory receptors. These receptors are found in all areas of the body, such as the skin, fascia, muscles, tendons, joints, blood vessels, visceral organs, and the structures of the special senses. Receptors act as transducers to convert chemical, thermal, mechanical, or photic forms of energy into action potentials, which then travel along sensory fibers to the central nervous system. By determining the frequency of action potentials, the type of receptor firing, and the location of the receptor the CNS can then recognize, interpret, and make value judgments as to the nature of the stimulus. Such information can then be used to influence thought processes, motor output, and all other body systems.

The following areas of this chapter deserve special attention:

1. Functional, anatomic, and structural classifications of the types of receptors
2. Concepts of receptor field and receptor specificity as they relate to sensory discrimination
3. Steps involved in the production of a sensory action potential
4. Effects of stimulus strength, duration, and velocity and adaptation on the frequency of action potentials produced
5. Location of the various cutaneous receptors and their adequate stimulus
6. Structure of the muscle spindle
7. Influence of the muscle spindle, Golgi tendon organ, and joint receptors on muscle tone and movement
8. Effects of gamma stimulation on the muscle spindle and subsequent discharge of the Ia and II fibers into the spinal cord
9. CNS connections of the Ia and II fibers from the spindle, the Ib fibers, and the II, III, and IV fibers from the joint receptors
10. Role of the muscle spindle, GTO, and joint receptors in the maintenance of posture and in coordinated movement
11. Effects of length and tension servomechanisms and alpha-gamma coactivation on the alpha motoneurons
12. Adequate stimuli of the visceral receptors
13. Steps involved in the production of sensations from the special sense organs, including location of the receptor, adequate stimuli needed for activation, and course of afferent fibers
14. Influence of the vestibular system on posture and equilibrium

CLASSIFICATION OF RECEPTORS
Functional classification

Receptors can be classified in several ways, one of which is the method used by Sherrington. He divided receptors into (1) *exteroceptors*, which relay data from the external environment, (2) *interoceptors*, which relay data from the viscera, and (3) *proprio-ceptors*, which relay data from the labyrinthine mechanism, muscles, and tendons. Another method based somewhat on a developmental scale divides receptors into those that produce *protopathic* sensations and those that produce *epicritic* sensations. The former are poorly localized sensations such as crude touch, crude temperature, and awareness of pain. These sensations are considered to be phylogenetically old and are used for protective reactions from potentially harmful stimuli. The latter involve an ability to localize and finely discriminate sensations such as touch, vibration, pain, temperature, and joint position. This ability is phylogenetically new and enables the organism to explore and learn from the environment.

Anatomic classification

Receptors can also be classified according to their general anatomic location: (1) cutaneous receptors, which are found in the superficial and deep layers of the skin, measure pain, touch, pressure, vibration, and temperature; (2) muscle, tendon, and joint receptors, which include the muscle spindle and Golgi tendon organs, measure muscle length, muscle tension, joint angle, and deep pain; (3) visceral receptors basically measure pressure (distention) and pain from the internal organs and blood vessels; and (4) receptors of the special senses including vision, hearing, taste, olfaction, and vestibular sense.

Structural classification

Receptors can be further classified according to their structural design, which allows them to respond to a particular kind of stimulus: (1) mechanical receptors respond to a deformation of the receptor ending or to the area around the receptor and measure touch, pressure, vibration, muscle length and tension, hearing, vestibular sense, and distention of an organ or vessel; (2) thermal receptors respond to changes in temperature gra-

dients from their normal surrounding temperature to give sensations of cold and warm; (3) chemical receptors respond to dissolved chemical substances to produce the sensations of gustation (taste) and olfaction and evaluate the carbon dioxide level of the blood; and (4) photic receptors, which are the rods and cones of the retina, respond to a particular wavelength of radient energy within the visible spectrum.

It is important to mention that the sense of pain is difficult to place in just one of these categories because extreme intensities of mechanical, thermal, chemical, or photic energy can cause pain to be experienced. The sense of pain is very complex, as it is difficult to quantify. Pain is discussed further in Chapter 8.

GENERAL PROPERTIES OF RECEPTORS

Receptor fields and two-point discrimination

The peripheral process of a unipolar or bipolar neuron usually terminates by branching into a varying number of processes. These branches are in a sense naked, since they lack the connective tissue and neurilemmal coverings of the parent fiber. These bare endings may themselves become receptors, they may become encapsulated with specialized connective tissue, or they may become associated with specialized receptor cells. The area supplied by the terminal branches (receptors) of one neuron is known as the *receptor field*. Receptor fields vary in size and density within a given area, and several fields may overlap. The concept of receptor fields is important to the understanding of sensory discrimination. If two pins are held about 4 to 5 mm apart and placed on different areas of the body, they will be felt as one or two points, depending on the size and density of the receptive fields (Fig. 4-1). The skin of the lips, fingers, palms, and soles has small receptor fields in close

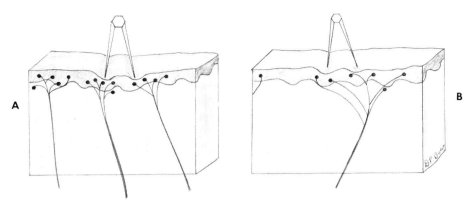

Fig. 4-1. Receptive fields and two-point discrimination. **A,** Schema of receptive fields in skin of fingertips. Note that area of skin supplied by single afferent fiber is small and that there is high density of receptor fields. There is greater chance that two points held close together will activate two receptor fields. **B,** Schema of receptive fields in skin of thigh. These receptor fields are large and farther apart and much less useful in tactual discrimination.

proximity to each other. Thus two points can be detected. Other areas such as the skin of the arm, thigh, or back have large receptor fields that are farther apart (less density). Two points placed 5 mm apart on these areas will be felt as only one point, since only one receptor field will be excited. Those areas of the body that are able to discriminate two points close together are essential to exploring and learning about the environment. For example, have a friend close his or her eyes. Take a paper clip and place it on the forearm. Move it around on the skin and ask the person to identify the object. Now place the paper clip between the person's thumb and index finger and allow a few seconds for manipulation. Identification will be quick and easy.

Receptor potentials

A given receptor will generally respond to only one type of stimulus. For example, the touch receptor around a hair follicle will re-

spond only to mechanical movement of the hair. Vibration or heat will not excite that receptor. This receptor specificity is known as *adequate stimulus.* It implies that only one stimulus is adequate to activate a particular receptor. Adequate stimulus also means that stimulus to which a receptor is most sensitive. A number of factors may contribute to receptor specificity, including the position of the receptor in the body, the cellular organization of the receptor, and the sensitivity of the receptor membrane.

The production of the *receptor (generator) potential* is easy to comprehend. The beginning of receptor activity involves either the terminal portion of the afferent fiber or a specialized receptor cell such as the hair cells associated with hearing in the cochlear duct. In either case there will be a resting membrane potential of about -70 mV. When a stimulus is applied the membrane of the receptor cell or afferent fiber will depolarize. In other words, the stimulus causes the membrane to alter its permeability, allowing ions such as sodium to enter (Fig. 4-2). This depolarization is known as a receptor potential (RP). It is different from an action potential (AP) in the following ways:

1. The amplitude of the RP will vary with the strength of the stimulus and rate of stimulation. A 1 lb weight placed on your hand will cause less receptor activity than a 2 lb weight dropped from 10 ft. An AP is all or none.

2. The RP does not propagate; it spreads passively with decrement for 1 or 2 mm along the membrane.

3. There are usually no refractory periods. In many receptors the RP will last as long as the stimulus is applied.

4. The RP can be summated. The RPs from two stimuli applied successively can be added together.

Obviously the membrane of the terminal portion of an afferent fiber is quite different from the rest of the fiber, as no action potential is produced here. Neither is one pro-

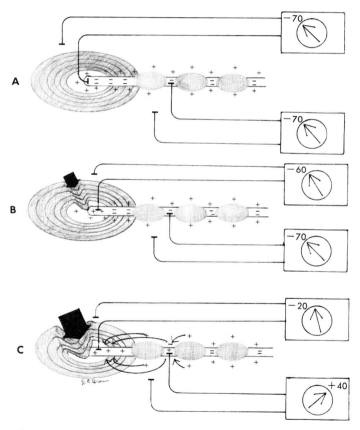

Fig. 4-2. Production of receptor potential and sensory action potential. **A,** Voltmeter reading from terminal portion of afferent fiber and adjacent node of Ranvier. **B,** Weak stimulus depolarizes terminal portion to −60 mV (receptor potential of 10 mV). This is insufficient to generate action potential. **C,** Stronger stimulus depolarizes terminal portion to −20 mV. This causes sufficient local current flow to generate action potential at node of Ranvier.

duced in a receptor cell. What happens is that the receptor potentials they generate attract positive ions from the immediate extracellular area of the afferent fiber. In the case of a myelinated fiber this might be at the first node of Ranvier. As with other neural currents, there is also an intracellular flow of positive ions to depolarize the adjacent area of the fiber to its critical firing level. An AP is produced and then propagated to the CNS (Fig. 4-2). With a weak stimulus there can be a receptor potential but no AP. With a strong but brief stimulus several APs may be produced. With a prolonged stimulus a maintained frequency of APs may be produced

(Fig. 4-3). In other words, the CNS can determine (1) the type of stimulus from the receptors that are responding to that form of energy, (2) the location of the stimulus by the area of the body being activated, and (3) the strength and velocity of the stimulus by the frequency of APs.

Receptor adaptation

In some receptors a maintained stimulus will cause a reduction in receptor sensitivity. This causes a lower frequency of APs as a result of a reduced receptor potential. This process is known as receptor adaptation and *may* be caused by changes in the mechanical

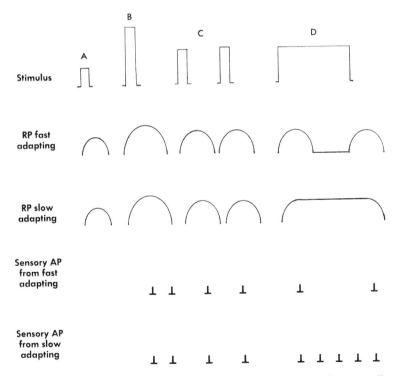

Fig. 4-3. Receptor activity from different stimuli. **A,** Weak stimulus produces small RP and no sensory APs. **B,** Strong stimulus produces big RP and several sensory APs. **C,** Two stimuli in succession produce separate RPs and several sensory APs. **D,** Prolonged stimulus produces RPs at "make" and "break" in fast-adapting receptors and prolonged RP in slow-adapting receptors. Note difference in sensory APs.

properties of the tissue about the sensory ending or a refractory period of the receptor membrane. Other possibilities are accommodation of the afferent fiber or of the synapse within the CNS.

Those receptors that produce a maintained level of APs to a prolonged stimulus are known as slowly adapting or tonic receptors. These include some touch receptors, temperature receptors, muscle spindle, vestibular receptors, pain receptors, position sense (joint) receptors, and receptors in the lung and carotid sinus. Some of these receptors do show some decline in AP production such as when one goes swimming in unheated water. Initially the water may be very cold, but within a few minutes there is some receptor adaptation. Other receptors such as the muscle spindle show a maintained output as long as there is a stretch on the muscle.

Those receptors that show a quick reduction or cessation of AP production are known as fast-adapting or phasic receptors. These include some touch receptors, pressure receptors, and some labyrinthine and joint receptors. While tonic receptors provide continuous data about the environment, phasic receptors respond very quickly to changes in the environment. For example, when you put a heavy coat on you feel its weight, but you soon are unaware of it until the coat is removed. Working together, tonic and phasic receptors provide a background level of sensory data that can indicate changes above and below this level as well as the rate of change. This background activity of electrical input to the CNS is also probably important in maintaining an active level of cortical arousal.

STRUCTURE AND FUNCTION OF CUTANEOUS RECEPTORS

Cutaneous receptors are the terminal branches of an afferent fiber and may or may not be encapsulated by organized connective tissue capsules. Keep in mind that the area of skin supplied by one afferent fiber is a receptive field and that one sensory neuron and all of the terminal branches make up a *sensory unit*. Fig. 4-4 is a composite diagram of the cutaneous receptors.

Free nerve endings

Free nerve endings are the naked axon cylinders that branch from unmyelinated and lightly myelinated fibers. They are found in the dermis and epidermis of the skin, cornea of the eye, mucous membranes, pulp of the tooth, intermuscular connective tissue, and the viscera. They are sensitive to touch, pain, or temperature, depending on the chemical nature of the membrane of the terminal branches. In first- and second-degree burns of the skin these receptors become highly sensitive to any stimulus, and they may even be exposed to the surface. Thus the weight of bed sheets or placement in a whirlpool can be quite painful to some patients with burns. In third-degree burn (destruction of the dermis) the receptor endings are lost, and there is much less sensitivity to pain.

Hair follicle endings

The base of each hair in the dermis is innervated by lightly myelinated fibers that end in a plexus in the connective tissue about the neck of the hair follicle. Thus each hair acts as a sense organ that is very sensitive to mechanical movement (touch). Pain afferents also end about the hair follicle.

Ruffini endings

Ruffini endings are encapsulated whorls of fine terminal branches of myelinated fibers. These receptors are found in the dermis and in joints. They were thought to be termperature receptors but are now generally believed to be touch receptors.

Merkel's discs (touch corpuscles)

These receptors are actually specialized cells in the skin called Merkel's cells found at the bottom of the epidermis. These cells are

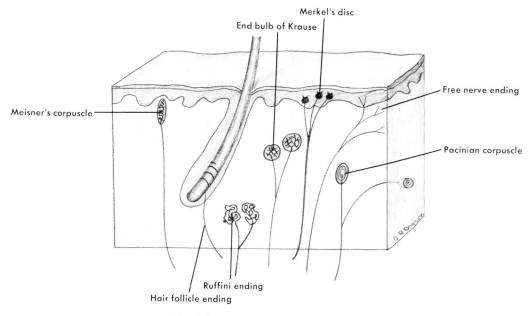

Fig. 4-4. Cutaneous receptors of skin.

found in hairy and glabrous skin and innervated by branches of a large myelinated fiber. The epidermis above the Merkel cells supplied by one sensory fiber is slightly thickened and rounded, giving the receptor field a discrete anatomic area with a rough diameter of several hundred micrometers (0.008 in). The touch corpuscle is sensitive to low-intensity touch as well as to the velocity of touch. Because of this and its discrete area these receptors are important in localizing touch on the skin and in two-point discrimination.

Krause's end-bulb

In these receptors the bare terminations of a myelinated fiber are surrounded by a round capsule (bulb) of connective tissue. These receptors probably detect cold sensations, but they are not the only ones that do, since cold can be felt in areas of skin void of these receptors.

Meissner's corpuscles

Meissner's corpuscles are encapsulated endings found in the dermis of the palms and soles. They are especially numerous in the toes and fingertips. Meissner's corpuscles are supplied by large myelinated fibers. They are very sensitive to touch. They undoubtedly subserve discriminative touch sensations. As many as 20 to 30 corpuscles may be concentrated in a square millimeter.

Pacinian corpuscle

This receptor was first described by Vater and Pacini, and because of its size (up to several millimeters in length) it has been a frequent subject of receptor physiology studies. The corpuscle is usually ovoid and composed of concentric layers of connective tissue (like an onion). These layers are loosely organized around the bare termination of a single large myelinated fiber. The loose arrangement allows for plenty of extracellular fluid to bathe the nerve fiber. Pacinian corpuscles are found in subcutaneous tissue, fascia around joints and tendons, and in the mesentery. They respond to pressure (deep touch) and to vibration. They are rapidly adapting, responding to the onset and termination of mechanical pressure. It is presumed that these receptors play an important role in our sensory environment. It has been found that the stimulation of a single corpuscle can excite an area of the postcentral gyrus.

STRUCTURE AND ORGANIZATION OF MUSCLE, TENDON, AND JOINT RECEPTORS

The information produced by muscle, tendon, and joint receptors is primarily concerned with posture, muscle tone, the awareness of the body's position in space, and the coordination of movement in terms of the force and velocity of contraction.

These receptors are often grouped together and called the proprioceptors. The sense of proprioception is used by the CNS for all aspects of conscious (willed) and unconscious movements. A distinction should be made between conscious and unconscious proprioception. Conscious proprioception is primarily concerned with joint position sense. Data traveling to the parietal lobe for conscious awareness is sometimes referred to as kinesthesia. Unconscious proprioception is primarily concerned with muscle length, tension, and velocity of movement. These data travel to the cerebellum. Although it may alter the output of the motor cortex, this information does not reach consciousness.

Pacinian corpuscles

The pacinian corpuscles in muscle are the same as those in the skin. They are found in

the facia within a muscle. They respond to deep pressure and vibration and signal this information to the cerebrum.

Pressure-pain endings

Pressure-pain endings are free nerve endings located in the fascia within a muscle. They respond to both mechanical and noxious stimuli and thus relay pressure and muscle pain to the brain. The pain of a muscle cramp or fatigue is detected by these endings.

Muscle spindle

Muscle spindles are among the most complex and widely researched receptors in the body. They are located between the individual muscle fibers. Muscle fibers are usually long, thin, cylindrical cells that often run most of the length of the muscle. These cells are known as *extrafusal fibers*. They usually vary between 0.01 to 0.1 mm in diameter and from 1 to 40 mm (rarely 120 mm) in length. Lying parallel to the extrafusal fibers are a group of shorter, thinner fibers known as *intrafusal fibers*. The intrafusal fibers can vary from a few millimeters to over 10 mm in length and are attached at either end to the connective tissue sheath that surrounds the extrafusal fiber. The intrafusal fibers are thus "parallel" to and attached to the muscle fibers. Any change in length of the muscle will also alter the length of the intrafusal fibers. A change in length as well as the rate of change are the adequate stimuli for this receptor.

Surrounding the middle third of the intrafusal fibers is a fluid-filled connective tissue sac that is about 100 μm thick. Penetrating the sac and innervating the intrafusal fibers are two kinds of sensory fibers and one kind of motor fiber. The sac, intrafusal fibers, afferent fibers, and efferent fibers are collectively known as the *muscle spindle* (Fig. 4-5). Since the shape of this receptor is long, thin, and slightly wider in the middle, it is often

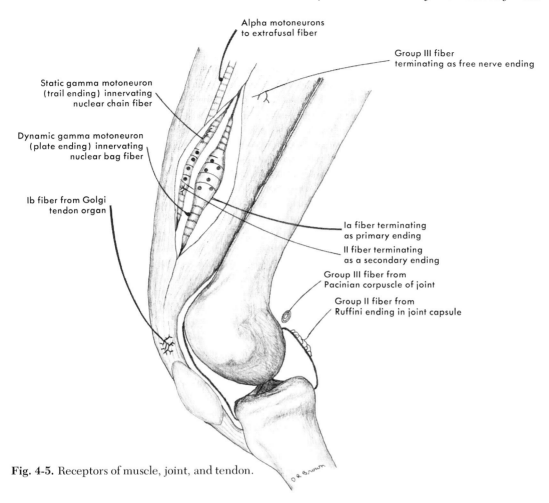

Alpha motoneurons
to extrafusal fiber

Group III fiber
terminating as free nerve ending

Static gamma motoneuron
(trail ending) innervating
nuclear chain fiber

Dynamic gamma motoneuron
(plate ending) innervating
nuclear bag fiber

Ib fiber from Golgi
tendon organ

Ia fiber terminating
as primary ending

II fiber terminating
as a secondary ending

Group III fiber from
Pacinian corpuscle of joint

Group II fiber from
Ruffini ending in joint capsule

Fig. 4-5. Receptors of muscle, joint, and tendon.

described as being fusiform. In general a muscle spindle contains each of these components, although some variations exist.

Except for some muscles innervated by cranial nerves, muscle spindles have been found in every striated muscle in every species of mammal studied. The muscles in which spindles have not been found in some species are the extraocular, facial, and some of the throat muscles, although in humans muscle spindles have been found in the extraocular and facial muscles. In fact, the number of

spindle-free muscles found is decreasing with more refined techniques.

What is interesting about the distribution of muscle spindles is not the number per muscle but the density (number of spindles per weight of the muscle). The muscles with the highest density of spindles are those such as the small muscles in the neck and hand that are involved in fine, highly skilled movements. For example, the abductor pollicis brevis contain 80 spindles and weighs 2.6 g. The density of spindles per gram is

29.3. The triceps brachii, on the other hand, contains 520 spindles but weighs 304 g. Its density is only 1.4 spindles per gram.

The individual components of the spindle reveal some fascinating differences. First to be considered are the *intrafusal fibers.* There are generally from two to 12 intrafusal fibers per spindle. Each intrafusal fiber has an arrangement of nuclei in the central third and the contractile proteins actin and myosin in each outer third. Each "polar" region appears striated (like the extrafusal fibers) and is capable of shortening (contracting). The "equatorial" region is mostly nonstriated.

Other histologic differences between the polar and equatorial regions influence the mechanical properties. The polar regions are more viscous, while the equatorial regions are more elastic. The significance of this difference is discussed briefly.

Two kinds of intrafusal fibers can be distinguished anatomically and functionally. Some intrafusal fibers are longer than the others, and the nuclei in the equatorial region are stacked two and three abreast, giving a slight bulge to this area. These fibers are called the *nuclear bag fibers,* although there is no "bag" around the nuclei other than the cell membrane. The other intrafusal fibers are called *nuclear chain fibers.* These are shorter and thinner, and their nuclei lie single file in the equatorial region. Functionally they differ in their response to stretch. The bag fibers respond not only to a change in length but also to the velocity of stretch. This difference results partly because the center region of the bag has fewer contractile proteins than the center region of the chain. This allows the center region of the bag to be less viscous (more elastic) and to show greater distortion to stretch. Other factors such as the position of the afferent endings and the contractile properties of the different polar regions also contribute to the velocity sensitivity of the bag fiber.

As a unit the muscle spindle is actually two receptors in one and has two afferent endings for this purpose. One ending is called the *primary ending.* It is the termination of a large myelinated fiber. Each spindle has only one of these fibers, but each intrafusal fiber receives a terminal branch (primary ending). This ending wraps itself around the equatorial region of the bag and chain fibers in a spiral fashion. This ending was formally called the annulospiral ending. The other ending is called the *secondary ending.* It terminates by branching or spiraling on one side or the other of the equatorial region, predominately on the chain fiber. This ending was formally called the flower-spray ending and is supplied by a medium sized afferent fiber. There may be from 0 to 5 secondary endings per spindle. As a rule the total number of primary endings for a given muscle is close to the total number of secondary endings.

Both the primary and secondary endings respond to mechanical deformation of the underlying membrane of the intrafusal fiber. Since the muscle spindle lies parallel to the extrafusal fiber and is attached to it, any stretch of the muscle will stretch the intrafusal fibers and activate the receptor endings. When either the bag or chain fiber is stretched the center region, with little contractile proteins, will show greater deformity as a result of a lesser viscosity than in the polar regions. The change in shape of the equatorial regions of the bag and chain causes the primary and secondary endings to depolarize. Since the primary ending is directly around the center region, it will show a greater sensitivity to stretch than the secondary ending, which is juxtaequatorial in position.

To summarize the above:

1. The center region of the bag is more elastic. It is distorted by even small stretches and shows proportionally greater distortion to equal stretches of different velocities.

2. The center region of the chain is less elastic and has a more linear distortion to the rate of stretch.

3. The secondary ending is located over or

very near the polar region. This region is more viscous and shows less distortion to stretch.

To help explain the above, assume the following position. Hold one arm at your side with the elbow flexed to 90°. The primary and secondary endings of the elbow flexor spindles will fire at a particular frequency related to the precise length of those muscles (all other skeletal muscles will do the same). Now slowly lower your arm to 180°. Both the primary and secondary endings will show an increased rate of firing in response to the new length. The primary endings (primarily from the bag), however, will show an excess activity initially in response to the velocity (Fig. 4-6). If the elbow is returned to 90° and quickly lowered to 180° there will be an even larger initial burst of activity from the primary endings. This type of stretch is often known as a "ramp" stretch because there is a steady increase in length. The faster the stretch the steeper the ramp (Fig. 4-6). At the end of the stretch a higher level of firing

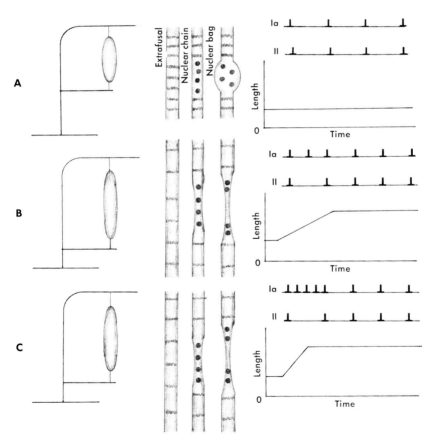

Fig. 4-6. Response of intrafusal fibers to ramp stretch. **A,** For any length of muscle, there will be resting discharge from primary (Ia) and secondary (II) endings. **B,** If muscle is stretched to new length and maintained there, there will be higher level of discharge on both Ia and II fibers and dynamic response on Ia fiber indicating velocity of stretch. **C,** With faster rate of stretch (steeper ramp) there will be bigger dynamic response on Ia fiber.

can be found. The initial excess activity seen during the stretch is known as the *dynamic response*. The secondary ending also shows a dynamic response only much smaller. For example, in a ramp stretch with a velocity of 10 mm/sec the dynamic response of the primary ending can range from 40 to 120 impulses per second, while that for the secondary ending can range from 5 to 21 impulses per second. This ten times greater dynamic (velocity) sensitivity allows the nuclear bag—primary ending to be sensitive to small-amplitude stretch as in a tendon tap and to sinosoidal stretches as from vibration (Fig. 4-7).

The muscle spindle is one of the few recep-

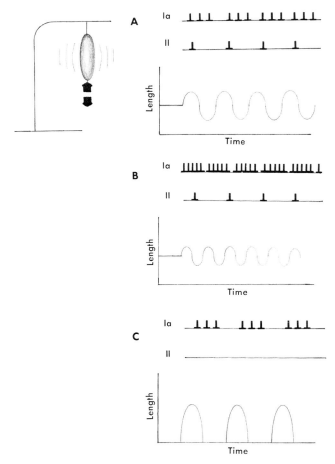

Fig. 4-7. Response from intrafusal fibers to sinusoidal and small-amplitude stretches. **A,** Applying sinusoidal stretch to muscle (like pulling light chain) will produce bursts of activity on Ia fiber related to amplitude and rate of stretches. II fiber will show steady response related to average length. **B,** Increasing frequency of stretches will cause bigger dynamic from Ia fiber but little change from II fiber. **C,** Applying small-amplitude stretches (like tendon tap) will cause bursts of activity on Ia fiber related to amplitude and velocity of stretch. Group II fiber activity will depend on original length. In above examples amount of static gamma activity will influence frequency of action potentials on Ia and II fibers.

tors receiving motor innervation. The motor supply to the intrafusal fibers is primarily derived from thinly myelinated fibers known as *gamma (fusimotor) fibers* (Figs. 4-5 and 4-8). The cell bodies of these motoneurons are found in the ventral horn of the spinal cord and some cranial nerve nuclei along with the alpha (extrafusal) motoneuron. Two types of gamma motoneurons have been found that differ in the way they terminate and function. One type of gamma fiber terminates as a discrete ending predominantly on the polar region of the nuclear bag fibers. These endings are similar to the discrete termination of the alpha motoneuron on the extrafusal muscle cell (the neuromuscular junction) and are referred to as "plate" endings. These gamma neurons are known as the gamma$_1$ fiber. The other type of gamma fiber terminates as a diffuse multibranching ending predominantly on the nuclear chain fiber adjacent to the equatorial region. These endings are referred to as "trail" endings and the motoneuron as the gamma$_2$ fiber.

Some researchers have found plate endings on the chain fiber and trial endings on the bag fiber, but the significance of this is uncertain. It is generally accepted that in spite of this occurrence the bag and chain are innervated by separate motoneurons that can be independently operated by the CNS.

In some muscle spindles a third ending has been found. It is a discrete termination and can be called a plate ending. The terminal fiber is not from a gamma motoneuron but is a branch from an alpha motoneuron. This branch is commonly called a *beta fiber* and ends predominantly on the nuclear bag fiber. The plate endings from beta fibers are known as p$_1$ endings, while those from gamma fibers are known as p$_2$ endings.

There have been reports of sympathetic innervation to the muscle spindle. These motor fibers have been found to innervate only the intracapsular blood vessels and not the intrafusal fibers.

The effects of fusimotor stimulation are interesting. When either gamma fiber is activated the polar regions of the nuclear bag or nuclear chain fiber contract as the contractile proteins form their actinmyosin bonds. This shortening of the polar regions effectively lengthens (distorts) the equatorial regions and can cause the primary and secondary endings to fire (Fig. 4-8). Although the overall intrafusal fiber length does not change and no contractile tension develops in the muscle, gamma activation will cause the following:

1. It will raise the level of output from the spindle for any given length. This is known as *gamma biasing*.
2. When the gamma fibers to the nuclear bag (gamma plate) are activated there will be a marked increase in the dynamic response (sensitivity). For this reason gamma plate fibers are often called *dynamic gamma fibers*.
3. When the gamma fibers to the nuclear chain (gamma trail) are activated there is a decrease in the dynamic response, since the primary and secondary endings are firing at a higher base level. Gamma trail fibers are often called *static gamma fibers*.
4. Static gamma activation maintains a level of output from the spindle even when the muscle is shortened. Under conditions of no fusimotor activity the primary and secondary endings do respond to ramp and small-amplitude stretches but at a lower frequency. If the elbow is at 90° and flexed to 130° the output from the spindle will fall to zero. The spindle is said to be "slack" and in a "silent period." With activation of the static gamma fibers, however, this slack of the equatorial region can be taken up by the shortening polar regions (Fig. 4-9).

This different effect from static gamma and dynamic gamma stimulation can be explained

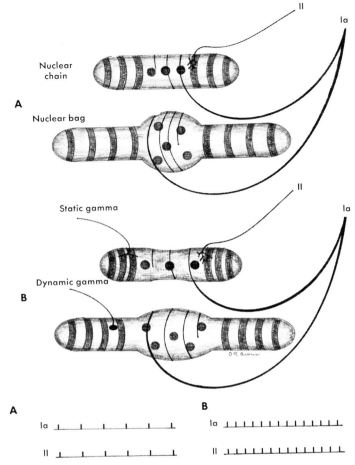

Fig. 4-8. Effects of fusimotor stimulation on intrafusal fibers. **A,** Recordings of activity on Ia and II fibers from muscle at maintained stretch with no gamma activity. **B,** Recordings of activity on Ia and II fibers from same muscle at same length with gamma stimulation.

partly on the basis of the contractile properties of the bag and chain fibers and partly on the propagation of the AP from the plate and trail endings. The nuclear chain fiber is capable of shortening (contracting) its polar regions faster than the bag fiber. The chain fiber shows a higher density of mitochondria (for ATP production) and a greater rate of ATPase activity. This allows for rapid forma-

tion and breaking of the actin-myosin bonds needed for shortening. The bag fiber has more myoglobin than the chain fibers, allowing for more sustained contraction. Thus the chain fiber contracts like a fast twitch (phasic) muscle fiber, and the bag fiber contracts like a slow twitch (tonic) muscle fiber. Static gamma activation causes a propagated AP to be quickly delivered to the entire polar re-

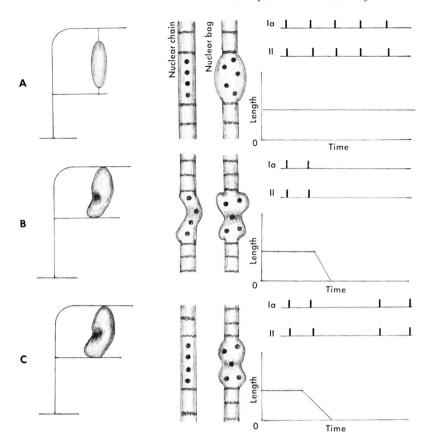

Fig. 4-9. Effects of fusimotor stimulation on "slack" spindle. **A,** Recordings from Ia and II fibers during maintained stretch of muscle. **B,** Placing muscle in shortened position (release of stretch) will cause silent period from spindle as result of lack of tension on equatorial regions of intrafusal fibers and no activity from gamma motoneurons. **C,** Adding static and dynamic gamma stimulation to polar regions will cause primary and secondary endings to fire. Note that dynamic gamma is much less effective in removing slack on bag fiber.

gion. Dynamic gamma activation causes a local nonpropagated response along the polar region. The overall effect is that when the polar regions of the nuclear bag are activated the polar regions maintain their bonds longer, thereby maintaining an increased polar viscosity. The equatorial region of the bag (and the primary ending) has a longer period of sensitivity to the velocity of a ramp stretch

(dynamic response) and to the displacement of even small sinosoidal stretches. Stimulation of dynamic gammas increases the dynamic sensitivity of the bag fiber by prolonging the period of polar viscosity. In the nuclear chain fiber activation of the polar regions by the static gammas causes the rapid formation of actin-myosin bonds, which are then quickly broken. Even with prolonged static gamma stimulation these bonds are not held long enough to raise the polar viscosity. Thus if a muscle is lengthened and both gammas are active, the equatorial region of the bag absorbs all of the stretch, while the equatorial *and* polar regions of the chain absorb the stretch. The primary ending of the bag experiences greater mechanical disturbance than the primary or secondary ending of the chain. Keep in mind that the faster the stretch the greater the response from the primary ending of the bag.

The activity of the gamma motoneurons is regulated by several sources. The static gammas are regulated by the supraspinal motor centers such as the reticular formation, vestibular nuclei, and the basal nuclei. The dynamic gammas are regulated by these supraspinal centers and also segmentally by cutaneous afferents. A patient with an injured spinal cord, lacking higher motor control, can still have an increased response to quick stretch (spasticity) because of excess dynamic gamma facilitation. In an intact nervous system the supraspinal centers function to activate the two gammas independently in a reciprocal manner (static gamma during a shortening contraction and dynamic gamma when a muscle is being lengthened by its antagonist).

The question arises as to the necessity of both a nuclear chain fiber and a static gamma motoneuron. The muscle spindle is a receptor that detects muscle length, and the nuclear bag–primary ending is capable of assessing both the amplitude and velocity of any stretch. Stimulation of the dynamic gamma increases this response. The justification for the nuclear chain fiber is that it provides a continuous output during the shortening of a muscle. When a muscle contracts and shortens the muscle spindle becomes slack. If there is no mechanical tension on the center region of the bag or chain, there will be no output from the primary or secondary ending. The CNS requires a continuous output

from the muscle spindle for the maintenance of muscle tone and posture and for the coordination of movement. If the static gammas are activated during the shortening, the quick formation of actin-myosin bonds can absorb the slack and keep the center region of the chain taut (Fig. 4-9). Thus both the primary and secondary endings can maintain a level of discharge.

The CNS therefore possesses two related but independent length receptors that can be very finely sensitized (biased) by two separate gamma motoneurons. One receptor (nuclear bag) responds best to quick changes in length, while the other is able to maintain a level of spindle output during any shortening of the muscle. The CNS can operate either system independently.

The afferent impulses from the primary and secondary endings are conveyed over two kinds of myelinated fibers. The primary endings are the terminal branches of moderately myelinated fibers known as Ia fibers. Sensory fibers are grouped according to diameter into four categories designated by Roman numerals I, II, III, and IV. Group I has two subgroups, Ia and Ib. As with all sensory fibers, the Ia and II afferents from the muscle spindle have their unipolar cell bodies in the dorsal root or cranial nerve ganglia. The only exceptions are those from the muscles of mastication (see Chapter 5). The central process from the dorsal root or cranial nerve ganglia enters the spinal cord or brainstem and give off several branches, which do the following:

1. They travel in a fiber tract to the cerebellum to provide continuous data as to the length of their muscle.
2. Ia fibers send a branch directly to an alpha motoneuron of the same muscle and powerfully excite those motoneurons (autogenic excitation).
3. Ia fibers send a branch to excite an internuncial, which in turn excites synergistic muscles. Activation of Ia fibers

from the quadriceps excites not only alpha motoneuorns to the quadriceps but also the soleus and adductors as well. This synergistic response is much smaller.

4. Ia fibers send a branch to excite an internuncial, which in turn inhibits the antagonistic muscle.

5. II fibers have the same effect and course as the Ia fibers, except that the autogenic excitation from the II fibers involves an internuncial.

It should be mentioned that between the late 1950s and 1970 the group II fibers from the muscle spindle were though to always excite flexor motorneurons and inhibit extensor motoneurons regardless of whether the II fibers originated in a flexor or extensor. Thus in a stretch of the quadriceps the Ia fibers would excite the quadriceps and inhibit the hamstrings, while the II fibers would inhibit the quadriceps and excite the hamstrings. These effects are no longer considered valid, as careful review of the experiments revealed serious flaws in the methodology.

The above discussion of the muscle spindle is summarized in Table 4-1. An understanding of spindle characteristics is necessary to appreciate how the spindle functions in the body, as discussed in the last part of this section.

Golgi tendon organs

The structure and function of the Golgi tendon organ (GTO) is much less complicated than the muscle spindle. Tendon organs are found in most skeletal muscles at the junction of muscle cells and their connective tissue sheaths that are about to enter the tendon. Any individual GTO is thus directly pulled on only by a relatively small number of muscle fibers. The GTO is a long tubular capsule that may be up to 1 mm in length. Within this delicate capsule a spray of bare nerve endings form a large myelinated (Ib) fiber terminal on the tendon fascicles. Apparently the tendon material (collagen tissue) within the capsule is less compact and presumably weaker per unit of area than the outside. Tendon organs occur at both the origin and insertion of a muscle as well as at the intermuscular septa (Fig. 4-5).

Unlike the muscle spindle, which is said to lie parallel to the long axis of the muscle

Table 4-1. Characteristics of muscle spindle

Differences	Nuclear bag	Nuclear chain
Structural	Longer; nuclei lie two and three abreast in center region	Shorter; nuclei lie in chain in center region
Elastic properties	Center region shows large deformity to stretch (more elastic)	Center region shows less deformity to stretch
Histologic	Polar regions are similar to "slow" twitch muscle fibers	Polar regions are similar to "fast" twitch muscle fibers
Afferent innervation	Ia fiber, primary ending around equator	Ia fiber, primary ending around equator; II fiber, secondary ending on either side of equator
Efferent innervation	Dynamic gamma (plate ending)	Static gamma (trail ending)
Conduction of intrafusal fiber AP	Slow, nonpropagated from plate ending	Fast, propagated from diffuse trail ending
Contractile properties	Slow formation and splitting of actin-myosin bonds	Quick formation and splitting of actin-myosin bonds

Continued.

Table 4-1. Characteristics of muscle spindle—cont'd

Differences	Nuclear bag	Nuclear chain
Response to ramp stretch	High sensitivity	Low sensitivity
At beginning	Marked increase in firing rate from primary ending; center region shows marked deformity	Linear increase in firing rate from primary and secondary endings; center region shows less deformity
At completion	Higher level of discharge than at beginning length	Higher level of discharge than at beginning length
On release	Center region becomes slack, silent period from primary ending	Center region becomes slack, silent period from primary ending and progressive decrease in firing to zero from secondary ending
Effect of dynamic gamma stimulation	Marked increase in rate sensitivity and overall discharge or primary ending; polar regions have prolonged increase in viscosity with slower rate of shortening; no effect on silent period	No effect
Effect of static gamma stimulation	No effect	Increased output from primary and secondary endings; eliminates silent period on release of stretch; less polar viscosity but faster rate of shortening
Response to tendon tap	Low threshold, less than 50 μm needed; large response on Ia fiber	Higher threshold, more than 500 μm needed
Response to vibration applied to tendon	Low threshold, less than 50 μm needed for "driving" of primary ending at 100 to 300 Hz	High threshold, more than 250 μm for response at 100 Hz
Sensitivity to small displacement with gamma biasing (as in decerebrate)	High, 100 impulses per second for 1 mm stretch	Low, 7 impulses per second for 1 mm stretch.
Effect on alpha motoneuron	Autogenic and synergistic excitation, antagonistic inhibition	Same
Functional application	Biased by gamma$_D$ during muscle lengthening to facilitate a reciprocal motion	Biased by gamma$_S$ during muscle shortening to maintain input to CNS over Ia and II fibers
Source of biasing	Higher centers, cutaneous input	Higher centers

fibers, GTOs are said to be located in series. Therefore they do not change length in tandem with muscle fiber and are much less sensitive to passive stretch of a muscle than the spindles. Tendon organs are, however, very sensitive to any contractile tension developed in the muscle. The sensitivity of the GTOs is very high. A single motor unit of only a few muscle fibers exerting a force of 0.1 g is sufficient to excite one GTO.

The traditional picture of the reflex action of the Ib fibers is that of autogenic inhibition via an internuncial. Thus contraction of a flexor will excite the Ib endings in the tendon organ, which in turn leads to the activation of an internuncial that is inhibitory to the alpha motonuerons of the same muscle. At the same time a branch of this Ib fiber will excite another internuncial, which then excites the antagonist muscle (Fig. 4-10). This is the opposite effect from Ia fiber activation. Recent evidence has indicated a wider, less con-

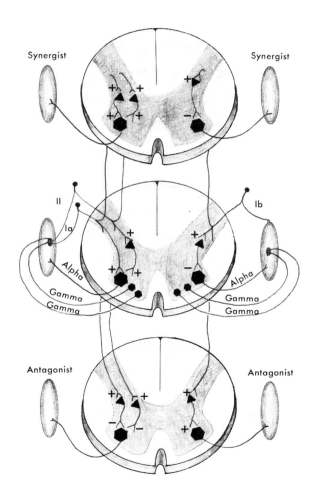

Fig. 4-10. Reflex connections of afferents from muscle spindle and tendon organ. Note opposite effects caused by Ib fiber compared with Ia and II fibers. Not included in diagram are branches from these afferents that ascend to cerebellum.

sistent range of effects from Ib activation, including presynaptic inhibition of Ia and Ib afferents, and similar effects on synergistic muscles regardless of whether the Ib is from a flexor or extensor.

Like the Ia fibers, the Ib fibers also join with the Ia fibers in the same fiber tracts that travel to the cerebellum, providing data as to the precise tension acting on a muscle. There is no evidence to indicate that data from the GTO reach conscious levels of the cerebrum.

Joint receptors

The joint receptors are located mainly in the connective tissue of the joint capsule (Fig. 4-5) and in the surrounding ligaments. There are few, if any, endings in the synovial membrane or the cartilaginous articular surfaces. The capsule contains free nerve endings, which possibly respond to pain, Ruffini endings similar in structure to those in the skin, and paciniform endings. The latter are similar to pacinian corpuscles only smaller. The receptors in the ligaments include Golgi endings similar to those found in tendons, Ruffini endings, and free nerve endings. All of the joint receptors are innervated by separate fiber types that fall in groups II, III, and IV.

The receptors primarily responsible for measuring joint movement are the Ruffini and paciniform endings. The former are slowly adapting receptors, while the latter are rapidly adapting. Originally each cluster of receptors was thought to respond over only a very restricted range of joint motion. There would be optimum angles for maximum output from each receptive field. In this way the CNS knew the exact angle of every joint. These endings, along with the paciniform corpuscles, also have a dynamic component that allows for the detection of velocity of motion as well. The Golgi endings in the ligaments were thought to behave in a manner similar to the Ruffini endings without the dynamic component.

Recent work has cast some doubt on these views. Of primary concern is the action of the Ruffini endings. One study found these receptors to be active at only the extreme ranges of movement and relatively silent at ranges in between. In addition a large number of them fired equally at both full flexion and full extension. Another study of the costovertebral joint has proved more helpful. The receptors here were maximally excited at only one extreme range and showed a discharge frequency proportional to the angle moving toward that extreme.

Whatever the mechanism, the output along the afferent fibers from the joints travels to the spinal cord and medulla. Here it is relayed to the postcentral gyrus for a conscious awareness of joint position sense. There have as yet been no established reflexes with motoneurons or synaptic involvement with muscle afferents.

FUNCTIONAL CONSIDERATIONS

Although the muscle spindle, Golgi tendon organs, and joint receptors are quite independent in terms of structure, location, and neural innervation, they function together to help produce skilled patterns of movement. In experiments on monkeys where the cervical dorsal roots of one extremity have been severed (deafferation), some interesting results occur following recovery:

1. There is loss of all afferent data from the limb, causing it to be ignored by the animal.
2. There is a marked loss of muscle tone and reflexes. The extremity is not paralyzed, as the cortical pathways and motoneurons are intact, but it is not used and hangs limply at the side.
3. If the unaffected extremity is restrained, the deafferented limb is used but only for crude grasping and reaching movements.

It is important to realize that data from these peripheral proprioceptors are continuously traveling to the CNS. These data are used consciously and subconsciously to monitor environmental influences acting on the body and to modify cortical commands to correlate with the data. For example, if you were to pick up a medium sized box of unknown contents, you have no idea of how much muscle force (alpha motoneuron activity) to produce. The instant you pick the box up, the GTO and pressure receptors in the skin signal the weight of the object and the muscle spindle and joint receptors signal the velocity of contraction. These environmental data are then used to modify alpha motoneuron activity. If the box contains, nothing and the arms are flexed rapidly, alpha activity will be reduced. If the box contains bricks there may be no arm flexion, and alpha activity will be increased until the weight is overcome.

Each of these receptors has its own specific function.

Muscle spindle

The muscle spindle is involved with the following.

Muscle tone through the reflex pathways of the Ia and II fibers. Muscle tone is related to the normal consistency of muscle tissue and the periodic asynchronous discharge of alpha motoneurons. This gives a muscle a certain consistency (viscosity), which is best assessed by palpation. Muscle tone varies between different muscles and is greatly influenced by the degree of stretch and the level of cortical arousal. Keep in mind that activating the spindles through gamma biasing can alter muscle tone.

Myotatic (deep-tendon) reflex. A firm tap on any major muscle tendon, such as the patellar tendon, causes a quick stretch to the muscle and the muscle spindle. The primary ending of the nuclear bag is excited sufficiently to cause a reflex movement (contraction) of the same muscle as well as a reflex inhibition of the antagonist. Tapping a very weak or hypotonic muscle can sometimes create enough facilitation on the alpha motoneurons to cause some movement.

Posture. The ability of the spindle to respond to small displacements and its autogenic excitation connections are important factors in the maintenance of posture. Have a friend stand sideways in front of you. Watch the small swaying back-and-forth movements. As he or she sways slightly forward the calf muscles become slightly stretched. The increased spindle output causes the alpha motoneurons to the gastrocnemius-soleus muscles to fire. This causes the calf muscles to contract, bringing your friend back to a more vertical position.

Length servomechanism. This concept is probably the fundamental role of the spindle and actually incorporates the above three functions. A servomechanism is an automatic closed-loop error-signaling device that possesses power amplification to maintain a specific factor. The length servomechanism of the spindle can be explained as follows: tension on muscle $\rightarrow$ stretch of muscle $\rightarrow$ excitation of spindle endings $\rightarrow$ excitation of alpha motoneurons of the same muscle $\rightarrow$ contraction of muscle $\rightarrow$ increase in muscle tension $\rightarrow$ return to resting length. The length will also be maintained when the load on a muscle is reduced. Implied in this length servomechanism is the fact that the muscle spindle assists in the velocity of contraction. If the load is too light for the force of contraction there will be a rapid decrease in output from the spindle and less facilitation of the alpha motoneurons. This will reduce the force of contraction and the velocity. The reverse (heavy load) would also be true.

Coordinated movement–alpha-gamma coactivation. The importance of muscle length to skilled, coordinated movements is obvious. Persons with lesions of the cerebellum can no longer integrate information from the spindles or GTOs. Among other problems, these persons have great difficulty reaching for objects. To keep a high level of accurate spindle data coming in the higher motor centers generally activate the static gammas (along with the alphas) of the contracting muscle and the dynamic gammas of the lengthening (antagonist) muscle. The static gammas keep the spindle firing in a shortened position to aid in muscle tension. The dynamic gammas increase spindle output from the lengthening muscle, possibly to put those alpha motoneurons closer to critical firing level for a reciprocal movement. Reciprocal movements are alternating contractions of agonist and antagonist (as in walking) and are the basis of coordination.

Golgi tendon organ

The Golgi tendon organ is involved with a tension servomechanism used to maintain the appropriate tension from the muscle. In this sense the GTO, like the muscle spindle, is vitally important for regulating the velocity of contraction. The length servomechanism produces less output with increases in velocity (the more shortening the less Ia and II activity). This reduces alpha motoneuron output to the contracting muscle and thus reduces velocity. With the tension servomechanism the velocity is regulated by comparing the weight of the load with the force of contraction. For example, hold a bucket under a faucet with your elbow at 90° of flexion. As the bucket fills, the arm is maintained at 90° because (1) cortical commands are given to elbow flexor alphas for muscle tension, (2) increased load stretches the elbow flexors, increasing Ia and II output on the alphas, and (3) increased muscle tension from cortical commands and increased load causes increased GTO output. This acts to inhibit the alphas and prevent any shortening (velocity) of the elbow flexors by inhibiting the alpha motoneuron output.

The GTO is also thought to play a role in the lengthening reactions. In instances where the load on the muscle is excessive and there is a potential danger of rupture, the inhibition from the Ib fibers reduces muscle tension (increasing velocity) to allow lengthening (putting down the load). The velocity of lengthening is usually regulated to prevent dropping the load. Lengthening reactions are also used in many nonstress daily activities. A good example is sitting down or walking downstairs. The release of tension in the hip and knee extensors is regulated in part by the tendon organs.

This lengthening reaction is sometimes referred to as the clasp-knife reflex. This reflex was originally described by Sherrington as a sudden release of tension and a rapid lengthening of a muscle following the buildup of rather high contractile tension. This reflex is best seen in patients with CVA with involvement of the corticospinal (pyramidal) motor system. Such patients usually have increased tone and spasticity (increased resistance to stretch) in the antigravity muscles (arm flexors, adductors, and internal rotators and leg extensors, adductors, and internal rotators). A strong quick stretch applied to one of these muscles will initially cause marked increase in resistance to movement, followed by a sudden release of tension (similar to closing a pocketknife). The sudden release of tension was thought to be due to a buildup of GTO inhibition overriding Ia and II facilitation on the alpha motoneurons. Such a simple explanation is no longer considered valid. Although GTO inhibition must be a factor, any such buildup would cause a gradual release of tension. Although nothing conclusive has been established, one study on decerebrate cats found a reduction in spindle discharge immediately prior to the release of tension. This indicates that some fusimotor inhibiton may be present that when combined with GTO inhibition can cause a sudden turning off of the alpha motoneurons.

Golgi tendon organs may also be involved with reciprocal movements. Contraction of a muscle causes GTO excitation of the antagonist. This may assist the dynamic gamma-Ia excitation to keep the antagonist "ready" for a contraction.

Joint afferents

The perception of the position of the extremities in space is often referred to as conscious proprioception or kinesthesia. This ability is frequently tested in neurologically impaired patients by having them duplicate, on the uninvolved side, a position of the involved extremity (e.g., thumb abduction, wrist extension). The patients's eyes, of course, must be closed. Testing for kinesthesia is an easy way to determine the integrity of the epicritic (discriminative) sensory system. An awareness of joint position is an important component of eye-hand coordination and bilateral extremity activities.

The joint receptors along with muscle spindles and GTOs are also involved in maintaining the upright position. There are three primary systems for upright control: the optic, the vestibular, and the proprioceptive. These systems operate together to keep the head and body vertical and the eyes facing the horizon. They project their sensory data to cortical and subcortical motor centers. These motor centers work with the spinal reflex mechanisms to alter motoneuron activity appropriately to maintain the upright position. One can function quite well with only two of these systems (e.g., a blind person). In patients with lesions of the dorsal funiculi and dorsal roots (tabes dorsalis), which interrupt the incoming proprioceptive afferents, there is usually no severe problem until they close their eyes or try to walk in darkness. The loss of optic and kinesthetic input cannot be overcome by the vestibular system.

STRUCTURE AND FUNCTION OF VISCERAL RECEPTORS
Location and types of receptors

The visceral sensory system is less complex than the somatic system primarily because of fewer kinds of receptors and sensations. The visceral receptors include free nerve endings, chemoreceptors (which may actually be free nerve endings), and stretch receptors (which may be free nerve endings or pacinian-type receptors). Free nerve endings are found in the wall of viscus such as the stomach and pelvic organs and in the large arteries. These receptors respond to distention and contact, producing sensations of pain, fullness, touch, and pressure (as in blood pressure and sexual feelings). Chemoreceptors are found mainly in the carotid and aortic bodies. They measure the amount of carbon dioxide in the blood. The stretch receptors are in the bronchi of the lungs and in the walls of the bowel and bladder. They respond to tension and distention.

Functional consideration

Visceral sensations are often described as being poorly localized. Many of them never reach consciousness. Visceral afferent data are transmitted to the CNS via lightly myelinated and unmyelinated fibers. These fibers travel with somatic afferents in the spinal and some of the cranial nerves. These data are used primarily by the autonomic nervous system to regulate blood pressure, heart and respiratory activity, swallowing, gagging and vomiting, micturition, defecation, and sexual functioning. Some of these visceral activities involve coordination with the somatic system via the brainstem and spinal cord.

In addition to reflex activity some visceral data are relayed to the parietal lobe and the limbic cortex. Perceived visceral sensations include pain, fullness, satiation, well-being, and malaise. For additional information on the visceral system see Chapter 5.

STRUCTURE AND FUNCTION OF SPECIAL SENSE RECEPTORS

The senses of smell, taste, vision, hearing, and equilibrium are considered special for several reasons. The receptors are in specific locations, not distributed generally throughout the body; the afferent fibers are involved only with cranial nerves; and loss of these senses (especially vision and hearing) can severely alter an individual's life style. Each of the special senses is discussed separately in terms of location and mode of excitation of the receptors. The course of the afferent fibers to the CNS and functional considerations are briefly mentioned. Further information in these areas is discussed in other chapters.

Olfaction

The olfactory receptors are actually the branches of the peripheral process of bipolar neurons. These branches (which are slightly motile cilia) and the soma are located high in each nostril in the nasal mucosa. Within the mucosa are glands that secrete a fluid that bathes the receptor sites on the cilia. This allows chemical particles (odors) in the air to dissolve. For a substance to be smelled it must not only be volatile but must also be able to be dissolved in water and lipids in order to be taken up by this fluid. Once dissolved the odoriferous molecules cause ionic changes on the receptor sites, which may or may not generate a receptor potential. The precise mechanism of excitation of these bipolar cells is complex and not completely understood. Some factors that determine sensitivity to an odor are (1) area of the mucous membrane, (2) size of odoriferous molecules, (3) ultrastructure of the cilia, and (4) age of the individual. Once the receptor cell has been activated the action potentials are conveyed along the central process of the bipolar cells. These processes are collectively known as the first cranial nerve, *the olfactory nerve*.

The olfactory nerve fibers pass through the foramina of the cribiform plate of the ethnoid bone to the *olfactory bulb*. The olfactory bulb is located on the inferior surface of the frontal lobe just lateral to the medial longitudinal fissure. The olfactory bulb contains four types of neurons that become involved in several closed-loop feedback circuits. These circuits act to process olfactory data by partly coding the quality of an odor in terms of frequency patterns of APs. Axons of the mitral cell (Fig. 4-11) project this information to the olfactory cortex (sometimes called the rhinencephalon). These axons collectively form the *olfactory tract*. The olfactory cortex provides awareness of smell and projects this information to other centers such as the amygdala for emotional and reflex reactions.

In humans the sense of smell is less sensitive and of less functional importance than in other animals. Olfaction is, however, a part of our daily lives. Smell has a large influence on taste (try eating an apple while smelling a peach), it can warn us of danger, and a smell can evoke a wide variety of feelings from nostalgia to nausea. Older persons are much less sensitive to odors. Some estimates have indicated that there is an annual decrease of about 1% of the receptors throughout life. This loss is reflected as an increasing blandness in the taste of food, which is often counteracted with an increased use of salt and other spices.

Gustation

The primary receptors for taste are known as the *taste buds*. These buds are oval structures found mainly just below the surface of the tongue. Taste buds contain neuroepithelial taste cells, which act as taste receptors, and less differentiated sustentacular cells, which act as reserves to replace the taste cells as they wear out. The taste bud has an opening on the surface of the tongue. The distal tip of the cells in each bud extends toward this pore (Fig. 4-12). Saliva, which is constantly secreted into the oral cavity, acts as a solvent to dissolve an ingested substance into individual ions or molecules. These ions or molecules can then enter the pore and

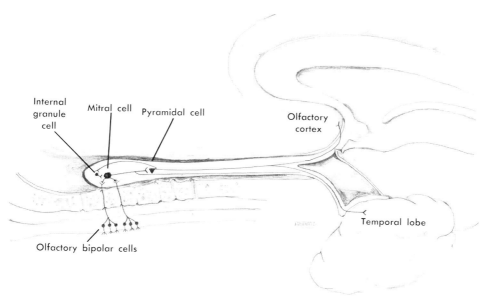

Fig. 4-11. Olfactory system. Not included are efferents from olfactory cortex to olfactory bulb and external granule cells and tufted cells within olfactory bulb that are involved with processing olfactory data from receptor cells.

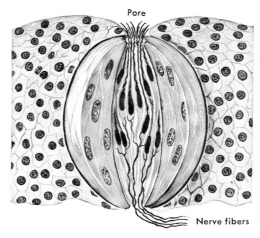

Fig. 4-12. Taste bud showing pore and nerve fibers. (From Schottelius, B. A., and Schottelius, D. D.: Textbook of physiology, ed. 18, St. Louis, 1978, The C. V. Mosby Co.)

alter the taste cell membrane to generate a receptor potential. At the base of the taste cells are the peripheral processes of afferent fibers, which respond to the receptor potentials.

The exact mechanism of how taste cells are activated to differentiate the many varieties of tastes is unknown. There are no apparent structural differences in the 10,000 buds or the taste cells in a young adult. Perhaps the patterns of discharge from taste cells of varying sensitivities can indicate not only the intensity of the taste but also the quality. Four basic taste modalities have been identified and localized on the tongue. Salt is perceived maximally on the sides of the tongue, probably from stimulation by chloride and sulfate ions. Sour is maximally perceived partway back on the tongue from stimulation by the hydrogen ions of acids. Sweets are best sensed on the tip of the tongue, where the sugar reacts with fatty substance in the tip of the taste cell. Bitter is best perceived on the back of the tongue, probably from stimulation by hydroxide ions. Taste sensations, especially bitter and sour, can also be sensed on the palate and pharynx. A wine taster swirls the wine in the mouth and swallows slowly to make use of these receptors.

From these four modalities a great variety of tastes can be described. In addition to the unexplainable neural mechanisms, many factors influence what we interpret as taste or flavor. These factors include smell, texture, temperature, sight, tanginess (cheese), chemical heat (chili peppers), and even the sound (frying bacon).

The afferent fibers from the taste buds in the anterior two thirds of the tongue form the corda tympani branch of the *seventh (facial) cranial nerve*. Afferents from the taste buds of the posterior third of the tongue and pharynx form components of the *ninth (glossopharyngeal) and tenth (vagus) nerves*. All taste afferents terminate in the upper lateral portion of the solitary nucleus of the medulla.

Secondary neurons then relay this data to other areas for reflex action and sensory awareness.

Taste is intimately related to smell, as is apparent when one has a cold. Taste declines with age. An elderly person, for example, may require three times the concentration of sugar for normal perception of sweetness. Of interest, the sense of taste can alter diet. Rats deprived of calcium will seek food with high amounts of calcium. It is supposed that infants will select from a tray of foods those that provide a balanced diet.

Vision

The receptors for vision are the rods and cones of the retina. They are activated by electromagnetic waves of energy that have wavelengths between 400 and 700 nm (0.0004 to 0.0007 mm). The retina, which is embryonically derived from the diencephalon, is the inner of three layers of the eyeball.

The eyeball is a spherical structure about 2.5 cm in diameter. Its components can be best seen in a horizontal section (Fig. 4-13). The outer coat is made of dense connective tissue and is known as the *sclera*, or the white of the eye. The anterior portion of the sclera is modified so as to be transparent and is called the *cornea*. Lying over the cornea is a thin mucous membrane, the *conjunctiva*. The middle layer is the *choroid*. It is a richly black pigmented layer that extends about two thirds of the way forward. It is coexistent with the back of the retina and functions to absorb excess light. The internal structures are the muscular *iris*, the *lens*, the *anterior* and *posterior chambers*, and the *vitreous body*. The chambers contain a clear fluid known as *aqueous humor*. The vitreous body contains a clear gelatinous material known as *vitreous humor*. The opening made by the iris is the *pupil* and is like the f-stop on a camera.

Light waves that enter the eye are refracted by the cornea, humors, and lens, are

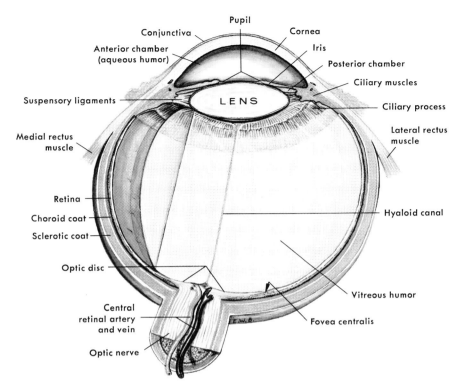

Pupil
Conjunctiva
Cornea
Anterior chamber
(aqueous humor)
Iris
Posterior chamber
Ciliary muscles
Suspensory ligaments
LENS
Ciliary process
Medial rectus
muscle
Lateral rectus
muscle
Retina
Choroid coat
Sclerotic coat
Hyaloid canal
Optic disc
Central
retinal artery
and vein
Vitreous humor
Fovea centralis
Optic nerve

Fig. 4-13. Eyeball in cross-section showing important anatomic structures. (From Schottelius, B. A., and Schottelius, D. D.: Textbook of physiology, ed. 18, St. Louis, 1978, The C. V. Mosby Co.)

reduced (or increased) in amount by the iris, and are focused to a pinpoint by the lens. The regulation of the muscles of the pupil and iris is under the reflex control of cortical and subcortical structures. In a normal eye the lens forms an inverted image on an area of the retina known as the *fovea centralis.* The fovea is only 0.4 mm wide and is actually the center area of the macula of the retina. The fovea is the area of greatest visual activity and contains only cone cells.

The retina is structured into ten layers with the receptor cells near the outer (back) layer. Light must pass through the eye and the inner eight layers before reaching the *rods* and *cones* (Fig. 4-14). Rods are thin

cylindrical cells. The outer (back) part of the rod cells contains a stack of layered membrane with a visual pigment called rhodopsin. Rhodopsin reacts to all frequencies of the visible spectrum. The inner (front) segment of the rod cell gives rise to one or more axons. Cones are similar in structure to rods only shorter and broader. The outer segment of a cone cell has one of three visual pigments. Each has a particular sensitivity for blue light waves (cyanolabe), green light waves (chlorolabe), or red light waves (erythrolabe). When the radiant energy of the light waves reaches the rods and cones a chemical reaction occurs with the visual pigment, resulting in a receptor potential. This potential is con-

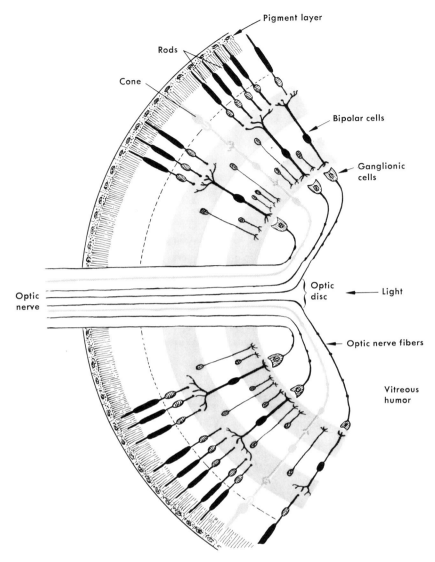

Fig. 4-14. Section of human retina showing principal neuron layers and their connections. (From Schottelius, B. A., and Schottelius, D. D.: Textbook of physiology, ed. 18, St. Louis, 1978, The C. V. Mosby Co.)

verted into action potentials by other neurons in the retina and is finally carried along the inner surface of the retina via axons that converge to form the optic disc. The fibers of the optic disc travel dorsally as the second cranial nerve, *the optic nerve*, to the *optic chiasm. The optic disc* is often called

the blind spot because there are no photoreceptors in this area. In the optic chiasm the fibers from the nasal portion of each retina cross to join with the temporal fibers of the opposite retina, forming the *optic tract* (Fig. 4-15). Since the pupil projects the visual image like a camera (inverted), each optic

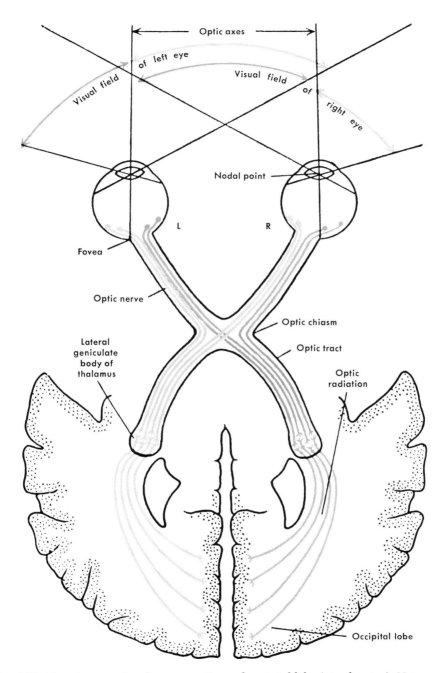

Fig. 4-15. Neural connections between retina and occipital lobe (visual cortex). Note crossing of optic nerve fibers from nasal portions of retinae of left and right eyes at optic chiasm. Visual fields and corresponding nerve tracts are shown in shades of gray. (From Schottelius, B. A., and Schottelius, D. D.: Textbook of physiology, ed. 18, St. Louis, 1978, The C. V. Mosby Co.)

tract contains fibers from the opposite visual field of each eye.

The frequency of the light waves determines the particular cones activated. Lower frequencies activate more erythrolabe, while higher frequencies activate more cyanolabe. The amplitude (strength) of the light waves reaching the retina is also important. In low intensities of light only the rods are activated. The impulses generated are interpreted as shades of gray. Thus at night houses, trees, and cars can be seen but their colors undetected. In higher intensities of light the cones become activated. The varying level of activity from the three different cones produces different frequency patterns of action potentials. When appropriately processed in the CNS this information yields the many colors we enjoy.

The point of focus on the retina is constantly changing as the result of minute movements of the eyeball. These movements are accomplished by the six extraocular eye muscles. They are necessary to prevent a bleaching out of the visual pigments. When not activated by a focal point of light the visual pigments are quickly regenerated. This bleaching effect can be easily experienced by walking from daylight into a dimly lit theater. Rhodopsin becomes bleached in bright light and is not replenished until the light intensity is low. Usually when you first walk into a dark room you can seen nothing. Within a few minutes enough rhodopsin can be reformed to respond to low-intensity light. Vitamin A is necessary for the reformation of rhodopsin.

A more detailed discussion of the anatomy

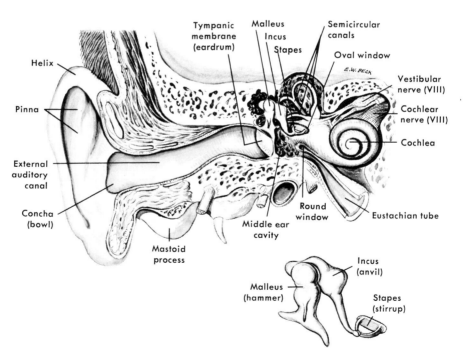

Fig. 4-16. Anatomy of external, middle, and inner ear. Auditory ossicles are shown enlarged below. (From Schottelius, B. A., and Schottelius, D. D.: Textbook of physiology, ed. 18, St. Louis, 1978, The C. V. Mosby Co.)

of the eye, the mechanics of vision such as depth and motion perception, and disorders of the eye are beyond the intent of this book. The visual reflexes and neurologic visual defects are discussed in other chapters.

Audition

The receptors for hearing are located in the *cochlear duct* of the inner ear (Fig. 4-16). The inner ear, which also includes the vestibular system, is actually a complex system of tubes and vesicles that is often called the *membranous labyrinth.* The entire complex is encased by the bony labyrinth.

The *cochlear duct* is part of a structure called the cochlea. The cochlea is made of three parallel ducts that spiral 2³/₄ times (Fig. 4-17). The three ducts are the *scala*

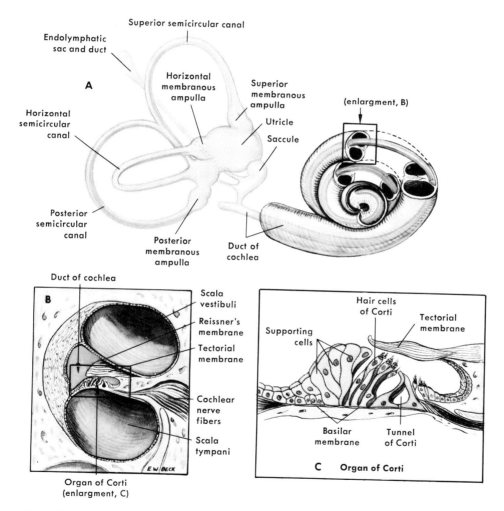

Fig. 4-17. Relationships of anatomic structures of inner ear. **B** and **C**, Successive enlargements of areas shown in **A** and **B**. (From Schottelius, B. A., and Schottelius, D. D.: Textbook of physiology, ed. 18, St. Louis, 1978, The C. V. Mosby Co.)

vestibuli on top, the *scala tympani* on the bottom, and the *cochlear duct* in the middle. Within the cochlear duct is the *organ of Corti*, which contains the receptors for hearing. The activation of these receptors is complex and interesting. The process begins with the collection of vibrations in the air or a liquid (we can hear underwater) by the *auricle* (ear). These vibrations are transmitted through the *external auditory meatus* (canal), which acts as a resonator to increase the amplitude of the vibrations. The external auditory canal and ear make up the outer ear.

At the end of the canal is the *tympanic membrane*, which is shaped like a conical loudspeaker. The vibrations in the canal cause the tympanic membrane to vibrate. So sensitive is this membrane that ordinary conversation produces inward and outward movements of about the diameter of a hydrogen ion. Attached to the tympanic membrane is the first of three ear ossicles (bones), the *malleus* (hammer). The tympanic membrane and these ear bones make up the middle ear. Vibrations on the tympanic membrane are relayed to the malleus, then to the *incus* (anvil) and then to the *stapes* (stirrup). The transfer of vibrations through the middle ear by these structures is extremely effective, operating at 99.9% efficiency. Vibrations from the stapes are transmitted directly to the oval window of the scala vestibuli of the cochlea. There is a fluid in the scala vestibuli called perilymph, which responds to the vibrations from the stapes on the oval window. If the ossicles become fused and immobile the transfer of sounds is greatly reduced. A hearing aid, which sends sound vibrations through the skull bones to the inner ear, may be needed.

The pressure waves in the perilymph travel in the scala vestibuli to the apex of the cochlea and then back along the scala tympani (the two scalae are continuous) to the resilient membrane of the *round window*. Since fluids are noncompressible, the inward (or outward) movement of the oval window is accompanied by an outward (or inward) movement of the round window a fraction of a second later.

The organ of Corti is composed of a rigid *tectorial membrane* on top and a movable *basilar membrane* on the bottom. Attached to the basilar membrane are special hair cells with rigid tuffs of hair that project toward the tectorial membrane (Fig. 4-17).

Vibrations set up in the perilymph cause the basilar membrane to vibrate and cause a shearing (tangential) motion of the hair cells against the tectorial membrane. This mechanical deformation of the hair cells generates receptor potentials that then produce action potentials on the afferent fibers located at the base of the hair cells. The base of the cochlea near the stapes shows greater displacement for high-frequency sounds, while the apex responds better to low frequency sounds. Part of the explanation for this may be a stiffness gradient along the basilar membrane from the base to the apex. The apex is said to be 100 times more flexible than the base.

Loudness (intensity) of the sound waves is related to the length of basilar membrane set into maximal motion (amplitude of vibration). Musical cords and harmonies are the result of several frequencies vibrating at once in simple numeric oscillations. Noises are several frequencies not in periodic oscillation. Like the visual system, the frequency and amplitude are coded into patterns of action potentials that are transmitted via fibers that make up the cochlear portion of the eighth cranial nerve. These fibers terminate in the cochlear nuclei of the upper medulla. From here secondary fibers are projected to other centers for reflexes and awareness of sounds.

The auditory system receives some efferent fibers from the CNS that alter the output from the organ of Corti. Projections from the auditory cortex in the temporal lobe terminate on the receptor hair cells via the *cochlear nerve*. The medial geniculate body and inferior colliculus are important relay nuclei in this pathway. These efferents act to inhibit those hair cells transmitting unwanted sounds, thereby sharpening those sounds to which the cortex is "listening." There are also efferent fibers from the trigeminal and facial nerves that innervate two tiny muscles attached to two of the ear ossicles. These muscles are the *tensor tympani*, which is attached to the malleus (trigeminal), and the *stapedius*, which is attached to the stapes (facial). Their contractions exert tension on the malleus and stapes that dampens the intensity of sounds. This acts as a source of protection against very loud noises. Prolonged exposure to excessive noises, however, can produce permanent damage to the auditory system. Relaxation of these muscles can amplify weak sounds as much as 50 times by removing the tension on the malleus and stapes. This auditory reflex is similar to the pupillary reflex that regulates the amount of light entering the eye.

One final mechanism influences the conversion of sound vibrations into action potentials, the eustachian tube. This tube connects the middle ear with the pharynx and acts to equalize the pressure on either side of the tympanic membrane in response to changes in atmospheric conditions (as in the ascent or descent of an elevator). The act of swallowing allows air to enter or leave the middle ear to equalize the pressure.

Equilibrium

The receptors for the vestibular system are located in the vestibular portion of the membranous labyrinth (Fig. 4-16). These receptors detect the angular and linear acceleration as well as the static position of the head. This information is used by the CNS to direct the gaze of the eyes, to maintain equilibrium (primarily by modifying extensor muscle tone), and to maintain the head in a vertical position so that the eyes can look forward, parallel to the horizon.

The components of the vestibular system are the three semicircular canals, the ampulla of each semicircular canal, the utricle and saccule, and the maculae of the utricle and saccule.

The *semicircular canals* are ducts about 0.1 mm in diameter and 10 to 15 mm long. They are arranged at right angles to each other like the three sides of a cube. The lateral (horizontal) canal is in the horizontal plane when the head is tilted forward 30°. The posterior (vertical) canal is in the vertical (sagittal) plane when the head is laterally flexed 55°. The inferior (anterior) canal is at right angles to the posterior canal. Each canal is filled with endolymph and has a tiny bulge at its base called an ampulla (Fig. 4-17). The *ampulla*, which is only 1 mm in diameter, contains *the crista ampulla*. The crista contains receptor hair cells embedded in a gelatinous wedge, the cupula. The cupula is attached on one side to the ampulla and acts like a swinging door (Fig. 4-18). Any movement of the endolymph pushes the cupula and bends the projecting hairs from the hair cells. This mechanical deformation generates receptor potentials that are converted into action potentials by the sensory fibers at the base of the hair cells. These receptors detect changing movements that involve acceleration and deceleration. With acceleration the inertia of the endolymph bends the cupula in one direction, causing an increase in the normal resting discharge frequency. This is like being pushed back in the seat of an accelerating car. When the endolymph attains the velocity of the head the cupula returns to its neutral position and the frequency of action potentials returns to resting levels. With deceleration the endolymph pushes the cupula the other way, causing a decrease in the resting frequency. This bidirectionality of response is important because it allows for the detection of acceleration as well as deceleration of the head. The resting frequency from the hair cells may be the result of synaptic-

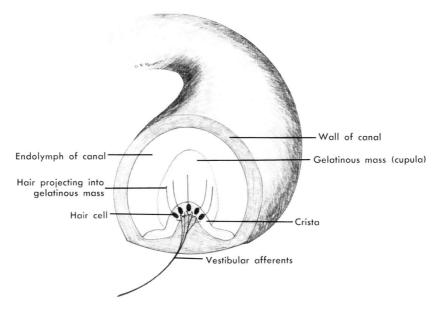

Endolymph of canal

Hair projecting into gelatinous mass

Hair cell

Wall of canal

Gelatinous mass (cupula)

Crista

Vestibular afferents

Fig. 4-18. Ampulla of semicircular canal.

like release of a transmitter substance from the hair cells causing a periodic firing of the vestibular afferents. The increase and decrease from resting levels can be partly explained by the structure of the hair cells (see below and Fig. 4-20).

The *utricle* and *saccule* contain a ridge of hair cells embedded in a gelatinous mass known as the *macula*. Above the hair projections of the hair cells are minute grains of calcium carbonate and protein called *otoliths*. The long axis of the macula of the utricle is oriented in the horizontal plane, while the macula of the saccule is oriented in the vertical plane. The weight of the otoliths responds to the pull of gravity. This will mechanically bend the projecting hairs of the receptor cells in varying amounts depending on the position of the head (Fig. 4-19). The maximum effect from the macula of the utricle is between the supine and inverted positions; the minimum effect is from the prone to the up-

right position. The effect from the utricle is to facilitate extensors. The saccule is aligned with its long axis in the vertical plane and is said to be maximally active when the head is bent to the side. The effects of saccule activation are obscure.

In the upright position the otoliths of the utricle are in greatest contact with the hairs, yet this is when there is minimal effect on extensors. In fact, going up in an elevator, which increases the otolith contact, causes the knees to flex. Going down in an elevator, producing less otolith contact (like the inverted position), causes the knees to extend. This contradiction and the fact that the face-up (supine) postition facilitates extensors, while the face-down (prone) position removes this facilitation, may be explained by the structure of the hair cells. The hair cells of the macula and crista contain two kinds of hair projection *kinocilia* and *stereocilia* (Fig. 4-20). Each hair cell has about 70 stereocilia

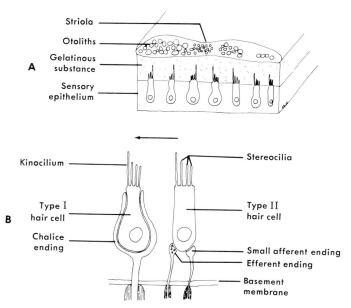

Fig. 4-19. Structure of utricular macula. **A,** Sensory epithelium and its associated otolithic membrane. Otoliths vary in size, smallest tending to be in region of striola. Sensory hair cells also show regional distribution. Free margins of cells are largest, and concentration of type I cells is highest at striola. **B,** Expanded views of type I and type II hair cells. Type I cells are pear shaped, and they receive chalice endings from large afferent fibers. Type II cells are cylindrical, and they receive small terminals of afferent and efferent fibers. There are also efferents to type I cells (not shown). Sensory hairs are cilia, including kinocilium and several stereocilia. Cilia are cause of functional polarization of hair cells. When cilia are bent in direction of kinocilium, afferent fiber is caused to discharge. (From Willis, W. D., Jr., and Grossman, R. G.: Medical neurobiology, ed. 2, St. Louis, 1977, The C. V. Mosby Co.; after Lindeman, 1969.)

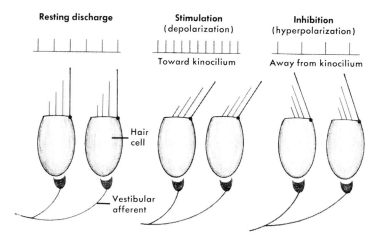

Fig. 4-20. Action of vestibular hair cells. Movements of endolymph in semicircular canals or pressure of otoliths will alter displacement of hairs and change discharge frequency on vestibular afferents.

and only one kinocilium at one end. The kinocilium acts as a polarized surface such that deviation of the stereocilia toward it causes the hair cell to depolarize, while deviation away causes hyperpolarization. This effect alters the resting discharge frequency from the hair cells. In the prone to upright position the otoliths presumably bend the kinocilium away from the stereocilia.

The hair cells also receive efferents from the CNS. Little is known about their origin or mode of action. They probably allow some degree of central control over the sensitivity of the hair cells.

The afferents from the ampulla and the macula join the auditory fibers from the cochlea to form the eighth cranial nerve (vestibulocochlear). The vestibular fibers terminate in the vestibular nuclei of the upper medulla and lower pons. Secondary neurons then relay this data to other centers for vital balance, posture, and ocular reactions.

TEST QUESTIONS

1. Match the following receptors with the correct classification:
   - ____ a. Free nerve ending
   - ____ b. Pacinian corpuscle
   - ____ c. Taste cells
   - ____ d. Muscle spindle
   - ____ e. Vestibular hair cells
   - ____ f. Rod cells
     - (1) Mechanoreceptor
     - (2) Protopathic
     - (3) Thermoreceptor
     - (4) Visceral receptor
     - (5) Proprioceptor
     - (6) Photic
     - (7) Special sense
     - (8) Chemoreceptor

2. In comparison with light (crude) touch, discriminating touch:
   a. Is signaled by different receptors.
   b. Is detected by areas of skin with large receptive fields.
   c. Allows for the recognition of objects by touch.
   d. Is best detected by the fingers and lips.

3. Match the following receptors with their correct adequate stimulus:
   - ____ a. Muscle spindle
   - ____ b. Semicircular canal
     - (1) Rotation of head
     - (2) Cold
     - (3) Low-intensity touch

   - ____ c. Pacinian corpuscle
   - ____ d. Merkel's disc
   - ____ e. Krause's end-bulb
     - (4) Muscle tension
     - (5) Deep touch

4. Increasing the stimulus strength to a particular receptor:
   a. Increases the amplitude of the receptor potential.
   b. Alters the permeability of a larger area of the receptor membrane.
   c. Causes a greater flow of positive ions from adjacent areas of the afferent fiber.
   d. Causes the receptor to adapt.

5. In rapidly adapting receptors:
   a. Velocity of stimulation is the primary cause.
   b. Frequency of APs generated is reduced.
   c. There may be a response to the removal of the stimulus.
   d. Amplitude of the RP is reduced.

6. A muscle tone is partly determined by the discharge frequency of the alpha motoneurons to the extrafusal fibers. Which of the following could increase the firing frequency of a flexor alpha motoneuron?
   a. Activation of extensor GTOs.
   b. Quick stretch of an extensor.
   c. Gamma biasing of the flexor.
   d. Lying in the supine position.

7. During an isotonic contraction:
   a. GTOs of the contracting muscle act to slow the velocity of contraction.
   b. Muscle spindles of the lengthening muscle act to slow the velocity of the contraction.
   c. Static gammas of the contracting muscle are activated.
   d. Dynamic gammas of the lengthening muscle are activated.

8. Vibration of an extensor:
   a. Causes a large increase in APs on the Ia and II fibers of the extensor.
   b. Acts to inhibit the antagonistic flexor.
   c. Causes the GTOs of the extensor to fire.
   d. Causes the muscle spindle to adapt after a few seconds.

9. The afferents from the joint receptors:
   a. Reflexly excite the contracting alpha motoneurons during movement.
   b. Provide an awareness of the position of a joint.

c. Are one of the proprioceptors that assist in maintaining upright posture.

d. Are one of the epicritic sensations.

10. During postural sway the calf and pretibial muscles are periodically stretched by gravity. As the soleus is being stretched (a slight forward lean) the following events occur:

a. Ia fibers from the soleus cause the soleus to contract.

b. Ib fibers from the soleus dampen (reduce) the force of soleus contraction.

c. Pretibial muscles contract.

d. Ia fibers from the pretibial muscles reduce the force of soleus contraction.

Place these events in the correct order:

_____ _____ _____ _____.

11. During any coordinated movement:

a. Muscle spindles of the lengthening muscle decrease the force of contraction of the shortening muscle.

b. GTOs from the contracting muscle indicate whether the tension produced is appropriate for the velocity intended.

c. Lengthening muscle is kept "ready" to contract by its muscle spindle and GTOs of the shortening muscle.

d. Velocity of movement is controlled by input from the joint afferent.

12. The vestibular system:

a. Is activated by the position of the head and acceleration-deceleration movements of the head.

b. Is located in the middle ear.

c. Has terminal branches of afferent fibers that act like cilia.

d. Reduces extensor tone in the prone position.

13. In the visual system:

a. Fovea of the retina is known as the blind spot.

b. Amount of light entering the eye is regulated by the lens.

c. Lesion of the left optic tract causes loss of vision in the right half of each eye.

d. Lesion of the optic chaism interferes with peripheral vision.

14. Impulses from the auditory system:

a. Originate from afferent fibers supplying the hair cells of the organ of Corti.

b. Travel to the brainstem by the cochlear division of the eighth nerve.

c. Are the result of a chemical reaction between the endolymph and the vibrating hair cells.

d. Can be modified by the stapedius and tensor tympanic muscles.

SUGGESTED READINGS

CIBA Foundation Symposium: Myotatic, kinesthetic, and vestibular mechanisms. section II. Vestibular mechanisms: fine structure, Boston, 1976, Little, Brown & Co.

Noback, C., and Demarest, R.: The human nervous system, New York, 1975, McGraw-Hill Book Co.

Schottelius, B. A., and Schottelius, D. D.: Textbook of physiology, ed. 18, St. Louis, 1978, The C. V. Mosby Co.

Willis, W. D., Jr., and Grossman, R. G.: Medical neurobiology: neuroanatomical and neurophysiological principles basic to clinical neuroscience, ed. 2, St. Louis, 1977, The C. V. Mosby Co.

CHAPTER 5

Peripheral nervous system

The peripheral nervous system (PNS) is simply any neuron cell body or process outside the CNS. This includes the dorsal and ventral roots, spinal nerves, peripheral nerves, cranial nerves, dorsal root ganglia, cranial nerve ganglia, and autonomic nervous system ganglia. The anatomy of the PNS is much less complex than that of the CNS, and the symptoms of PNS lesions are quite different from those seen with CNS lesions. In this chapter are presented the different kinds of peripheral fibers, synaptic and neuromuscular transmission, the anatomy and pharmacology of the autonomic nervous system, and the location, function, and distribution of the cranial nerves. Related functional and clinical considerations are also presented. The following areas deserve special attention:

1. Types of fibers in a peripheral nerve and their different functions
2. Properties of large-diameter fibers
3. Differences between segmental innervation and peripheral nerve innervation
4. Types of fibers in a peripheral nerve, their functions, and their properties
5. Characteristics of cranial nerves, including the kinds of fibers they carry
6. Methods of recording nerve conduction velocity
7. Anatomy and functional characteristics of synaptic and neuromuscular transmission
8. Mechanisms of producing excitatory and inhibitory synapses

9. Effects of denervation on the muscle cells and the electrical properties of peripheral nerves
10. Anatomy of the two divisions of the autonomic nervous system and their specific effects on the eye, glands, and blood vessels and the cardiac, respiratory, digestive, genital, urinary, and rectal systems
11. Types of cholinergic and adrenergic receptors
12. Clinical significance of causalgia, referred pain, stellate ganglion block, and denervation hypersensitivity
13. Reflex pathways for bowel, bladder, and sexual function
14. Location of the cranial nerve nuclei, distribution of the cranial nerves, and their function

CHARACTERISTICS OF PERIPHERAL NERVES
Types of fibers

A nerve trunk is composed of many individual nerve fibers surrounded by several layers of connective tissue. These fibers may be grouped functionally into the following areas:

1. Somatic afferents from a receptor in the skin, muscle, tendon, joint capsule, or ligament going to the CNS
2. Somatic effects from the motoneurons of the cranial nerve nuclei or ventral horn going to extrafusal or intrafusal skeletal muscle fibers

Table 5-1. Types of nerve fibers on the basis of diameter

Type	Group	Diameter	Conduction velocity	Sub-group	Function
Afferent fibers					
A	I	12-20	70-120	Ia	Primary ending of muscle spindle
				Ib	Golgi tendon organ afferent
	II	6-12	36-70	Muscle	Secondary ending, joint receptor
				Skin	Pressure and touch afferents
	III	1-6	6-36	Muscle	Pressure-pain, joint afferents
				Skin	Touch, temperature, pain
				Viscera	Pressure-pain
C	IV	1	5-2	Muscle	Pain afferents
				Skin	Touch, temperature, pain
				Viscera	Pressure-pain
Efferent fibers					
A	Alpha	12-20	70-120		Extrafusal skeletal muscle fiber
	Beta	5-12	30-70		Muscle spindle (?)
	Gamma	2-8	12-48		Intrafusal muscle fiber
B		3	3-15		Preganglionic, autonomic
C		1	5-2		Postganglionic, autonomic

3. Visceral efferents from the autonomic ganglia or cranial nerve nuclei to cardiac muscle, skeletal (brachiometric) muscle, smooth muscle, or a gland
4. Visceral afferents from the internal organs going to the CNS (not all peripheral nerves contain visceral afferents)

The individual nerve fibers can also be classified according to their diameter and conduction velocity of action potentials. There are three types of fibers, designated by the letters A, B, and C. Types A and B are myelinated, type C is not. Table 5-1 shows how motor and sensory fibers are grouped within these types. Keep in mind that the larger the diameter of the axon (or dendrite) the more myelin, the faster the conduction velocity, the lower the threshold to electrical stimulation, and the more sensitivity to anoxia. Also note that afferent fibers are designated by Roman numerals I to IV.

Characteristics of spinal nerves

Spinal nerves occur as 31 pairs of nerves that begin at the junction of the ventral root and dorsal root ganglia. There are eight cervical, 12 thoracic, five lumbar, five sacral, and one pair of coccygeal spinal nerves. On exiting the vertebral column via the intervertebral foramina the spinal nerves give off the following branches:

1. Dorsal ramus, which innervates the skin and intrinsic muscles of the back
2. Meningeal ramus, which supplies the meninges and blood vessels of the spinal cord
3. White ramus communicans (only spinal nerves T_1 through L_2), which is associated with the autonomic nervous system
4. Ventral ramus to the neck, chest, abdominal and perineal regions, and the limbs

Many of the ventral rami join together to

form a plexus (cervical, brachial, lumbar, or sacral) where fibers from several spinal nerve segments combine to form a peripheral nerve such as the median, obturator, or sciatic. A peripheral nerve therefore contains fibers from two or more cord segments. A distinc-

tion can thus be made between a dermatome and a peripheral nerve field and between the muscles supplied by one ventral root as opposed to those supplied by a peripheral nerve (Fig. 5-1). A lesion of a peripheral nerve will cause similar symptoms as a lesion of a

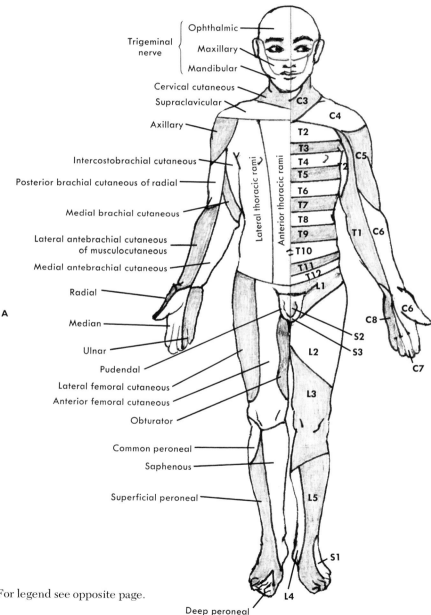

Fig. 5-1. For legend see opposite page.

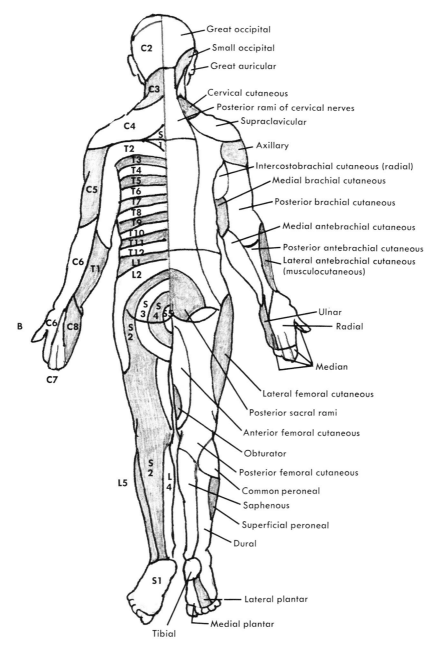

Fig. 5-1. Comparison of dermatomal and peripheral nerve innervation. It is important to realize that areas indicated vary among different authors. (Adapted from Chusid, J. C.: Correlative neuroanatomy and functional neurology, Los Altos, Calif., 1976, Lange Medical Publications.)

spinal nerve only the distribution will be different.

Characteristics of cranial nerves

Later in this chapter a more detailed discussion of the individual cranial nerves is presented. The general characteristics are as follows:

1. There are 12 pairs of cranial nerves.
2. There are no dorsal and ventral root attachments.
3. In addition to somatic afferent and efferent innervation, the cranial nerves convey data from the special senses and innervate the skeletal muscle derived from the embryonic brachial arches. The special senses are equilibrium, vision and hearing (special somatic afferents), and taste and smell (special vis-ceral afferents). The brachial muscles are skeletal muscles derived from mesoderm around the foregut rather than from somites. These muscles include those for mastication, facial expression, swallowing, and phonation. Motoneurons to these muscles are called special visceral efferents.
4. There is no segmental distribution, and the cranial nerves do not form plexuses.

Electrical responses

The characteristics of an action potential of an individual axon are presented in Chapter 3. When a peripheral nerve is stimulated the summated responses of the individual fibers can be recorded. This is known as the *compound action potential.* Since fibers of different diameters have different thresholds, the compound AP will increase in amplitude as the stimulus strength increases. The shape of the compound AP will also vary depending on the distance between the recording and stimulating sites (Fig. 5-2). As the recording site is moved farther away, the slower fibers will take longer to reach the electrode. This causes the compound AP to spread out.

The compound AP can be useful in study-ing the effects of certain drugs. For example, local anesthetics such as procaine hydro-chloride (Novacain) block Na^+ and K^+ activation primarily on small-diameter fibers. Also certain disorders such as compression of a nerve trunk can produce an altered compound AP.

The conduction velocity of the APs in a peripheral nerve is easy to determine and can be a very useful diagnostic aid. To measure *nerve conduction velocity (NCV)* apply surface-stimulating electrodes to an area along the nerve. Place recording electrodes on the surface of a muscle innervated by the nerve. Apply a moderately strong stimulus (to activate many A alpha fibers) and measure the latency between the time of stimulation and the time of response. Measure the distance between recording and stimulating electrodes. Velocity equals distance divided by time. This procedure measures the conduction velocity of the large motor fibers. Normal values range from 70 to 120 m/sec. Some authors report 50 to 70 m/sec as normal.

In conditions of nerve compression such as carpal tunnel syndrome (compression of the median nerve at the wrist, usually by swelling of the flexor tendons) the conduction velocity will drop. The drop in velocity is caused by the higher sensitivity of the larger fibers to the anoxic conditions resulting from the compression. In severe compression of a nerve all of the fibers can be affected, and there may even be axonal degeneration. To localize the compression the nerve should be stimulated twice in different places (Fig. 5-3). The conduction velocity can be determined between S_1 and R, S_2 and R, and S_1 and S_2 (the latency for S_2 to R minus that for S_1 to R divided by distance between S_1 and S_2). In carpal tunnel syndrome the NCV between S_1 and S_2 will be normal, while that between S_1 and R will be much slower. As a general rule velocities below 50 m/sec are considered abnormal.

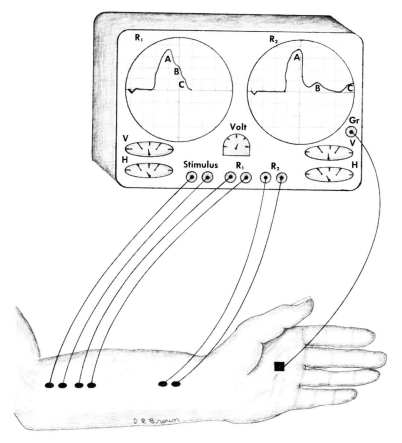

Fig. 5-2. Recording compound action potential at different distances. Since any nerve trunk contains fibers of different diameters, recording at some distance from source of stimulation (R_2) will show spreading out of recorded potentials. This would be like looking at horses during a race. Keep in mind that compound AP represents summation of all individual axon APs.

Both age and temperature have a strong influence on NCV. The conduction velocity of infants, young children, and elderly persons is significantly lower. Cold will slow conduction velocity to the extent that each drop in temperature of 1 C will lower the velocity 2 to 4 m/sec. These factors must be considered when interpreting NCV results.

The conduction velocity of sensory fibers can be assessed by stimulating a fingertip or toe and recording from the nerve trunk. The normal velocities will be similar to those for motor fibers. This procedure is more painful than that for motor fiber conduction and is thus not used as frequently (Fig. 5-3).

It is important to mention that whenever a nerve (like an axon) is stimulated somewhere along its course, APs are propagated in both directions. Those APs that travel in the normal direction are conducted *dromically*. Those that travel in the opposite from normal direction are conducted *antidromically*.

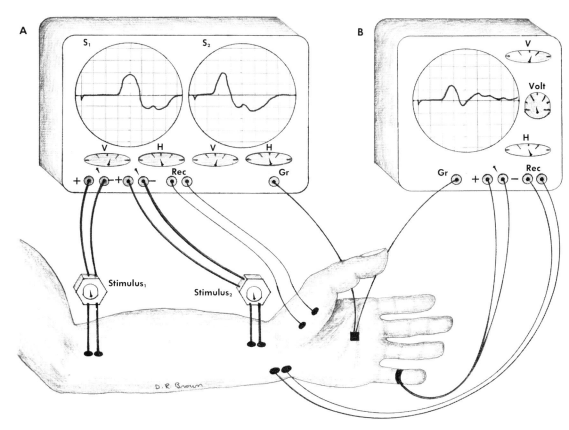

Fig. 5-3. Recording nerve conduction velocity. **A,** Motor axon conduction velocity using median nerve. Median nerve can be stimulated below elbow (S_1) medial to biceps tendon and brachial artery and at wrist (S_2) between palmaris longis and flexor carpi radialis. Recordings are made from abductor pollicis brevis. Note difference in latency between S_1 and S_2. Each horizontal division equals 2 msec, and each vertical division equals 5 mV. The NCV between S_1 and R or S_2 and R is equal to the distance divided by the latency. NCV between S_1 and S_2 is equal to distance divided by difference in latencies. **B,** Sensory axon conduction using ulnar nerve. Sensory nerves are stimulated by ring electrode on little finger. Recordings are made from ulnar nerve at wrist lateral to flexor carpi ulnaris tendon. Same calibrations are used.

Thus if the median nerve is stimulated at the elbow there will be dromic APs going to the thenar muscles (on A alpha fibers) and to the dorsal horn (on sensory fibers) and antidromic APs going to the ventral horn (on A alpha fibers) and to sensory receptors (on sensory fibers).

SYNAPTIC ACTIVITY OF PERIPHERAL FIBERS IN CNS

The synapses of the CNS are one of the most intriguing elements in the body. A synapse is an area where two neurons communicate with each other. In this communication the action potentials of one neuron re-

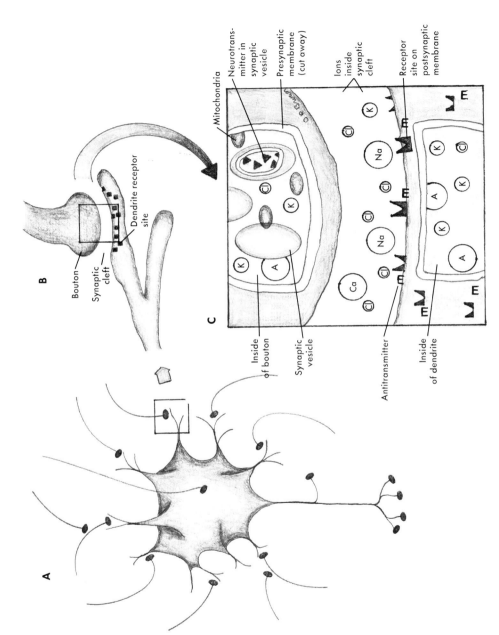

Fig. 5-4. Components of synapse. **A,** Typical neuron with proportion of its synapses. **B,** Closeup of one synapse. **C,** Closeup of presynaptic membrane, synaptic cleft, and postsynaptic membrane. Portions of two membranes have been cut away to show internal components. (These membranes are actually transparent but have been shaded to provide depth.) Receptor sites are probably protein molecules that lie over opening in postsynaptic membrane. Their odd shape is schematic and better understood in Fig. 5-5.

lease a chemical that can produce APs on the second neuron. The components and intricacies of synaptic activity are discussed below.

Structure of synapse

The synapse has only three components: *the presynaptic knob or bouton*, *the synaptic cleft*, and *the postsynaptic membrane* (Fig. 5-4). The bouton is a bulbous termination of a branch of an axon. The number of boutons per axon may vary from a few to 50,000, but they usually average around 1000. Within each bouton are tiny *synaptic vesicles* that store *neurotransmitter chemicals*. Mitochondria are also present.

Between the bouton and postsynaptic membrane is a junction averaging 20 nm that is known as the synaptic cleft. The neurotransmitter diffuses across this cleft to the postsynaptic membrane. The extracellular fluid of the cleft contains the normal concentration of ions (high Na^+, low K^+), which become involved in the transmission of the impulse.

The postsynaptic membrane may be part of a dendrite (such as the spine), a cell body, or another axon. Of interest, the dendrites and cell body of a single neuron may receive from several hundred to as many as 100,000 synapses. The postsynaptic membrane, like any other area of a neuronal membrane, separates the intracellular fluid from the extracellular fluid. There are voltage and concentration gradients on either side of the postsynaptic membrane, a resting membrane potential (RMP) of -70 mV, and a sodium-potassium pump. Associated with the postsynaptic membrane is an antitransmitter substance that acts to break down the neurotransmitter and receptor sites that "lock in" to the neurotransmitter.

Synaptic activation

The initial event in synaptic activity is the release of the transmitter substance from the synaptic vesicles into the synaptic cleft. The release of the neurotransmitter is accomplished by the depolarization of the bouton membrane by an action potential. The depolarization allows sodium and calcium ions in the extracellular fluid of the cleft to enter the bouton. In some manner the calcium ions trigger the synaptic vesicles to release their transmitter substance into the cleft. The amount of transmitter substance released is directly related to the amount of calcium ions that enter the bouton. The amount of calcium that enters is directly related to the amplitude of the AP (from the resting level to the peak of the spike) and the frequency of APs.

When the neurotransmitter is released it diffuses across the cleft in a fraction of a millisecond and attaches to *receptor sites* on the postsynaptic membrane. On the postsynaptic membrane the transmitter will do one of two things, either "open the gates" (alter membrane permeability) to Na^+ or open "other gates" to K^+. Keep in mind that the synaptic cleft contains extracellular fluid (Fig. 5-5).

There are many kinds of transmitter substances, including acetycholine, norepinephrine, serotonin, dopamine, gamma-aminobutyric acid (GABA), glycerine, glutamic acid, and others still unknown. In spite of this variety neurotransmitters cause only two responses, either the increase in Na^+ or K^+ permeability mentioned above. In the former Na^+ flows into the postsynaptic neuron, causing a local depolarization. This depolarization is known as an *excitatory postsynaptic potential* (EPSP). The EPSP represents a depletion of positive ions from the synaptic cleft. This creates a flow of positive ions from the adjacent areas of the neuron. The EPSP is similar to a receptor potential in that it is a local nonpropagating depolarization that can be summated.

If enough excitatory transmitter is released the EPSP will cause sufficient positive ions to flow from the axon hillock (that part of the

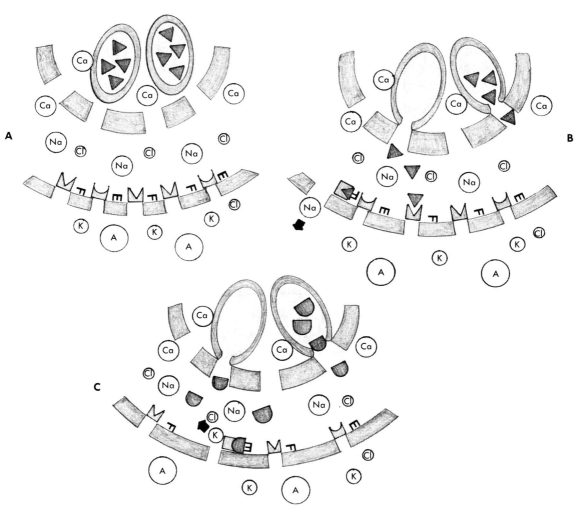

Fig. 5-5. "Unlocking of gates" by transmitter substance. **A,** Bouton has been depolarized and Ca^{++} ions have diffused inside. Note shape of different structures over "gates" (pores) of postsynaptic membrane. Also note antitransmitter substances on postsynaptic membrane. **B,** Calcium has mobilized transmitter substance in synaptic vesicles. This allows transmitter to leave bouton and diffuse across cleft. Excitatory transmitter will "unlock" larger gates and allow Na^+ ions to enter dendrite. This depolarizes area (EPSP) and decreases number of positive ions in extracellular fluid. **C,** Inhibitory transmitter will "unlock" smaller gates, allowing K^+ ions to leave dendrite. This hyperpolarizes area (IPSP) and increases number of positive ions in extracellular fluid.

neuron with the lowest threshold to excitation). This will depolarize the axon hillock. If the flow of positive ions is large enough the axon hillock will be depolarized to its critical firing level (CFL) and postsynaptic APs will be produced (Fig. 5-6).

The other transmitter response is to in-crease K⁺ permeability. This causes K⁺ to diffuse out into the synaptic cleft. The result is a hyperpolarization of the postsynaptic membrane known as an *inhibitory postsynaptic potential* (IPSP). The increase in K⁺ in the cleft causes positive ions to flow to adjacent areas along the neuron. If enough inhibitory

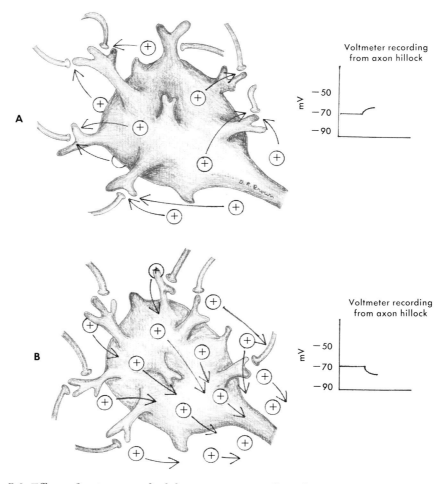

Fig. 5-6. Effects of excitatory and inhibitory synapses on flow of positive ions around soma. **A,** Excitatory synapses cause Na⁺ to enter neuron, creating concentration gradient in synaptic clefts. This causes positive ions along outside of soma to flow toward clefts. Loss of positive ions around axon hillock will depolarize that area. If enough positive ions flow to clefts, axon hillock will depolarize to critical firing level. **B,** Inhibitory synapses cause K⁺ to be released into synaptic cleft, creating concentration gradient in opposite direction. Positive ions flow toward axon hillock, causing hyperpolarization of that area.

transmitter is released the IPSP causes positive ions to flow to the axons hillock, resulting in hyperpolarization of that area.

Several points about synaptic activation are important:

1. All the boutons of a single neuron contain the same neurotransmitter. Thus a given neuron is either inhibitory or excitatory but never both.
2. Synaptic transmission is unidirectional. Stimulating the postsynaptic neuron will not produce APs on the bouton.
3. The time for the transmitter to be released, diffuse across the cleft, and alter the permeability of the postsynaptic membrane is known as synaptic delay.
4. The effect on the axon hillock of several thousand excitatory and inhibitory synapses is a simple algebraic summation of positive ions leaving or entering the area. If there is a net loss to depolarize the hillock to the CFL, then APs will be produced.

The neurotransmitter that attaches to the postsynaptic membrane is broken down (hydrolyzed) by the antitransmitter on the postsynaptic membrane. The speed at which this occurs varies (average, 1 to 2 msec). The end result is the splitting of the transmitter into simpler components, which break away from the postsynaptic membrane. The postsynaptic membrane can then repolarize to resting levels. The transmitter components that diffuse back into the cleft are probably absorbed back into the bouton for resynthesis.

The concepts of spacial and temporal summation have an important influence on the output from the axon hillock. In the former many synapses on a neuron are activated, while in the latter a given synapse has a higher frequency of stimulation. These effects are summated at the axon hillock. For example, the alpha motoneurons to the triceps brachii can be activated with a tendon tap. To employ spacial summation the vestibular system can be utilized by lying supine (extensor facilitation). This increases the number of excitatory synapses on the motoneurons. To employ temporal summation the triceps tendon can be tapped repeatedly (or the muscle vibrated). In both instances there will be greater flow of positive ions away from the axon hillock. It is important to keep in mind that many therapeutic techniques such as sensory stimulation and posturing can be used to depolarize or hyperpolarize a particular group of motoneurons. When used individually there may be no response visible in the muscle. The synaptic events, however, still occur. Lying supine may not cause extension, but the extensor motoneurons are being depolarized. If another technique is utilized or conscious effort is employed an active response may occur.

Some neurons show an interesting response to synaptic stimulation known as *repetitive discharge*. A single volley on the bouton will cause the postsynaptic neuron to fire repeatedly. The cause may be an excess amount of transmitter released or a slow breakdown by the antitransmitter. Related to this is *afterdischarge* where the postsynaptic neuron shows a prolonged discharge after the initial volley has ceased. Both of these mechanisms act to amplify the original signal with a higher frequency output.

Other neurons have a response known as *posttetanic potentiation (PTP)*. PTP is an increase in size of an EPSP to a single stimulus that follows a period of repetitive stimulation. Although the exact mechanism responsible for PTP is not clear, it may be that the initial volley of stimuli helps mobilize the transmitter in the bouton. Thus successive stimuli will release greater amounts of the transmitter. PTP may last for minutes from a single initial volley. It functions to make more effective those synapses that are used the most (such as highly developed motor skills).

Of all areas of a neuron the synapse is most susceptible to insufficient levels of oxygen (hypoxia). Synaptic transmission will start to fail within 45 sec of anoxia; the neuron, however, can remain alive. This effect is seen when a person stands up quickly and feels faint or passes out. It is possible that in some vascular lesions neurons may cease to function as a result of low oxygen levels. They can remain alive, however, and when conditions improve sudden recovery may occur.

Numerous drugs and compounds affect synaptic activity. The major ones are as follows:

1. Curare, atropine, and snake venom compete with acetylcholine for postsynaptic receptor sites, reducing or preventing an EPSP. Acetylcholine is used at the neuromuscular junction of skeletal muscle. If the dose is large enough it will cause paralysis of the respiratory muscles.
2. Prostigmine and some nerve gases and pesticides act as an anticholinesterase. Cholinesterase is the antitransmitter for acetylcholine. If cholinesterase is prevented from acting, acetylcholine will have an enhanced effect. If the dose is large enough it will cause spasm of the respiratory muscles.
3. Tetanus and strychnine act to block inhibitory synapses. This can produce lockjaw and generalized muscle spasm including the respiratory muscles.
4. Botulism acts to block the release of acetylcholine.

If the presynaptic ending is stimulated repeatedly the EPSP (or IPSP) may decline in amplitude. This effect is termed synaptic fatigue and may be caused by a depletion of stored neurotransmitter. It is uncertain whether synaptic fatigue occurs normally in the body. It has been mentioned as a possible cause of muscle fatigue (at the neuromuscular junction).

The above discussion of synaptic activity concerns chemical transmission. There are electric synapses that occur between two neurons. In such cases the synaptic cleft is so narrow (around 2 nm) that the presynaptic AP can cause local current flow directly on the postsynaptic neuron. Such synapses are not uncommon and have been found in several areas of the mammalian nervous system. Electrical synapses also are found between smooth muscle cells and cardiac muscle cells.

SYNAPTIC ACTIVITY OF PERIPHERAL FIBERS WITH MUSCLE CELLS

In this section the neuromuscular synapse between alpha motoneurons and skeletal muscle cells (extrafusal fibers) is discussed. The synaptic activity between gamma motoneurons and intrafusal fibers, and between the autonomic system efferents and smooth and cardiac muscle is similar. Not discussed is the activity between autonomic system efferents and gland cells in which neural impulses initiate events to cause the release of a hormone, enzyme, or other secretion.

Structure of neuromuscular synapse

The structure of the neuromuscular synapse is similar to that between two neurons. There is a synaptic knob, an area of the muscle cell that is the postsynaptic membrane, and a synaptic cleft known as the *neuromuscular junction*. Keep in mind that all of the muscle cells innervated by one neuron represent a *motor unit*. The bouton contains synaptic vesicles that contain the neurotransmitters. All alpha motoneurons to skeletal muscles use acetylcholine (ACh) as their transmitter. The neuromuscular junction, which is about 5 nm wide, contains extracellular fluid. The membrane of the muscle cell under the junction is folded at regular intervals and is known as the *motor endplate*. Associated with it is the antitransmitter cholinesterase and receptor sites for ACh.

Neuromuscular activation

The method of producing a muscle cell action potential is the same as that for producing one with another neuron. APs on the motor axon depolarize the bouton membrane. This allows calcium to enter to trigger the release of acetylcholine into the junction. ACh attaches to receptor sites on the motor end-plate and increases the membrane permeability to Na^+ ions. A local depolarization results, known as the *end-plate potential (EPP)*. The EPP causes local current flow of positive ions from adjacent areas, resulting in the production of two action potentials. These APs are propagated along the *sarcolemma* (muscle cell membrane). At regular intervals the sarcolemma projects into the muscle cell (like poking a finger into a balloon). This forms a network of tiny tubes known as the *T (transverse) system*. The T system functions to conduct the AP to the contractile proteins actin and myosin inside the muscle cell.

Several interesting features of neuromuscular transmission deserve mention:

1. All synapses between motoneurons and skeletal muscle cells are excitatory.
2. A single motor axon AP releases enough ACh to produce a muscle cell AP.
3. There is an apparent leakage of ACh from the bouton even when the motoneuron is inactive. This leakage results in miniature EPPs, which do not cause a muscle cell AP but may act to keep the muscle fiber "ready" to fire.
4. The ionic events for the production and propagation of the muscle cell AP are over in a few milliseconds (because of the action of cholinesterase). The contractile elements then produce a twitch response that may take several hundred milliseconds. A motor unit can be stimulated at a particular frequency so that the tension of individual twitches can be fused into a tetanic contraction.

Neuromuscular pharmacology

Many of the drugs and compounds that alter synaptic activity also affect neuromuscular transmission. Of clinical significance are the drugs that block cholinesterase and thus prolong the effect of ACh. These drugs have been found useful in myasthenia gravis. Patients with this condition are thought to have either too much cholinesterase, an insufficient supply of ACh in the bouton, or an insufficient release of ACh. The symptoms are marked fatigue, weakness, or paralysis. Administering an *anticholinesterase* such as prostigmine can prolong the action of the available ACh, allowing some degree of relief from the symptoms.

EFFECTS OF DENERVATION
Motor and sensory symptoms

Involvement of the fibers of a peripheral nerve produce motor and sensory symptoms as well as changes in the effector cell. The motor symptoms basically involve paralysis of the somatic muscles in the distribution of that nerve. Sensory symptoms involve some degree of anesthesia. The area of anesthesia is usually less than the spinal (dermatomal) or peripheral nerve field because of overlap from adjacent fields. Related sensorimotor symptoms can include hypotonus and decreased or absent deep-tendon reflexes. Occasionally with compression of a nerve trunk there may be a dissociated sensory loss. The large fibers carrying epicritic sensations are initially involved, leaving the smaller fibers carrying protopathic sensations intact.

Of interest clinically is "glove" or "stocking" anesthesia in which the motor or sensory loss is similar to the area of a hand or foot in a glove or sock. Since this does not correspond to any dermatome or peripheral nerve field, patients with this complaint are often suspected of having some type of psychosomatic problem. Certain metabolic disorders, however, such as polyneuritis from chronic diabetes can produce glove or stocking anesthesia, since the most distal segments of the peripheral nerves are involved first.

Trophic effects

During periods of nonactivation neurons are thought to secrete a substance that diffuses across the cleft and into the effector cell. This substance, which is actually formed in the neuron soma, is often referred to as a neurotrophic factor. It is apparently essential for the physiologic maintenance of the effector cell. The best example of this is the cross-innervation experiments. In these experiments the nerves to a muscle of predominantly red fibers (slow twitch) and to a muscle of predominantly white fibers (fast twitch) were severed. The nerves were then crossed, sutured to the distal stump, and allowed to regenerate. When reinnervation occurred biochemical changes were noted involving glycolytic and oxidative enzyme activity. These changes caused a physiologic reversal of roles such that the red fibers showed a faster twitch time and the white fibers showed a slower twitch time.

One of the main tropic effects is a hypersensitive reaction to chemical substances (circulating in the body) that resemble transmitter substances. For example, a denervated skeletal muscle is 1000 times more sensitive to ACh (or ACh-like compounds). The individual muscle cells show a sporadic firing known as *fibrillation potentials* in response to denervation. These fibrillation potentials begin between 1 and 3 weeks after the denervation and usually persist until reinnervation occurs. They do not occur synchronously, so no contractile tension develops in the muscle. Fibrillations cannot be felt or seen through the skin; they can only be detected by electromyography (EMG).

Another important trophic reaction to denervation is *atrophy;* Denervation atrophy is more severe than disuse atrophy and is most prominent in skeletal muscle. Although the muscle cells remain alive after denervation, the lack of natural stimulation and the neurotrophic factor cause the cell to shrink in size from a loss of the contractile proteins. There is also a proliferation of connective tissue around the muscle. A denervated muscle cell can survive well over a year but will eventually die unless reinnervation occurs.

Other trophic effects include degeneration of some receptor cells such as the taste cells, brittleness of nails, loss of hair, dry and flaky skin, lysis of bones, and changes in skin temperature. These are discussed again later with the autonomic nervous system.

Changes in electrical responsiveness

A normally innervated muscle will show a brisk response to electrical stimulation using galvanic (DC) or faradic (AC) current. Of importance is the fact that an innervated muscle will respond to currents of brief duration because of the inherent properties of the motor axons. In denervation only the muscle cells remain. They are incapable of responding to currents of short duration. This inability to respond to certain types of current can be tested by performing a *strength-duration test*. The procedure is as follows:

1. Locate the motor point of a muscle (the area where the motor axons enter the muscle) by finding the area that offers the best response to a suprathreshold stimulus.

2. Determine the smallest amount of current with a prolonged duration (300 msec) that is needed to produce a barely visible response. This is the *rheobase current*.

3. Using a current strength of double the rheobase current, determine the shortest duration of current needed to produce a barely visible response. This is the *chronaxie value*.

4. Plot the rheobase and chronaxie values on the strength-duration graph. Additional values between 300 msec and the chronaxie value can be found by adjusting the amplitude and duration of the current (Fig. 5-7).

Normal values for rheobase are around 30 mV and for chronaxie between 0.1 and 0.7 msec. In denervation the rheobase has not been found to show consistent change (although some authors report a consistent slight drop in value). Chronaxie values, however, show a consistent marked increase of around 100 times normal. In other words, with doubling of the rheobase current a muscle cell must be exposed to that current 100 times longer than a motor axon. The change in chronaxie with denervation causes the SD curve to shift upward and to the right (Fig. 5-7). With reinnervation there may be a rise in the rheobase, "kink" may appear in the curve indicating that both muscle and nerve are responding, and the curve will shift down and to the left.

The change in chronaxie following denervation is one of several responses known as *reactions of degeneration*. These reactions occur 1 to 2 weeks after denervation and also include changes in the following:

1. Erb's polarity formula. An innervated muscle gives a brisk response to galvanic current. The response from the cathode current is greater than from the

anode current. In other words, cathode closing contraction (CCC) is greater than anode closing contraction (ACC). This is written as CCC > ACC. A denervated muscle gives a slower writhing contraction, and the anode response is better than that from the cathode (ACC > CCC).

2. Galvanic-faradic response. A normal muscle responds to both AC and DC current. A denervated muscle responds only to DC current.

3. Galvanic-tetanus ratio. The current required for tetanus in a normal muscle is about four times that needed for a twitch (a 1:4 ratio). For a denervated muscle the ratio is 1:1.

The most reliable method of determining denervation is electromyography. However, this requires expensive equipment and a high degree of skill. The SD curve is an acceptable alternate and requires less expensive equipment. The other reactions to denervation are less reliable but can provide a quick assessment of the situation.

AUTONOMIC NERVOUS SYSTEM

This division of the nervous system can be referred to as the department of internal affairs, for it functions to maintain a delicate balance of homeostasis in the internal organs of the body. Included in this are the regulation of blood pressure, temperature, and digestion. This self-governing system operates primarily on a subcortical level involving both segmental spinal reflexes and regulatory centers in the brainstem and cerebrum. There are two divisions of the autonomic nervous system (ANS): the *sympathetic* and *parasympathetic*. Each has two efferent neurons, called *preganglionics* and *postganglionic* neurons, connecting the CNS with the effector organ. The cell bodies of the preganglionic neurons are found in the CNS, while those of the postganglionic neurons are found in the periphery. The preganglionic neurons receive input from vis-

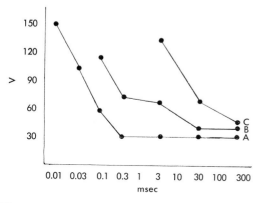

Fig. 5-7. Strength-duration curve. By plotting voltage required for threshold response to stimulus of decreasing duration, strength-duration curve can be determined. When voltage required is double that needed for rheobase, threshold duration of that stimulus is known as chronaxie. *A,* Normal SD curve. Note that response is obtained at 0.01 msec. *B,* Regeneration curve showing "kinks" and shift downward and to left. *C,* Denervation curve showing marked rise in chronaxie and shift upward and to right.

ceral afferents and from higher centers such as the hypothalamus and limbic system. The postganglionic neurons terminate on smooth or cardiac muscle or on glandular (secretory) cells.

Sympathetic division

The preganglionics for the sympathetic system are located in the intermediolateral gray horn of the spinal cord between segments T1 and L2. Because of this it is sometimes referred to as the *thoracolumbar division*. The axons from the preganglionic neurons leave the spinal cord via the ventral roots. Immediately after leaving the verte-

bral canal these axons leave the spinal nerve as the *white rami communicantes* (Fig. 5-8). These rami are white because the preganglionic axons are lightly myelinated B fibers. Keep in mind that the white rami are only found between spinal nerves T_1 to L_2 (Fig. 5-9). The white rami connect to a long chain of ganglia called the *paravertebral chain ganglia*. There are two chains, which extend parallel to and the full length of the cord on either side of the body of the vertebrae (Fig. 5-10). These two chains of ganglia are one location of the postganglionic neurons. The other location is the *collateral (prevertebral) ganglia* in the abdomen. The

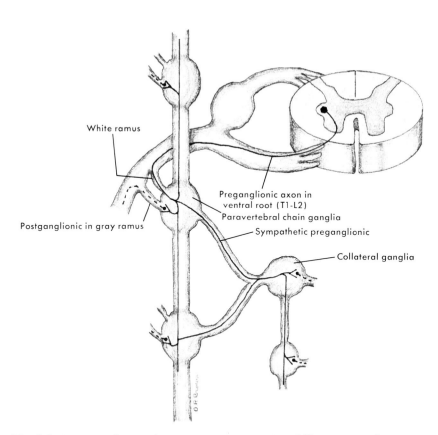

White ramus

Postganglionic in gray ramus

Preganglionic axon in ventral root (T1-L2)
Paravertebral chain ganglia

Sympathetic preganglionic

Collateral ganglia

Fig. 5-8. Anatomy of sympathetic nervous system. Note different areas of synapsing.

preganglionics that snyapse in the paravertebral chain may do so at that level or ascend or descend several segments. In any event, the postganglionic axons rejoin the spinal nerve, forming the *gray rami communicantes*.

These rami are gray because the axons are unmyelinated C fibers. The gray rami are attached to all of the spinal nerves. The postganglionic axons from the chain ganglia provide sympathetic input to the blood vessels of

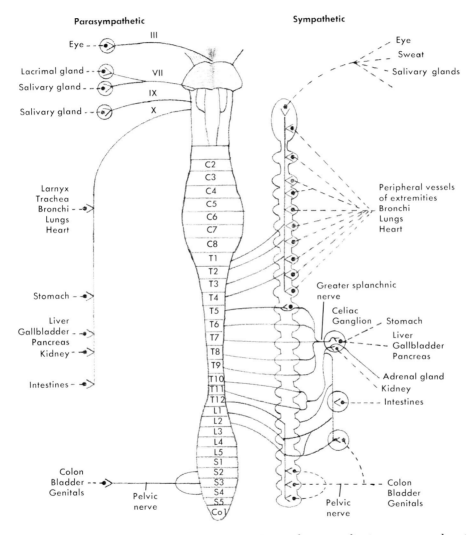

Fig. 5-9. Anatomy of autonomic nervous system. Pre- and postganglionic neurons and major distribution of each division are schematically shown. Realize that both divisions project to both sides of body. Not included are postganglionic sympathetic fibers that connect to all spinal nerves via gray rami. These fibers travel in spinal and peripheral nerves to innervate cutaneous vessels, sweat glands, and piloerector muscles.

Fig. 5-10. Location of paravertebral chain ganglia.

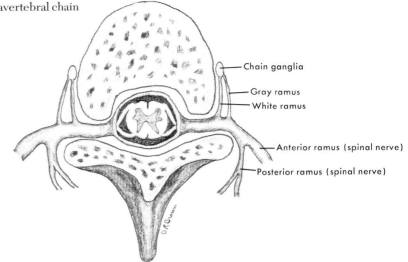

Chain ganglia

Gray ramus

White ramus

Anterior ramus (spinal nerve)

Posterior ramus (spinal nerve)

the head, neck, body wall, and extremities and to the visceral muscles and glands of the eye, lungs, heart, and skin.

Some preganglionic axons enter the chain and pass through it, forming the three splanchnic and other nerves. These preganglionics come from cord segments T5 to L2 and terminate in the *collateral ganglia*. These include the *ciliac, superior mesenteric, inferior mesenteric*, and others. The postganglionic neurons of these ganglia send axons that innervate the smooth muscle and glands of the abdominal and pelvic viscera and their blood vessels. Some of the preganglionics from T10 and T11 pass through the ciliac ganglia and synapse on the cells of the adrenal medulla on top of the kidney. These cells are actually modified postganglionic neurons and are involved with the release of epinephrine into the bloodstream.

The overall function of the sympathetic system is to mobilize the body during emergency or stress situations (fight, fright, and flight activities). This involves an increased cardiac output, an increase in blood pressure and blood sugar levels, and the shunting of blood from the viscera and skin to skeletal muscle. This system operates continually at varying levels of activity. When needed, however, the sympathetic system provides a fast diffuse reaction involving the entire body. The fast diffuse response results from the release of epinephrine into the bloodstream from the adrenal medulla and the extensive synapsing of one preganglionic with many postganglionics. This sympathetic reaction is also long-lasting because of the slow breakdown of norepinephrine by the antitransmitter.

Parasympathetic system

The preganglionics for the parasympathetic system are found in the brainstem and spinal cord segments S2 to S4. Because of this it is often referred to as the *craniosacral division*. In the brainstem the preganglionics are found in the *Edinger-Westphal nucleus, the superior and inferior salivatory nuclei*, and the *dorsal vagal nucleus*. The axons of these preganglionics travel out from the brainstem via the third, seventh, ninth, and tenth cranial nerves. These axons travel to *terminal ganglia* located near the effector organ. The preganglionic axons synapse with the postganglionic neurons in these ganglia. The postganglionic axons then travel a short distance to the effector cell. These axons provide parasympathetic input to the head, thoracic viscera, and most of the abdominal viscera.

The preganglionics from the sacral cord leave the cord via the ventral roots and help form the pelvic spinal nerves. They travel to terminal ganglia in the lower abdominal and pelvic areas to synapse with the postganglionic neurons. These postganglionic axons supply parasympathetic input to the lower gastrointestinal tract, urinary system, and genitals. Like the sympathetic system, the preganglionic axons are myelinated B fibers and the postganglionic axons are unmyelinated C fibers.

The overall function of the parasympathetic system is the restoration of energy stores used during sympathetic activity. Activities involved include slowing of the heart rate, reduced force of heart contraction, stimulation of the digestive system, and increased elimination of ingested food and water. The parasympathetic reaction is slower and more localized than the sympathetic reaction. Both divisions of the ANS at times work independently and at other times have antagonistic effects.

Transmitters of autonomic nervous system

In all of the ganglia of the ANS the transmitter substance is *ACh*. A synapse using ACh is often referred to as a *cholinergic synapse*. The synapse at the effector cells uses two transmitters. Most parasympathetic postganglionic axons and sympathetic postganglionics to sweat glands (except hands and feet) and skeletal muscle blood vessels liberate ACh. All other sympathetic postganglionics liberate *norepinephrine*. These synapses are often referred to as *adrenergic synapses*.

Cholinergic and adrenergic synapses can be divided into two subgroups on the basis of varying sensitivities to different compounds. These sensitivities are apparently the result of the receptor sites on the postsynaptic membrane. These sites, of which there may be thousands on an effector cell, are large molecules of protein. The response of the protein molecule to the transmitter can vary depending on the protein molecule. Cholinergic synapses are classified as *nicotinic* or *muscarinic receptors*. Nicotinic receptors are excited by nicotine and are found at the neuromuscular junction of skeletal muscle and in the autonomic ganglia. Muscarinic receptors are excited by muscarine (a substance derived from mushrooms) and are found in the effector synapses of smooth and cardiac muscles and glands.

Adrenergic synapses can be classified as *alpha* or *beta receptors*. Alpha receptors have an affinity for norepinephrine and epinephrine. Epinephrine is released by the adrenal medulla during sympathetic activity. Beta receptors have a selectively higher sensitivity to epinephrine. Alpha receptors generally cause excitatory response of smooth muscle contraction or glandular secretion. An exception is the smooth muscle of the gastrointestinal tract, where the effect is inhibition (relaxation). Beta receptors generally cause an inhibitory response involving relaxation of the smooth muscle of the viscera and blood vessels. Exceptions are the beta receptors of the sinoatrial node, atria, and ventricles, where the effect is increased heart rate and force of contraction.

Visceral afferent fibers

Like the somatic and proprioceptive systems, the visceral nervous system has afferent fibers. These fibers are found in the somatic nerves and in the facial, glossopharyngeal, and vagus nerves with their cell bodies in the dorsal root, geniculate, or inferior ganglia. These fibers are mostly un-

Table 5-2. Specific actions of sympathetic and parasympathetic systems

Structure	Sympathetic	Receptor type	Parasympathetic
Eye			
Radial muscles of iris	Dilates pupil (mydriasis)	A	
Sphincter muscles of iris			Contracts pupil (miosis)
Ciliary muscle of lens	Relaxes (far vision)	B	Contracts (near vision)
Glands of head			
Lacrimal			Stimulates secretion
Salivary	Thick, viscous secretion caused by vasoconstriction	A	Profuse, watery secretion caused by direct vasodilation
Nasal	Same as salivary		Same as salivary
Heart			
Rate	Increases	B	Decreases
Force of contraction	Increases	B	Decreases
Lungs			
Bronchial muscles	Relaxes (dilates)	B	Contracts (constricts)
Bronchial glands	Inhibits secretion		Stimulates secretion
Digestive tract			
Motility and tone	Reduces	A, B	Increases
Sphincters	Stimulates	A	Relaxes
Gallbladder and ducts	Relaxes		Contracts
Genital organs	Constriction of vas deferens, seminal vesicles, and prostate muscle (ejaculation)		Vasodilation (erection)
Urinary system			
Ureter (motility and tone)			Increases
Bladder (detrussor muscle)	Relaxes (usually)	B	Contracts
Trigone and extension into urethra (internal sphincter)	Contracts	A	Relaxes
Blood vessels			
Cutaneous and mucosa	Constricts	A	
Skeletal muscles	Dilates (relaxation of tone); postganglionics release ACh		
GI tract	Constricts	A	
	Dilates	B	
Pulmonary	Constricts	A	
Renal	Constricts	A	
Salivary gland	Constricts	A	Dilates
Coronary	Constricts	A	
	Dilates	B	
Skin			
Sweat glands of palms and soles	Stimulates	A	
Sweat glands of body	Stimulates (ACh is used by postganglionics)		
Pilomotor muscles	Contracts		
Liver	Glycolysis (increases blood sugar)	B	
Pancreas	Secretes insulin	B	Secretes insulin
	Inhibits insulin secretion	A	

myelinated and are concerned with the following:

1. Visceral sensations such as pain, cramps, fullness, and distention
2. Respiratory and cardiac reflexes from receptors in the aortic arch and carotid sinus that measure blood pressure and carbon dioxide in the blood
3. Peristaltic activity of the digestive system
4. Bowel and bladder reflexes for urination and defecation
5. Vomiting and coughing reflexes

The visceral afferents apparently enter the CNS and synapse on an internuncial, which in turn synapses on a preganglionic, autonomic, or somatic motoneuron to mediate the reflex. Autonomic reflexes may be simple segmental reactions or they may involve long chains of neurons in the spinal cord and brainstem. Autonomic reflexes are modified by input from many of the higher centers such as the hypothalamus, cerebral cortex, thalamus, and basal nuclei. Most of the regulatory control is conveyed to the preganglionics via the reticular formation.

Specific functions of autonomic nervous system

Table 5-2 presents the action of the sympathetic and parasympathetic divisions on specific areas of the body. When necessary a brief explanation of the mode of action is presented.

Table 5-3 presents the pharmacology of the peripheral nervous system.

Denervation hypersensitivity

When a somatic motoneuron is injured the skeletal muscle cells depolarize sporadically (fibrillate) as a result of hypersensitivity to ACh-like compounds in the blood. When the postganglionic axon or cell body in a ganglion is injured the smooth muscle, cardiac muscle, and gland cells also show hypersensitivity to their own transmitters. The hypersensitivity is less than for skeletal muscle but is readily visible in patients with peripheral or cranial nerve lesions. In brachial plexus injuries, for example, the arm is at times warm (no vasomotor tone) and at other times cold and cyanotic (vasospasm) as a result of hypersensitivity to circulating epinephrine.

Of interest is the fact that injury to the preganglionic neuron markedly reduces this hyperesensitivity as the postganglionic neuron remains intact. Although the normal regulatory control influences and reflex reactions are lost, there is little overresponse to circulating transmitter-like compounds.

Part of the explanation of denervation hypersensitivity may involve an increase in the number of postsynaptic receptor sites and a lack of normal reuptake of the transmitter by the bouton.

Trophic effects

The changes in the postsynaptic membrane and cell following denervation are known as trophic changes. They have been well documented in the effectors of the somatic and autonomic nervous systems. Since most neurons in the CNS receive input from many other neurons, loss of a given number usually will not cause much in the way of physiologic changes in the neuron. Thus trophic effects are seen only in lesions of peripheral nerves and cranial nerves.

The influence of the neurotrophic factors has already been mentioned. Those trophic effects attributed to the autonomic nervous system include the following:

1. Changes in skin temperature and color
2. Abnormal sweating
3. Brittle nails
4. Dry, flaky skin subject to ulceration
5. Loss of hair

Somatic-autonomic integration

In spite of the anatomic and functional differences between the somatic and autonomic nervous systems there are many activities where they interact. For example, somatic

Table 5-3. Pharmacology of peripheral nervous system

	Parasympathetic	Sympathetic	Somatic
Preganglionic transmitter	ACh	ACh	
Transmitter at effector	ACh	Norepinephrine ACh	ACh
Synaptic sensitivity	Muscarine at effector Nicotine at ganglia	Alpha receptors for both norepinephrine and epinephrine Beta receptors stronger affinity for epinephrine	Nicotine at N-M junction
Effect on postsynaptic	Muscarine stimulates Nicotine initially stimulates but later inhibits	Alphas excite except for relaxation of GI muscles Betas inhibit except for excitation at SA node, atria, and ventricles	Nicotine (same)
Transmitter degradation	Cholinesterase (AChE)	Monoamine oxidase (MAO) Catechol-O-methyl transferase (COMT)	Cholinesterase (AChE)
Inhibition of antitransmitter		Amphetamines Alpha blockers (dibenamines)	Physostigmine
Blocking agents	Curare Atropine Nicotine	Beta blocker (dichloroisoproterenol)	Curare Atropine Nicotine

sensory input from the retina (light) alters the output of the parasympathetic preganglionics in the Edinger-Westphal nucleus of the oculomotor nerve that regulates the opening of the pupil. Visceral afferents from chemoreceptors in the carotid sinus can alter the activity of the respiratory (somatic) muscles. Cutaneous stimuli can also be effective. Cold temperatures can activate the sympathetic system, while gentle rhythmic stroking can allow for more parasympathetic activity.

Central control of autonomic functions

The regulation of visceral functions by the sympathetic and parasympathetic systems is a complex process. Many visceral activities such as those involving cardiopulmonary functions employ complex reflex pathways that involve the "vital centers" in the reticular formation of the pons and medulla. Other visceral functions such as voiding involve cerebral, brainstem, and spinal cord centers. The central control of the major autonomic functions is discussed on the following pages.

Cardiac and respiratory systems

FUNCTION. The cardiac and respiratory systems function to deliver the appropriate amount of blood containing oxygen and vital nutrients to different areas at different times. The flow of blood is regulated by the diameter of the blood vessels, the rate of heart beat, and the force of contraction. Also important is the elimination of carbon dioxide and other waste products of metabolism.

AFFERENT INPUTS. The principal receptors are baroreceptors (pressure) and chemoreceptors in the carotid sinus and aortic arch and stretch receptors in the lung tissue and respiratory muscles. These receptors detect blood pressure, level of carbon dioxide in the blood, and expansion of the lungs.

CNS REFLEX CENTERS

Cardiac. In the medullary reticular formation are several clusters of neurons that produce either excitation or inhibition of the heart and varying levels of vasoconstriction.

The function of this center is to maintain a functional (safe) level of blood pressure (flow).

Respiratory. In the pontine and medullary reticular formations are inspiratory and expiratory centers that control the rate and depth of respiration. These centers are influenced by the amount of carbon dioxide in the blood.

HIGHER CENTER INPUT. Both cortical and diencephalic structures (such as the hypothalamus) project to these centers. Although not common, some individuals are able to consciously alter their heart rate. The higher centers, however, are not necessary for cardiorespiratory functions, as the output from the vital centers is primarily regulated by afferent input from the chemoreceptors and baroreceptors.

OUTPUT. Efferents from the vital centers travel via the vagus nerve, phrenic nerve, and spinal nerves to the heart, respiratory muscles, and blood vessels.

CLINICAL APPLICATION

Cushing response. An increase in intracranial pressure causes a reflex rise in blood pressure to maintain the cerebral perfusion pressure.

Essential hypertension. Although the exact cause of hypertension is uncertain, overactivity of the vasomotor control center in the medulla may be a factor.

Orthostatic hypotension. Orthostatic hypotension is an inability to maintain normal blood pressure when assuming a standing position, resulting in dizziness or fainting. Although common to a mild degree in most persons, it is most severe in lesions of the anterolateral portion of the cervical or upper thoracic cord that interfere with the descending fibers from the vasomotor center (reticulospinal tract).

Inflation reflex. In inflation reflex (one of the Hering-Breuer reflexes) afferent input from stretch receptors in the lungs causes an inhibition of the inspiratory neurons in the medulla and pons. This results in relaxation of respiratory muscles.

Hypoventilation. Many motoneuron diseases such as poliomyelitis and depressant drugs such as barbiturates lower the rate and depth of respiration or prevent respiration completely.

Cheyne-Stokes respiration. Extensive cerebral damage with depressed levels of consciousness can produce periods of rapid inspirations interspersed with periods of apnea.

Central neural hyperventilation. Patients with some midbrain lesions and who are comatose can have a regular increased rate of breathing (21 to 28 times per minute).

Apneustic breathing. In patients with lesions of the lower pons there may be inspiratory spasms. The breathing pattern is that of several prolonged inspirations per minute interspersed with periods of apnea.

Ataxic breathing. Some lesions of the medulla can cause an irregular rate and depth of breathing.

Other respiratory reflexes. Alterations in the normal breathing pattern occur in many reflex and voluntary acts such as sneezing, coughing, sniffing, vomiting, yawning, Valsalva's maneuver, and talking.

Digestive system

FUNCTION. The central control of the digestive system concerns the ingestion and swallowing of food and the protective response (vomiting) to noxious substances that have entered the bloodstream from the stomach.

AFFERENT INPUT

Swallowing. Tactile stimulation of the palate, pharynx, and epiglottis is conveyed to the medulla via the glossopharyngeal and vagus nerves.

Feeding. The contractions of an empty stomach produce hunger sensations. Receptors in the hypothalamus detect blood glucose and osmolarity of the blood. This input initiates the search for food and liquid.

Vomiting. Sensory receptors in the pharynx that travel over the glossopharyngeal and vagus nerves and chemoreceptors in the medulla that respond to chemical agents in the bloodstream project their input to the vomiting center in the lateral medulla.

CNS REFLEX CENTERS

Swallowing. The nucleus ambiguus of the medullary reticular formation sends fibers via the hypoglossal, glossopharyngeal, and vagus nerves to the muscles of the tongue and throat. The face and jaw muscles are more involved with chewing.

Feeding. The hunger and thirst centers of the hypothalamus provide the motivation to search for food and water. They also provide a sense of fullness or satiety.

Vomiting. The lateral medulla contains a vomiting center that integrates somatic and autonomic activities to cause a strong contraction of the abdominal muscles, which results in compression of the stomach. This empties the stomach through a relaxed esophagus. The epiglottis closes off the trachea to prevent any influx to the lungs.

HIGHER CENTER INPUT. There is little, if any, conscious control over vomiting, as the reflex response is quite strong. Swallowing can be initiated voluntarily. Once started, however, the reflex centers take over. In humans the search for food and drink is more under conscious control than in lower animals. This is obvious, since an animal usually does not overeat or diet.

OUTPUT

Swallowing. Output is primarily to the tongue (hypoglossal nerve) and throat (glossopharyngeal and vagus nerves). The epiglottis must seal off the trachea during swallowing.

Vomiting. Output is primarily to the abdominal muscles via the thoracic spinal nerves and the cardiac sphincter and esophagus.

Feeding. The somatic nervous system is essential for the search for food and drink.

CLINICAL APPLICATION

Swallowing. Difficulty in swallowing (dysphagia) can result from lesions of the medulla or of the cranial nerves involved. In addition to the loss of nourishment, food may get into the lungs, resulting in pneumonia.

Vomiting. In addition to being an important protective reflex, vomiting also is an important early indicator of increased intracranial pressure. Vomiting in a comatose or anesthetized patient may be fatal.

Feeding. It is possible that obesity may be caused by an inability of the hypothalamus to detect when satisfactory blood glucose levels are reached. Although physically full, obese persons are physiologically hungry. Drugs such as caffeine and amphetamines can suppress appetite by acting on the hypothalamus. Some hypothalamic lesions can cause the satiety centers to be inappropriately active. This could yield to starvation, as there is no drive to search for food. Anorexia nervosa is a condition of complete lack of appetite. It is generally thought to be psychogenic.

Regulation of temperature

FUNCTION. Warm-blooded animals, including humans, require a constant temperature, within a narrow range, for survival. In humans the core (heart blood) temperature averages 37 C (98.6 F) and may vary in healthy individuals between 36 C and 38 C (96.8 F to 100.4 F).

AFFERENT INPUT. The warm and cold receptors in the skin and thermoreceptors in the hypothalamus signal cutaneous and core temperature respectively.

CNS REFLEX CENTERS. The regulation of body temperature is the function of the hypothalamus. This includes centers for heat conservation and for heat loss.

HIGHER CENTER INPUT. There is no apparent conscious control over body temperature.

OUTPUT. The sympathetic system is utilized to lose excess heat by cutaneous vasodilation and sweating. The somatic system is activated to increase respiration, which also removes excess body heat. The sympathetic and somatic systems are activated to conserve heat by cutaneous vasoconstriction and piloerection and by an increase in skeletal muscle activity (shivering).

CLINICAL APPLICATION. Many infectious agents induce fever, presumably by inhibiting the heat loss center and activating the heat conservation center. The chills (shivering) associated with fever are an example of activating the heat conservation center, which then elevates the core temperature (fever).

Micturition

FUNCTION. Since the bladder (detrusor muscle) receives about 150 ml of urine per hour, periodic emptying is necessary. Of interest is the CNS control over this emptying process, which allows for voiding at the appropriate time.

AFFERENT INPUT. The bladder wall contains receptors that are located in series with the smooth muscle fibers. They therefore respond to stretch or contraction of the bladder. Afferents from these receptors enter the spinal cord at segments S2 to S4.

CNS REFLEX CENTERS. The parasympathetic center in cord segments S2 to S4 has neurons that when stimulated cause voiding by contraction of the detrussor.

The sympathetic center in cord segments T12, L1, and L2 has neurons that when activated cause urine retention by inhibition of the detrussor muscle and contraction of the internal sphincter. Note that some authors believe the internal sphincter may be regulated by the mechanical arrangement of the detrussor muscle fibers around the bladder neck (where the urethra and detrussor join). Contraction of the detrussor pulls on the bladder neck, allowing it to open for the voiding of urine.

The somatic center in cord segments S2 to S4 has neurons that when activated contract the muscles of the perineal floor including the external sphincter of the urethra.

EMPTYING OF BLADDER. At rest the sympathetic and somatic centers in the spinal cord are tonically active, causing relaxation of the detrussor and constriction of the internal and external sphincters. This prevents any dribbling of urine.

As the bladder fills, input from the stretch receptors acts to stimulate the parasympathetic centers and inhibit the sympathetic and somatic centers and ascends to the cerebrum to cause an awareness of the need to void.

Conscious control can prevent voiding (mainly by increased facilitation of the somatic spinal center) for a period of time.

CLINICAL APPLICATION

Urinary retention. In the early recovery period of some spinal cord lesions the patient is unable to void, presumably due to a flaccid bladder and spastic sphincters. Such patients require catheterization to remove urine. This phase often develops into a reflex (automatic) bladder.

Reflex bladder. In spinal cord lesions that do not disturb the reflex centers T12, L1, L2, and S2 to S4 (or their spinal nerves) the bladder may empty on a simple reflex basis. The filling and emptying of the bladder is never complete, and voiding often occurs inappropriately. In some cases catheterization may be necessary. In other cases voiding can be initiated at regular intervals following stimulation of the skin of the perineum, abdomen, or lower extremity.

Flaccid bladder. With lesions of the cauda equina that interrupt the efferents to the detrussor and sphincters there is incontinence, and catheterization is necessary. Lesions of the dorsal roots that interfere with the afferent input (S2 to S4) can also lead to retention.

Defecation. The rectal system is very similar to the urinary system. The spinal reflex centers are in the same segments and essentially do the same thing; parasympathetics induce contraction of the smooth muscle of the sigmoid colon and rectum, sympathetics induce relaxation of these muscles and con-

traction of the internal sphincter, and somatics cause contraction of the external anal sphincter. There are stretch receptors that respond to the filling of the rectum. Their afferents activate the parasympathetic centers and inhibit the sympathetic and somatic centers. These afferents also send branches to the cerebrum for awareness. Clinically bowel retention, a reflex bowel, and a flaccid bowel normally occur with the same lesions that produce the corresponding bladder problems. The use of suppositories, careful attention to diet, and a regular schedule can provide for less complication than are seen with loss of bladder control.

Sexual function. Although the pathways and reflex centers are similar in both sexes, greater emphasis is placed here on the male genital system because males show more serious symptoms with certain nervous system lesions.

STAGES. The stages involved in sexual activity are as follows:

1. Erection of the penis as a result of vascular engorgement of the corpora cavernosa and corpus spongiosum
2. Ejaculation of semen as a result of contraction of the vas deferens, seminal vesicles, prostate gland, bulbocavernosa, ischiocavernosa, and compressor urethra
3. Orgasm in which there are varying reactions and sensations in response to ejaculation (or in women in response to rhythmic contractions of muscles of the distended vagina)

AFFERENT INPUT. The cutaneous distribution to the genitals is from cord segments S3 and S4. Although stimulation of the genitals is an important contributor to the stages of sexual activity, hormonal and psychogenic factors are also important.

CNS REFLEX CENTERS

Erection. Parasympathetic neurons in cord segments S2 to S4 act to dilate the blood vessels of the penis. Somatic neurons in the same segments innervate the bulbocavernosa and ischiocavernosa. Contraction of these muscles aids in engorgement by compressing the venous flow from the penis. Stimulation of the genitals has a direct reflex effect on these centers. Sympathetic neurons in cord segments T12, L1, and L2 may participate in erection, but their role is uncertain.

Ejaculation. The sympathetic center in segments T12, L1, and L2 are involved in the emission of prostatic and vesicular fluid (semen) and the peristaltic-like contractions of the smooth muscle of the vas deferens, seminal vesicles, and prostate gland that propel the semen out. Also important is the contraction of the internal urinary sphincter by the sympathetic center to prevent a backflow of semen into the bladder. The somatic center in the sacral cord assists in ejaculation by contraction of the muscles of the pelvic floor.

Orgasm. The centers for orgasm are not well known but are in the cerebrum and limbic system.

HIGHER CENTER CONTROL. Electrical stimulation of the septal area in monkeys can produce penile erection, and bilateral ablation of the amygdala has caused increased sexual activity. The importance of these (and probably other) higher centers, however, is to produce a drive to seek sexual activity. The level of sex hormones in the blood is also important.

OUTPUT. The sympathetic fibers travel via the hypogastric plexus, while the parasympathetic fibers travel via the pudendal and pelvic nerves.

CLINICAL APPLICATION. Other than psychologic or emotional factors, sexual dysfunction is primarily seen with damage to the lower portion of the spinal cord or lesions of the cauda equina. These lesions interrupt the afferent or efferent limb of the reflex arc, thus preventing erection and ejaculation. With higher cord lesions manipulation of the genitals can produce an erection and ejaculation on a reflex basis. This can occur even with complete cord transection. The influence of psychogenic factors and the extent of orgasm will vary with the severity of the lesion. Although some sexual activity in men with spinal cord injury may be possible, there is frequently retrograde ejaculation of semen into the bladder. It is thought that this is caused by failure of the internal urinary sphincter to contract during ejaculation. Another complication in some patients is acute spontaneous erection, which may persist for several hours. It is not known if this is an irritative or release phenomenon.

Women with spinal cord injury suffer less sexual dysfunction and are generally able to become pregnant and bear healthy children.

Clinical application

In addition to the disorders mentioned above several items uniquely related to the ANS are of clinical significance.

Stellate ganglion block. In stellate ganglion block a local anesthetic such as procaine is injected into the area of the stellate ganglion (at the level of the C-6 vertebra). This temporarily disrupts the activity along the sympathetic postganglionics to the ipsilateral arm as well as the superior cervical ganglion. The result is a warm, dry upper limb and Horner's syndrome (see below). This procedure can be useful in diagnosing the cause of vascular insufficiency in the arm. In disorders with chronic vasospasm such as Raynaud's disease a stellate ganglion block provides immediate relief. In disorders that compress the subclavain artery such as cervical rib syndrome (scalenus anterior, thoracic outlet syndrome) there is little or no relief. Blocking of the sympathetic ganglia of the leg can also be done. Permanent effects by sympathectomy of the white rami can be performed if indicated.

Horner's syndrome. Any interruption of the sympathetic postganglionics from the superior cervical ganglion to the head, or the preganglionics from T1 to T4 to the superior cervical ganglion, or the descending regulatory fibers to the thoracic cord from brainstem centers will disrupt sympathetic activity in the head, resulting in Horner's syndrome. The symptoms are ipsilateral miosis (constricted pupil), partial ptosis, enophthalmos (sunken eyeball), and flushed, dry skin.

Referred pain. Ischemia or pressure in a visceral structure may be "felt" as a cutaneous pain. For example, a person suffering from myocardial ischemia or infarct may feel the pain in the left axila and left arm. Why visceral sensations can be referred to the skin is not completely known. It may be that the visceral afferents entering the cord may activate somatic afferents from a particular dermatome (using the same dorsal root). This interaction may occur in the dorsal horn, where both somatic and visceral afferents discharge into a common pool. Other areas of referred pain are the left shoulder blade from gallstone, the groin from a stone in a ureter, and the area around the umbilicus from the appendix.

Causalgia. In crushing or partial lesions of a peripheral nerve, burning and often excruciating pain sensations develop in the cutaneous distribution of the nerve. These sensations are known as causalgia and are seen more commonly with involvement of the median and tibial nerves than with other limb nerves. The cause is thought to involve stimulation of cutaneous afferents by tonically active sympathetic efferents in the nerve stump. This implies some kind of direct or close contact between the fibers causing a "short circuit" (the efferent APs cause local current flow and APs on the afferents). Causalgia can be relieved by sympathetic nerve block and by selective stimulation of large A fibers in the nerve trunk. The explanation for the latter may involve the *gate theory* (see Chapter 6).

Autonomic dysreflexia. In lesions of the spinal cord above T4 to T6 (above the sympathetic splanchnic outflow) acute paroxysmal hypertension can develop from a variety of stimuli to the viscera or the skin innervated by cord segments below T4 to T6. These stimuli include tactile, thermal, or painful stimulation of the skin and distention or spasm of a visceral structure. Distention of the bladder is the most common cause. The effect is a sudden rise in blood pressure (250 to 300 mm Hg systolic and 140 to 160 mm Hg diastolic). Secondary symptoms include a pounding headache. The mechanism behind autonomic dysreflexia is as follows:

1. Visceral or somatic afferents from the lower abdominal or pelvic organs or the skin of the lower part of the body enter the spinal cord and reflexly excite sympathetic preganglionics as the impulses ascend the cord stump.
2. This results in arteriolar spasm in the vessels of the skin and splanchnic bed, causing an elevation in blood pressure.
3. The increased blood pressure is detected by baroreceptors in the carotid sinus and aorta and conveyed to the CNS by the ninth and tenth nerves.
4. The vasomotor centers in the medulla try to reduce the pressure by slowing the heart rate but are unsuccessful in inhibiting the sympathetic preganglionics for vasodilation.

Of interest, patients with dysreflexia may experience postural hypotension when they assume a sitting-standing posture.

CRANIAL NERVES
Characteristics

The cranial nerves are the peripheral nerves of the head. There are 12 pairs of cranial nerves, often designated by Roman numerals I to XII. Unlike spinal nerves, there are no dorsal and ventral root attachments, and not every nerve has a ganglion. Cranial nerves transmit information from the special senses, somatic receptors, and proprioceptors. Their output is involved with movement of the eyes, face, jaw, tongue,

A

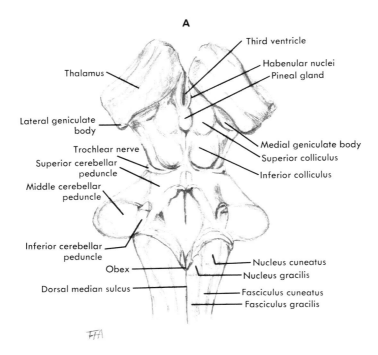

Third ventricle

Thalamus

Habenular nuclei

Pineal gland

Lateral geniculate body

Medial geniculate body

Trochlear nerve

Superior colliculus

Superior cerebellar peduncle

Inferior colliculus

Middle cerebellar peduncle

Inferior cerebellar peduncle

Nucleus cuneatus

Obex

Nucleus gracilis

Dorsal median sulcus

Fasciculus cuneatus

Fasciculus gracilis

B

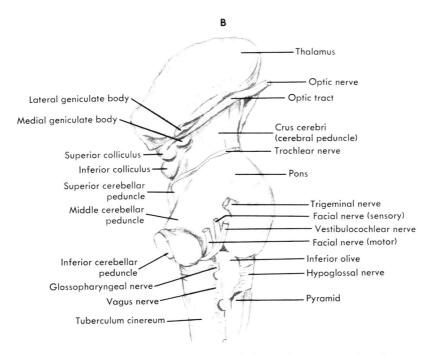

Thalamus

Optic nerve

Lateral geniculate body

Optic tract

Medial geniculate body

Crus cerebri (cerebral peduncle)

Superior colliculus

Trochlear nerve

Inferior colliculus

Pons

Superior cerebellar peduncle

Middle cerebellar peduncle

Trigeminal nerve

Facial nerve (sensory)

Vestibulocochlear nerve

Facial nerve (motor)

Inferior cerebellar peduncle

Inferior olive

Glossopharyngeal nerve

Hypoglossal nerve

Vagus nerve

Pyramid

Tuberculum cinereum

Fig. 5-11. External features of brainstem. Telencephalon and cerebellum have been removed to give better perspective. Cranial nerves are included here but are discussed in Chapter 2. **A,** Dorsal view. **B,** Lateral view.

Continued.

C

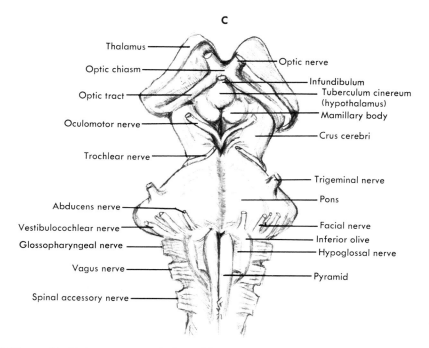

Thalamus —
Optic chiasm —
Optic tract —
Oculomotor nerve —
Trochlear nerve —
Abducens nerve —
Vestibulocochlear nerve —
Glossopharyngeal nerve —
Vagus nerve —
Spinal accessory nerve —

— Optic nerve
— Infundibulum
— Tuberculum cinereum (hypothalamus)
— Mamillary body
— Crus cerebri
— Trigeminal nerve
— Pons
— Facial nerve
— Inferior olive
— Hypoglossal nerve
— Pyramid

Fig. 5-11, cont'd. C, Anterior view. (Adapted from Everett, N.: Functional neuroanatomy, ed. 6, Philadelphia, 1971, Lea & Febiger.)

pharynx, and larynx and the parasympathetic supply to the viscera.

Each cranial nerve is discussed here in terms of its composition, associated ganglia and nuclei, and functional-clinical considerations. Fig. 5-11 shows the location of the cranial nerves on the brainstem. It is important to remember which nerves come off laterally for treatment of patients with bulbar (brainstem) injury (see also Chapter 10).

Since most of the motor nuclei for the cranial nerves receive bilateral cortical innervation, most unilateral supranuclear (upper motoneuron) lesions have little permanent effect. The facial nerve, however, presents an interesting exception, which is explained later.

As a group the cranial nerves contain the following types of fibers:

1. General somatic afferents from the skin, jaw, and muscles of the eyes and tongue

2. General somatic efferents to the muscles of the eyes and tongue
3. Special visceral afferents from the nose and tongue
4. Special somatic afferents from the eyes and ears
5. General visceral efferents to the thoracic and abdominal viscera
6. General visceral afferents from the viscera
7. Special visceral efferents to the jaw, face, pharynx, and larynx.

The openings in the floor of the cranial cavity for the cranial nerves are shown in Fig. 7-1, *B.*

Olfactory nerve (I)

The olfactory nerve is composed entirely of special visceral afferent fibers that convey the sense of smell. These fibers are the central processes of bipolar neurons in the nasal mucosa. They terminate on neurons in the olfactory bulb on the base of the frontal lobe. (The olfactory nerve is thus quite short.) Axons from these second neurons form the olfactory tract, which travels dorsally along the base of the frontal lobe (see Fig. 2-24). The olfactory fibers terminate in the rostral portion of the hippocampal formation (which is composed of the uncus and other areas), the anterior perforated substance of the frontal lobe (just lateral to the optic chiasm and tract), and the amygdala (see Fig. 4-11).

Lesions of one olfactory nerve or tract are often not noticeable, as the other side can compensate. Loss of sense of smell (anosmia) occurs if both sides are involved. Taste is also affected.

Optic nerve (II)

The optic nerve is composed entirely of special somatic afferent fibers that originate from cells in the ganglionic (innermost) layer of the retina. These axons converge toward the optic disc and then project dorsally, penetrating the sclera of the eyeball. The nerve continues dorsally, passing through the orbit and optic canal of the skull. It then travels along the base of the frontal lobe (lateral to the olfactory tract) to an area just in front of the infundibulum known as the optic chiasm. Here the axons from the nasal half of each retina cross and join with the axons from the temporal half of the retina to form the optic tract (Fig. 5-12). Since the eye is like a camera, the image falling on the retina is inverted and reversed. Thus the nasal half of the retina sees the temporal half of the visual field and vice versa. Each optic tract therefore transmits visual data from the contralateral visual field of each eye. From the optic chiasm the fibers of the optic tract travel dorsally along the lateral border of the midbrain to the *lateral geniculate body*. Some of the optic tract fibers terminate directly in the superior colliculus for visual reflexes. From the lateral geniculate body the visual data is

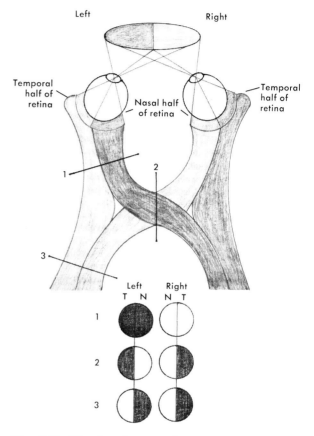

Fig. 5-12. Visual pathway and several lesions. Fibers from retina in optic nerve are arranged in temporal and nasal halves. Lesion of optic nerve *(1)* can produce varying degrees of visual field loss or blindness in that eye. In optic chiasm, nasal half from each retina crosses to opposite side. Since nasal part of retina "sees" temporal (lateral) part of visual field, lesion of optic chiasm *(2)* can result in bitemporal hemianopia. Optic tract contains ipsilateral temporal retinal fibers (nasal visual field) and contralateral nasal retinal fibers (temporal visual field). Lesion of optic tract *(3)* can cause loss of vision from opposite side of body. This is known as contralateral homonymous hemianopia.

relayed to the primary visual cortex of the occipital lobe via the posterior limb of the internal capsule.

Lesions of the optic nerve can cause varying degrees of loss of vision in one eye. Lesions of the optic chiasm can cause loss of vision of the temporal visual field of both eyes *(bitemporal hemianopia).* Lesions of the optic tract, lateral geniculate body, posterior limb of the internal capsule, or the occipital lobe can cause the loss of either the right or left halves of the visual field of each eye *(contralateral homonymous hemianopia).* For example, a lesion of the left optic tract causes loss of vision in the right half of the visual field of each eye.

Since the dura mater is continous with the sclera of the eyeball, any increased intracranial pressure can be transmitted to the back of the eyeball, causing the optic disc to bulge forward. This pressure can also cause the retinal veins to swell and pinkness of the disc. These signs are readily visible by looking through the pupil and are often referred to as *papilledema,* or a choked disc.

Oculomotor nerve (III)

The oculomotor nerve is composed of the following:

1. General somatic efferents to all of the extraocular muscles except the lateral rectus, superior oblique, and tarsalis
2. General somatic afferents from the muscle spindles of these muscles
3. General visceral efferents that innervate the smooth muscle of the eye

The general somatic efferent fibers arise from the *oculomotor nucleus* of the midbrain. This nucleus is located at the level of the superior colliculus in the medial tegmentum in front of the cerebral aqueduct. The fibers from these neurons travel anteriorly through the red nucleus and substantia nigra to emerge from the midbrain in the interpeduncular fossa. These fibers continue anteriorly in the interpeduncular cistern

through the inferior orbital fissure and orbit. These fibers innervate the medial, inferior, and superior recti, the inferior oblique, and the levator palpebrae muscles. The recti muscles cause the eye to look inward (converge), upward, or downward. The inferior oblique works with the superior oblique to help rotate the eye around a sagittal axis (making circles with the eye) and to look to the upper (inferior oblique) and lower (superior oblique) corners of the visual field. The levator palpebrae muscle helps to raise the eyelid. It works along with the tarsalis muscle (sympathetic innervation) to keep the eyelid (the palpebral fissure) open.

The general somatic afferents from the proprioceptors probably have their unipolar cell bodies in the *mesencephalic nucleus* of the trigeminal nerve (V). This nucleus is the only area of the CNS that seems to have unipolar neurons. The central processes of these neurons are involved with the reflex control of jaw movements and probably with the control of eye movements.

The general visceral efferents are preganglionic parasympathetics from the *Edinger-Westphal nucleus* (accessory oculomotor nucleus). This nucleus is located just posterior to the upper part of the oculomotor nucleus. These fibers travel in the oculomotor nerve to the ciliary ganglia just in back of the eye. From the ciliary ganglia, postganglionic parasympathetic fibers travel to the sphincter muscles of the iris for constriction (miosis) and to the ciliary muscles of the lens for contraction for near vision (Fig. 5-13).

Lesions of the third nerve interfere with the two basic patterns of eye movement: conjugate and convergent. In conjugate movements both eyes look to the same side of the body (like watching a Ping-Pong game). In convergence both eyes look medially (as in holding a book close). Loss of one oculomotor nerve will cause the following:

1. Inability to look medially (adduct) for conjugate (lateral) gaze

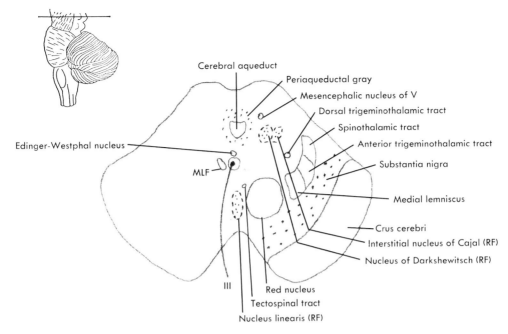

Fig. 5-13. Section through midbrain at level of superior colliculus.

2. Lack of convergence for near vision
3. Deviation of the eye (strabismus) laterally
4. Partial drooping of the eyelid (ptosis)
5. Dilated pupil (mydriasis)
6. Lack of accommodation for near vision (no focusing of the lens or constriction of the pupil)
7. Double vision (diplopia)

This collection of symptoms is often known as Weber's syndrome.

Trochlear nerve (IV)

The trochlear nerve is composed primarily of general somatic efferents. The cell bodies of these fibers are found in the *trochlear nucleus*. This nucleus is actually a caudal extension of the oculomotor nucleus and is found at the level of the inferior colliculus

(Fig. 5-14). The axons from this nucleus travel dorsally, cross midline, and exit from the back of the lower midbrain. The nerve then curves forward around the brainstem, through the superior orbital fissure and orbit, and innervates the superior oblique muscle. General somatic afferents from muscle spindles also travel in this nerve.

A lesion of one trochlear nerve causes diplopia, especially when the person attempts to look down or down to the side. This makes walking and especially walking downstairs difficult. The person may have to tilt his or her head to the shoulder on the side opposite the paralyzed muscle to reduce the diplopia.

Abducent nerve (VI)

The abducent nerve is composed primarily of general somatic efferents. The cell bodies

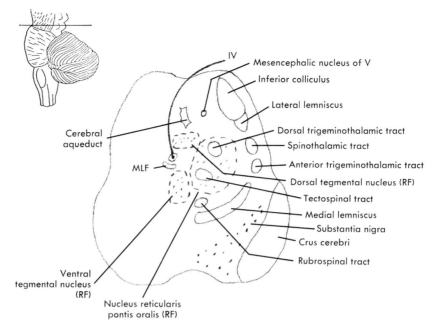

Fig. 5-14. Section through midbrain at level of inferior colliculus.

of these fibers are found in the *abducent nucleus* located near the midline of the lower pons in the floor of the fourth ventricle (Fig. 5-16). The axons travel anteriorly and inferiorly through the tegmentum to emerge at the pontomedullary junction just lateral to the pyramids. The nerve continues forward through the pontine cistern, superior orbital fissure, and orbit to innervate the lateral rectus muscle. General somatic afferents from muscle spindles also travel in this nerve.

A lesion of the abducent nerve causes internal strabismus, double vision, and loss of the abducting component of lateral gaze.

Trigeminal nerve (V)

The trigeminal nerve is composed of the following:

1. General somatic afferents from the skin of the anterior scalp and face, mucous membranes of the mouth, the gums, the tongue, the jaw, the teeth, nasal sinuses, meninges, the cornea, and the muscles of mastication. These afferents convey somatic and proprioceptive sensations to the CNS.
2. Special visceral efferents to the muscles of mastication, auditory tube, soft palate (to aid in swallowing), and the tensor tympani muscle.

The general somatic afferents have their unipolar cell bodies in the *trigeminal ganglion* (gasserian ganglia, semilunar ganglion), which is located in the floor of the middle fossa just lateral to the internal carotid artery. The peripheral processes of these neurons form three separate sensory branches: the ophthalmic, maxillary, and mandibular.

The ophthalmic branch supplies receptors in the dura (including the tentorium), the eye (including the cornea), upper eyelid, and skin of nose, forehead, and anterior scalp back to the ears.

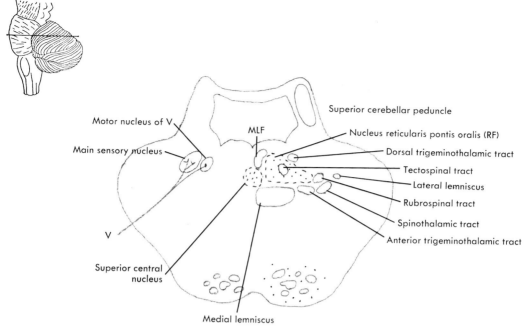

Fig. 5-15. Section through midpons.

The maxillary branch innervates part of the dura, lower eyelid, skin of the upper cheek, temple, and upper lip, mucous membranes of the upper mouth, nose, roof of the pharynx, and sinuses, and gums, teeth, and palate of the upper jaw.

The mandibular branch supplies part of the dura, teeth and gums of the lower jaw, mucosa of the cheek, floor of the mouth, anterior two thirds of the tongue, skin of the lower jaw and cheek, part of the ear, and the temporomandibular joint.

The central process of these unipolar cells form a large trunk that enters the lateral surface of the midpons (Fig. 5-15). On entering the pons the sensory fibers split into an ascending and descending branch. The ascending branch contains afferents relaying (1) position sense, vibration, pressure, and discriminatory touch to the *main sensory nucleus* (later portion of the upper pons) and (2) proprioceptive data to the *mesencephalic*

nucleus of the fifth nerve (a midbrain continuation of the main sensory nucleus). The mesencephalic nucleus contains the unipolar cell bodies for these afferents.

The descending branch contains afferents relaying pain, temperature, and crude touch to the *spinal nucleus* of the fifth nerve (a caudal continuation of the main sensory nucleus), which then descend to the first two cervical cord segments. As these afferents descend they form the *spinal tract* of the fifth nerve.

The continuation of the sensory data from these three nuclei is discussed in Chapter 8.

The special visceral efferent fibers arise from the motor nucleus of the fifth nerve (just lateral to the main sensory nucleus). They emerge from the lateral surface of the midpons adjacent to the sensory root. The motor root travels anteriorly with the mandibular branch to innervate the muscles of mastication (temporalis, masseter, and pterygoids), the mylohyoid muscle and anterior belly of the digastric muscle, and the tensor tympani.

Lesions of the trigeminal nerve can cause the following:

1. Ipsilateral anesthesia
2. Weakness in chewing and deviation toward the side of involvement as a result of paralysis of the muscles of mastication on the involved side and the action of the contracting pterygoids on the uninvolved side
3. Loss of the jaw jerk as a result of interruption of the proprioceptive afferents and lower motoneurons
4. Loss of ipsilateral corneal reflex (touching the cornea with a piece of cotton [V] will normally cause the eye to blink [VII])
5. Flaccidity of the ipsilateral half of the floor of the mouth as a result of paralysis of the mylohyoid muscle and anterior belly of the digastric muscle
6. Insensitivity of the ipsilateral nasal mucosa to ammonia and other volatile substances
7. In some instances, sharp agonizing pain localized to the cutaneous distribution of one of the branches (trigeminal neuralgia or tic douloureux)
8. More sensitivity to sounds (especially loud noise) as a result of paralysis of the tensor tympani muscle

The trigeminal nerve has three interesting features. First, sympathetic postganglionics from the superior cervical ganglia and parasympathetic postganglionics are conveyed to the smooth muscles and glands of the eye and head by the three sensory branches. Recall that the parasympathetic preganglionics travel via nerves VII, VIII, and IX. Second, each of the three sensory branches has a distinct dermatome with very little overlap (unlike spinal nerves). Third, proprioceptive information from the extraocular and facial muscles reach the brainstem via nerves III, IV, VI, and VII.

These fibers are presumed to have their cell bodies in the mesencephalic nucleus.

Facial nerve (VII)

The facial nerve is actually two nerves: the facial nerve proper, containing special visceral efferents to the muscles of facial expression, and the nervus intermedius, containing general visceral efferents and general and special visceral afferents and general somatic afferents. Both nerves emerge from the lateral aspect of the pontomedullary junction just anterior to the eighth nerve. All three nerves then continue laterally through the internal acoustic meatus of the floor of the cranial cavity. Distal to this the special visceral efferents travel through the stylomastoid process (facial canal) to the muscles of facial expression and the stapedius muscle. The cell bodies of these motoneurons are found in the *facial motor nucleus* in the dorsolateral tegmentum of the lower pons. Note in Fig. 5-16 that the motor axons curve around the abducens nucleus before joining the nervus intermedius.

The general visceral efferents are actually parasympathetic preganglionics from the *superior salivatory nucleus*, which is located just lateral to the abducens nucleus. These preganglionics synapse with postganglionics that supply the lacrimal and salivary glands.

The afferent fibers of the nervus intermedius have their cell bodies in the *geniculate ganglia*. The peripheral processes travel to the taste buds of the anterior two thirds of the tongue (special visceral), to the skin of the back of the external ear (general somatic), and to the glands of the face (general visceral). General somatic afferents from muscle spindles probably travel in this nerve. The central processes of these afferents enter the brainstem and go to the spinal tract of the fifth nerve (general somatic) to join those from the trigeminal nerve and to the rostral and lateral portions of the solitary nucleus (general and special visceral). This part of the

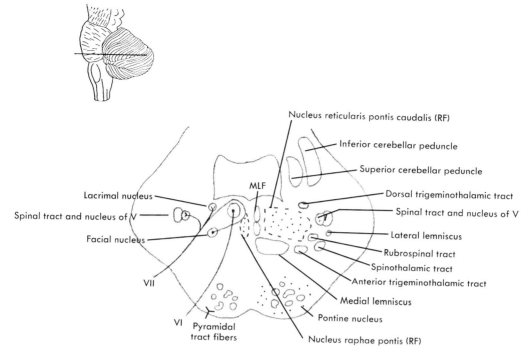

Fig. 5-16. Section through caudal pons.

solitary nucleus is called the gustatory nucleus.

A lesion of the seventh nerve can produce varying symptoms depending on the location of the lesion. The most common lesion is due to edema of the facial nerve proper in the facial canal of the petrosal bone as a result of an inflammatory reaction. These symptoms are associated with Bell's palsy and include the following:

1. Paralysis of the ipsilateral facial muscles with drooping of the corner of the mouth
2. Inability to close the eyelid, resulting in an irritated eye
3. Loss of the motor limb of the corneal reflex
4. Hyperacusis (increased sensitivity to sounds, especially low sounds) as a re-

sult of paralysis of the stapedius muscle, which normally dampens the vibration of the ear oscicles

A lesion of the seventh nerve at the brainstem can also include lack of tearing in the ipsilateral eye, decreased saliva production, and loss of taste on the ipsilateral anterior two thirds of the tongue.

A supranuclear lesion of the seventh nerve causes an interesting effect. The facial motor nucleus is divided into two portions, one for the upper facial muscles (above the eye) and the other for the lower facial muscles. The upper facial motoneurons receive bilateral hemispheric input, while the lower facial motoneurons receive only contralateral hemispheric input. A lesion of one motor cortex or internal capsule causes paralysis of only the contralateral lower facial muscles.

Vestibulocochlear nerve (VIII)

The vestibulocochlear nerve is actually two separate nerves from the inner ear that travel together. Both contain special somatic afferents.

The afferents of the vestibular nerve have their bipolar cell bodies in the *vestibular ganglion*, which is located within the distal part of the internal acoustic meatus. The peripheral processes travel laterally to terminate on the receptor cells of the ampullae of the semicircular canals or the receptor cells in the maculae of the utricle and saccule. The central processes travel medially through the internal acoustic meatus to the dorsolateral corner of the pontomedullary junction. These fibers then synapse in the vestibular nuclei (Fig. 5-17).

The afferents of the cochlear nerve have their bipolar cell bodies in the *spiral ganglion*, which is located in the modiolus bone of the cochlea (inner ear). The peripheral processes innervate the hair cells of the organ of Corti. The central processes join those of the vestibular nerve at the internal acoustic meatus and travel with them to the dorsolateral corner of the pontomedullary junction. Here they enter the brainstem to synapse in the cochlear nuclei.

A lesion of the eighth nerve can produce the following:

1. Deafness of the ipsilateral ear
2. Ringing in the ipsilateral ear (tinnitus) if the lesion is a slow compression as from a tumor (acoustic neuroma)
3. Nystagmus and vertigo from an imbalance in the input from the labyrinthine system, which may lead to nausea and vomiting and disturbances in gait
4. Possible decrease in protective extension responses to loss of balance
5. Possible changes in extensor tone on

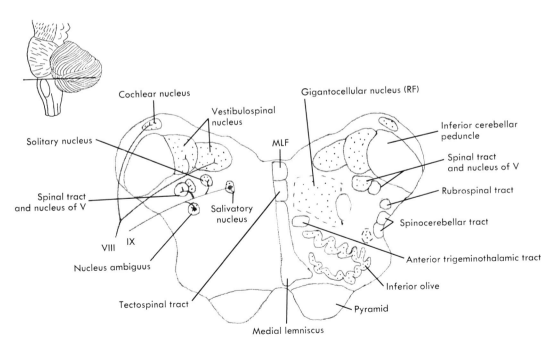

Fig. 5-17. Section through pontomedullary junction.

one side of the body (slight or transient)

Glossopharyngeal nerve (IX)

The glossopharyngeal nerve is composed of the following components:

1. Special visceral afferents conveying taste from the posterior third of the tongue and pharynx
2. General somatic afferents from the back of the external ear and external auditory meatus
3. General visceral afferents from the back of the tongue, upper pharynx, soft palate, tonsils, auditory tube, mucous membranes of the tympanic cavity, and the palatoglossal arches and from pressure receptors in the carotid sinus and chemoreceptors in the carotid body
4. General visceral efferents to the parotid gland
5. Special visceral efferents to the muscles of the upper pharynx involved with swallowing

The ninth nerve emerges from the lateral surface of the upper medulla as rootlets. These rootlets unite to form the nerve, which then travels with the tenth and eleventh nerves through the jugular foramen. The ninth nerve continues along the lateral side of the pharynx to the region of the tonsils and soft palate.

The cell bodies of the somatic afferents are located in the *superior* or *jugular ganglion* in the jugular foramen. The peripheral processes go to the skin of the ear, while the central processes enter the brainstem and join the spinal tract of the fifth nerve.

The cell bodies of the visceral afferents are found in the *inferior* or *petrosal ganglion*, which is located just distal to the jugular ganglion. The peripheral processes go to their respective areas, while the central processes enter the brainstem and terminate as follows. The taste afferents terminate in the rostral and lateral portion of the solitary nucleus; the afferents from the carotid sinus and body terminate in the medial and caudal part of the *solitary nucleus* for respiratory and cardiac reflexes; and the general visceral afferents from the back of the mouth and upper throat terminate in the spinal tract of the fifth nerve.

The visceral efferents to the parotid glands are parasympathetic preganglionics. Their cell bodies are in the *inferior salivatory nucleus* located in the dorsal part of the tegmentum of the lower pons.

The visceral efferents to the upper throat muscles for swallowing have their motoneurons in the *nucleus ambiguus* which is located in central tegmentum at the pons-medulla junction (Fig. 5-17).

Lesions of the ninth nerve include the following:

1. Loss of sensation and taste from the dorsal third of the tongue and adjacent areas (ipsilateral)
2. Unilateral loss of the gag reflex (sensory and motor limbs) with deviation of the uvula to the uninvolved side
3. Difficulty in swallowing (dysphagia)
4. Disturbance of the carotid sinus reflex, which may result in tachycardia.

Vagus nerve (X)

The vagus nerve is closely related to the glossopharyngeal nerve and is equally as complex. The vagus nerve is composed of the following:

1. Special visceral afferents from taste buds on the epiglottis
2. General visceral afferents from the structures of the throat, digestive system, and internal organs; pressure receptors in the aortic arch, heart, and lungs; and chemoreceptors in the aortic body and major thoracic arteries
3. General somatic afferents from the back of the external ear and external auditory meatus
4. Special visceral efferents to the muscles

of the soft palate, pharynx, and larynx for swallowing, regulating the laryngeal opening for respiration and phonation and regulating the opening of the auditory tube to equalize pressure in the middle ear

5. General visceral efferents (parasympathetic preganglionic) to the heart and smooth muscles and glands of the thoracic and abdominal viscera.

The vagus nerve emerges as rootlets from the lateral surface of the upper medulla just below the ninth nerve. It then travels laterally with the ninth and eleventh nerves through the jugular foramen into the neck. Branches are distributed to the ear, neck, thorax, and abdomen.

The somatic afferents have their cell bodies in the *jugular ganglion* of the jugular foramen (along with those from the ninth nerve). The central processes terminate in the spinal tract of the fifth nerve.

The visceral afferents have their cell bodies in the *inferior* or *nodose ganglion* located just distal to the superior ganglion. The special visceral afferents for taste enter the gustatory nucleus. The general visceral afferents terminate in the medial and caudal part of the solitary nucleus for cardiorespiratory and other visceral reflexes.

The general visceral efferents have their cell bodies in the *dorsal vagal (motor) nucleus*, which is located dorsally in the upper medulla between the solitary nucleus and hypoglossal nucleus (Fig. 5-18). These parasympathetic preganglionics travel to terminal ganglia associated with the larynx, esophagus, trachea, lungs, gastrointestinal tract (except bowels), heart, and associated glands.

The special visceral efferents for swallowing have their cell bodies in the *nucleus ambiguus*. These axons assist those from the ninth nerve in supplying the muscles of the throat.

A lesion of the vagus nerve can result in the following:

1. Flaccid soft palate resulting in a voice with a twang

2. Difficulty in swallowing
3. Deviation of the uvula to the uninvolved side during phonation
4. Transient tachycardia
5. Hoarse voice reduced to a whisper as the vocal cords become fixed near midline

Spinal accessory nerve (XI)

The spinal accessory nerve has two portions: a spinal root and a cranial root. Both roots contain special visceral efferents. The cranial root emerges from the lateral surface of the lower medulla below the vagus nerve. These axons are from the caudal portion of the *nucleus ambiguus*. The spinal root emerges from the lateral surface of the first five cervical cord segments. These axons have their cell bodies in the *ventral horns* (Fig. 5-19). After leaving the cord the axons angle upward to form a single root that passes through the foramen magnum to join with the cranial root. Both roots travel through the jugular foramen. Just distal to this foramen the cranial root joins with the vagus nerve and then forms the recurrent laryngeal nerve to innervate the intrinsic muscles of the larynx.

The spinal root descends in the neck in a dorsocaudal direction to innervate the ipsilateral sternocleidomastoid and upper trapezius muscles. General somatic afferents from muscle spindles probably travel in the upper cervical spinal nerves.

A lesion of the eleventh nerve can result in the following:

1. Weakness in contralateral head rotation (away from the side of the lesion)
2. Some weakness of neck (head) extension, flexion, and lateral bending
3. Weakness of ipsilateral shoulder shrugging
4. Some weakness in raising the ipsilateral arm above 90°
5. Ipsilateral shoulder sag
6. A hoarse weak voice caused by laryngeal paralysis on one side

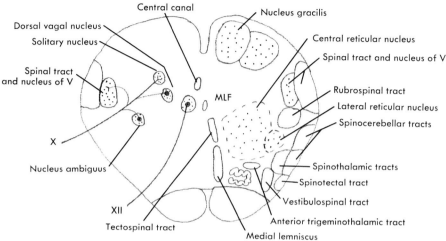

Fig. 5-18. Section through lower border of inferior olive.

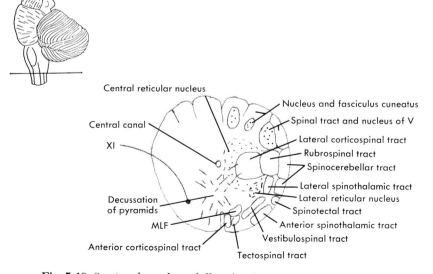

Fig. 5-19. Section through medulla at level of pyramidal decussation.

Table 5-4. Cranial nerves

Name	Nuclei	General location of nuclei	Function
I Olfactory	None	Inferior surface of frontal lobe near midline	Smell
II Optic	Lateral geniculate body	Posterolateral part of pulvinar	Sight
III Oculomotor	Edinger-Westphal nucleus	Dorsomedial part of midbrain at level of superior colliculus	Parasympathetic to eye-pupil constriction, lens accommodation
	Nucleus of III		Extraocular eye muscles: medial, inferior and superior recti, inferior oblique, and levator palpebrae
IV Trochlear	Nucleus of IV	Dorsomedial part of midbrain at level of inferior colliculus	Superior oblique muscle (causes eye to look down and out)
V Trigeminal	Motor nucleus	Dorsolateral part of the middle third of pons	Muscles of mastication, tensor tympani
	Mesencephalic nucleus	Dorsolateral part of lower half of midbrain	Proprioception, muscles of mastication
	Main sensory nucleus	Dorsolateral part of middle third of pons	Sensation of face, anterior part of scalp, mucosa, anterior two thirds of tongue, dura
	Spinal tract nuclei	Dorsolateral part of lower pons and medulla	Sensation of face, anterior part of scalp, mucosa, anterior two thirds of tongue, dura
VI Abducens	Nucleus of VI	Dorsomedial part of midpons	Lateral rectus (makes eye look laterally)
VII Facial	Motor nucleus	Dorsolateral part of lower pons	Muscles of facial expression, stapedius
	Salivatory nucleus	Dorsolateral part of lower pons	Production of saliva and tears
	Solitary nucleus	Dorsolateral part of lower pons	Taste, anterior two thirds of tongue
	Spinal tract and nucleus of V	Dorsolateral part of lower pons	Skins of ear

VIII	Vestibulocochlear	Nuclei of VIII	Dorsolateral part of lower pons and upper medulla	Balance and hearing
IX	Glossopharyngeal	Nucleus ambiguus	Dorsolateral part of upper medulla	Swallowing
		Solitary nucleus	Dorsal part of upper medulla	Taste, posterior third of tongue, blood pressure, regulation
		Salivatory nucleus	Dorsal part of upper medulla	Production of saliva
		Spinal tract and nucleus of V	Dorsal part of upper medulla	Touch, pain, temperature of posterior third of tongue, pharynx, and part of external ear; proprioception and facial muscles (?)
X	Vagus	Dorsal motor nucleus	Dorsolateral part of upper medulla	Parasympathetic to thoracic and abdominal viscera and heart
		Nucleus ambiguus	Dorsolateral part of upper medulla	Swallowing
		Solitary nucleus	Dorsolateral part of upper medulla	Respiratory, cardiovascular, and alimentary reflexes
		Spinal tract and nucleus of V	Dorsolateral part of upper medulla	Pain, temperature, and touch for pharynx, posterior tongue, and external ear
XI	Spinal accessory	Nucleus ambiguus	Dorsolateral part of midmedulla	Intrinsic muscles of larynx (not important)
		Ventral horn	C2 to C4	Sternocleidomastoid (C2) and upper trapezius (C3, C4), contralateral head rotation and ipsilateral shoulder elevation
XII	Hypoglossal	Nucleus of XII	Dorsomedial part of upper two thirds of medulla	Tongue movements

Spasticity may involve the upper trapezius and sternocleidomastoid muscles, resulting in torticollis (wryneck).

Hypoglossal nerve (XII)

The hypoglossal nerve is composed of general somatic efferents. Their cell bodies are found in the *hypoglossal nucleus*, which is a long column of cells in the dorsomedial portion of the medulla (between the dorsal vagal nuclus and midline) (Fig. 5-18). The axons travel ventrally along the lateral surface of the medial lemniscus. They emerge from the upper medulla between the inferior olive and pyramid. They pass through the hypoglossal canal and arc laterally in the pharynx to the muscles of the tongue. General somatic afferents from muscle spindles probably travel in this nerve.

A lesion of the twelfth nerve results in ipsilateral paralysis of the tongue. On protrusion the tip of the tongue deviates to the side of the lesion as a result of the unopposed action of the contralateral genioglossus muscle. Since many of the intrinsic tongue muscles cross midline, there is little functional disturbance. A unilateral upper motoneuron lesion may cause the tongue to temporarily deviate away from the side of the lesion.

Summary of cranial nerve function

Because of the added complexity of the special senses, the muscles of branchiomeric origin, and the many nuclei and ganglia, the cranial nerves can be somewhat overwhelming. A simplified summary is presented below. Reference to Table 5-4 is also helpful.

Afferent components. Conscious sensations from receptors in the skin of the face and scalp, eye, and mucous lining of the nose, mouth, and pharynx are relayed to the main sensory nucleus and spinal nuclei of the fifth nerve. Reflex connections are available for blinking, sneezing, coughing, gagging, swallowing, sucking, and rooting responses.

Visual data are relayed to the superior colliculus for reflex protective reactions of the head and body (via the tectospinal tract).

Gustatory data synapse in the upper portion of the solitary nucleus. Reflex connections to the salivatory nucleus are especially strong.

Unconscious visceral sensations in the vagus and glossopharyngeal nerves synapse in the lower portion of the solitary nucleus. Reflex connections are made with the "vital centers" and the vomit center.

Proprioceptive data from muscle spindles in the extraocular and mastication muscles have their unipolar cells bodies in the mesencephalic nucleus of the fifth nerve. Those from the sternocleidomastoid and upper trapezius have their cell bodies in dorsal root ganglia of C2 to C4. Information regarding spindles and their afferents from lingual and facial muscles is inconsistent.

Efferent components. The activities of eating involve salivation (inferior and superior salivatory nucleus), chewing (motor nucleus of the fifth nerve and hypoglossal nucleus), swallowing (nucleus ambiguus), and sucking (facial motor nucleus).

The occulomotor, trochlear, and abducens nuclei move the eye in its socket and open the eyelid.

The facial motor nucleus produces facial expression and closes the eyelid.

The motoneurons of C1 to C5 move the head in different planes.

The Edinger-Westphal nucleus constricts the opening of the pupil and focuses the lens for light and accommodation reactions.

The lacrimal nucleus causes the eye to tear.

TEST QUESTIONS

1. Early compression of a peripheral nerve may initially involve only the larger fibers. Which of the following is true?
   a. Threshold to electrical stimulation is raised.
   b. Position sense may be lost but pain sensibility retained.

c. Hyporeflexia occurs.

d. Compound AP is smaller in amplitude.

2. In conditions of anoxia:

a. Smaller fibers are affected initially.

b. Synaptic transmission is unaffected.

c. NCV decreases.

d. More specialized neurons are affected initially.

3. Anesthesia of both hands to the wrist could be caused by:

a. Lesion of the C6 to C8 dermatomes.

b. Hysterical reaction.

c. Polyneuropathy.

d. Lesion of the ulnar, median, and radial nerves.

4. Which of the following symptoms commonly occur with a complete lesion of a peripheral nerve?

a. Dissociated sensory loss.

b. Causalgia.

c. Absent DTRs.

d. Fasciculations.

5. Compared with spinal nerves, cranial nerves:

a. Have no dorsal or ventral roots.

b. Contain no general somatic efferent fibers.

c. Are unmyelinated.

d. Are distributed to structures only within the skull.

6. List the cranial nerves associated with the following functions:

a. Swallowing _____

b. Taste _____

c. Chewing _____

d. Salivation _____

e. Ocular movements _____

f. Convergence _____

g. Lateral gaze _____

h. Blood pressure _____

i. Tearing _____

7. If the latency between the time of stimulation and time of recording is 3 msec and the recorded velocity is 60 m/sec, what is the distance between stimulating and recording electrodes?

a. 14 cm

b. 10 cm

c. 22 cm

d. 18 cm

8. The size of the EPSP, EPP, or IPSP:

a. Is related to the amount of neurotransmitter released.

b. Is related to the amount of antitransmitter substance on the postsynaptic membrane.

c. Determines the frequency of firing from the axon hillock.

d. Varies with the frequency of presynaptic stimulation.

9. Place the following events in the appropriate order:

a. Opening the Na^+ gates.

b. Depolarization of the axon hillock.

c. Influx of Ca^{++}.

d. Production of an EPSP.

10. In denervation the involved muscle:

a. Has its strength-duration curve shift upward and to the left.

b. Responds briskly to direct current.

c. Shows fibrillation potentials.

d. Undergoes severe atrophy.

11. An overdose of epinephrine causes:

a. Increase in blood pressure.

b. Increase in blood flow to skeletal muscle.

c. Urine retention.

d. Dilation of the pupil.

12. If the parasympathetic system becomes too active, which of the following *could* result?

a. Ulcers.

b. Fainting.

c. Difficulty seeing in the dark.

d. Difficulty breathing.

13. Stimulation of the beta receptors:

a. Is accomplished by injections of muscarine.

b. Inhibits motility of the gastrointestinal tract.

c. Increases blood pressure.

d. Elevates blood glucose level and insulin secretion.

14. Damage to the sacral dorsal roots bilaterally could cause:

a. Autonomic dysreflexia.

b. Urine retention.

c. Impotence.

d. Tropic effects in the pelvic organs.

15. If the radial nerve is severely crushed:

a. Causalgia may result.

b. Referred pain may be felt over the sternum.

c. Tropic effects can occur.

d. Stellate ganglion block may help in the relief of pain.

16. If the lateral portion of the lower pons and upper medulla were compressed by a tumor, which of the following symptoms could result?
    a. Medial strabismus.
    b. Ipsilateral paralysis of the facial muscles.
    c. Nystagmus.
    d. Deviation of the tongue toward the side of the lesion.

17. The amount of transmitter released:
    a. Is determined by the amplitude of the pre-synaptic AP.
    b. Increases if the extracellular Na^+ is increased.
    c. Decreases if the presynaptic membrane has already been partially depolarized.
    d. Is constant because it follows the "all or none" law.

18. When a nerve is stimulated along its course:
    a. Dromic stimulation of the alpha axons may cause a muscle contraction.
    b. Dromic stimulation of the sensory fibers causes the receptors to depolarize.
    c. A second muscle response can be recorded as a result of dromic sensory stimulation.
    d. No potentials will be recorded from the ventral root.

19. Which of the following fibers are classified as type B?
    a. Myelinated somatic afferents and efferents.
    b. Unmyelinated somatic afferents.
    c. Myelinated autonomic preganglionics.
    d. Unmyelinated autonomic postganglionics.

20. The white rami communicantes:
    a. Connect spinal cord segments T1 to L2 with the sympathetic chain ganglia.
    b. Connect the chain ganglia with all spinal cord segments.
    c. Should be severed instead of the gray rami to avoid denervation hypersensitivity.
    d. Can cause Horner's syndrome if those to the C^1 to C^4 spinal nerves are severed.

21. Dermatomes are characterized by:
    a. Innervation of one spinal nerve.
    b. Overlap from adjacent dermatomes.
    c. Symmetric distribution on both sides of the body.
    d. Anesthesia if the dorsal root is severed.

22. Inhibitory synapses:
    a. Show an increased permeability to Na^+.
    b. Cause a hyperpolarization of the post-synaptic membrane.
    c. Are found only in the CNS.
    d. Reduce the frequency of production of APs.

23. All of the following structures receive only sympathetic innervation except:
    a. Adrenal medulla.
    b. Sweat glands.
    c. Lungs.
    d. Pilomotor muscles.

24. In the autonomic nervous system:
    a. Parasympathetic preganglionics to the bladder originate in segments T8 to T12.
    b. Postganglionic sympathetics to the arm come from the stellate ganglion.
    c. Epinephrine is liberated at all sympathetic effector sites.
    d. Horner's syndrome can result from a lesion of the T1 to T4 ventral roots.

25. Match the cranial nerve nuclei with their appropriate location or function:

____ a. Nucleus ambiguus	1.	Midpons, lateral portion
____ b. Upper solitary nucleus	2.	Midbrain
____ c. Main sensory nucleus	3.	Parasympathetic preganglionics
____ d. Edinger-Westphal nucleus	4.	Taste
	5.	Medulla, lateral portion
	6.	Swallowing

SUGGESTED READINGS

Chusid, J. G.: Correlative neuroanatomy and functional neurology, ed. 16, Los Altos, Calif., 1976, Lange Medical Publications.

Goodgold, J., and Eberstein, A.: Electrodiagnosis of neuromuscular diseases, Baltimore, 1972, The Williams & Wilkins Co.

Everett, N. B.: Functional neuroanatomy, ed. 6, Philadelphia, 1971, Lea & Febiger.

Eyzaguirre, C., and Fidone, S. J.: Physiology of the nervous system, ed. 2, Chicago, 1975, Year Book Medical Publishers, Inc.

Licht, S.: Electrodiagnosis and electromyography, ed. 2, Baltimore, 1961, Waverly Press.

Netter, F.: The CIBA collection of medical illustrations.

I. The nervous system, Summit, N.J., 1972, CIBA Chemical Co.

Noback, C., and Demarest, W.: The human nervous system, ed. 2, New York, 1975, McGraw-Hill Book Co.

Pierce, D., and Nickel, V.: The total care of spinal cord injuries, Boston, 1977, Little, Brown & Co.

Schmidt, R., editor: Fundamentals of neurophysiology, New York, 1975, Springer-Verlag.

Willis, W. D., Jr., and Grossman, R. G.: Medical neurobiology: neuroanatomical and neurophysiological principles basic to clinical neuroscience, ed. 2, St. Louis, 1977, The C. V. Mosby Co.

CHAPTER 6

Spinal cord

Compared with the other components of the central nervous system the spinal cord is less complex in structure. In addition the functions of the spinal cord have been well documented, in part because of less complexity and in part because of ease of access for experimental work. In this chapter are discussed the anatomic composition of the spinal cord including the vascular supply, the functional characteristics of the motoneurons, the activity of spinal internuncials including the Renshaw cell, and the spinal reflexes. Additional data about the fiber tracts of the spinal cord and clinical aspects are presented in greater detail in Chapters 8 to 10.

The following areas deserve special attention:

1. Types of information conveyed in the three funiculi
2. Functional arrangement of the gray matter of the cord in terms of nuclei, lamina, and motoneurons of the ventral horn
3. Relationship of the meninges to the spinal cord and spinal nerves
4. Arterial distribution to the white and gray matter of the cord
5. Initiation of an AP on the axon hillock and the influence of spacial and temporal summation
6. Effects of presynaptic depolarization on the output of the motoneuron
7. Comparison between motoneurons supplying phasic muscle fibers and those supplying tonic muscle fibers

8. Effects of repetitive discharge and posttetanic potentiation on CNS neurons
9. Method of excitation and influence of the Renshaw cell
10. Adequate stimulus, spinal pathways, and functional role of the somatic spinal reflexes
11. Description of Babinski's and Hoffman's signs and clonus
12. Expected pattern of movement seen with the overflow reactions

ANATOMIC COMPOSITION
External features

The spinal cord begins at the foramen magnum and travels caudally to vertebra L-1 or L-2. When viewed in full length the spinal cord has somewhat of an hourglass shape with bulges in the cervical and lumbosacral areas. The anterior and posterior surfaces are dented through their entire length by the anterior median fissure and dorsal median sulcus. In the upper portion of the dorsal surface is the dorsal intermediate sulcus. The terminal tip of the cord is known as the conus medullaris.

Associated with the spinal cord are 31 pairs of ventral roots, dorsal roots, and spinal nerves. The attachment of these rootlets breaks up the long axis of the cord into 31 segments. Spinal cord segmentation is related clinically to three factors.

First, the cutaneous distribution of one spinal nerve (dermatome) can be a useful evaluative aid in determining the level of a spinal cord lesion.

Second, afferent impulses from one dorsal root can enter the cord and reflexly influence activity of motoneurons of that same segment.

Third, since each spinal nerve must exit the vertebral canal via its corresponding vertebra, there is a downward angling of the spinal nerves beginning with the lower cervical segments. Thus many traumatic cord lesions can affect spinal nerves as well as the cord itself. For example, a bullet wound through the lower sacral cord (above the twelfth rib) may also damage the lumbar spinal nerves.

Internal organization

The white matter of the cord is divided into three pairs of *funiculi* (anterior, lateral, and dorsal) and a commissure area. Within the funiculi are the following:

1. Long ascending fiber tracts for conscious awareness of sensations and for relay of somatic visceral and proprioceptive sensory data to subcortical motor centers
2. Long descending fiber tracts from the higher (supraspinal) motor centers
3. Short descending and ascending tracts for intersegmental reflexes

The commissure area is in front of the central canal and is known as the anterior white commissure. It contains fibers traveling to the opposite side of the spinal cord for reflexes or to form a fiber tract.

It is important clinically to know the general location and function of the tracts in each of the funiculi. The location of these tracts is shown in Fig. 6-1. In the *anterior funiculus* the fiber tracts include the following:

1. Motor tracts from the vestibular nuclei, reticular formation, and tectum for the regulation of muscle tone and posture
2. A small motor tract from the ipsilateral motor cortex for voluntary skilled movements
3. A sensory tract for conscious awareness

of crude touch, tickle, and itch from the opposite side of the body and possibly for sexual sensations (orgasm)
4. An intersegmental pathway for spinal reflexes (propriospinal tract)

The *lateral funiculus* contains the following:

1. A large motor tract from the contralateral motor cortex for voluntary skilled movements
2. A tract from the red nucleus for flexor muscle tone
3. Two tracts (one crossed, one uncrossed) relaying muscle spindle and Golgi tendon organ data to the cerebellum
4. A tract relaying pain and temperature from the opposite side of the body to consciousness
5. A small tract conveying unconscious proprioceptive data to the inferior olive primarily from the cervical cord
6. Two tracts conveying exteroceptive data to the reticular formation and tectum
7. An intersegmental ascending and descending pathway for reflex connections (propriospinal tract)
8. An ascending visceral pathway conveying visceral sensations to the reticular formation and diencephalon (tract not well defined and not shown in Fig. 6-1)

The *dorsal funiculus* contains two ascending pathways, the fasciculus gracilis and the fasciculus cuneatus, which together are known as the dorsal columns. The fasciculus gracilis conveys sensations of pressure, discriminating touch, vibration, kinesthesia from the lower extremities and trunk to consciousness. The fasciculus cuneatus conveys the same sensations from the upper extremities to consciousness.

The dorsal columns also contain two small intersegmental pathways. The fasciculus septomarginalis, in the lower half of the cord, transmits collaterals from the fasciculus gracilis to internuncials and motoneurons in different segments. The fasciculus interfas-

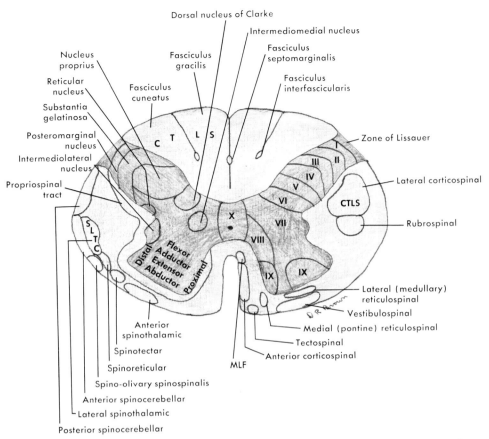

Fig. 6-1. Cross-section of spinal cord showing organization of gray and white matter. This is composite drawing showing composition of various cord levels. There would be some variations in structure at different levels. Note organization of dorsal columns, lateral spinothalamic tract, lateral corticospinal tract, and ventral horn. *C*, Cervical; *T*, thoracic; *L*, lumbar; *S*, sacral.

cicularis, in the upper half of the cord, does the same for the fasciculus cuneatus.

The gray matter of the spinal cord appears roughly as the letter *H* in transverse section. It is divided into a *dorsal horn* for afferent input, a *ventral horn* for somatic efferent output, an *intermediolateral horn* between segments T1 and L2 for sympathetic preganglionics (visceral efferents), and a central commissure around the central canal for crossed reflexes and fiber tracts.

The gray matter has also been organized on a more specific functional basis into *nuclei* and *lamina* (Figs. 2-3 and 6-1). These nuclei and lamina can be thought of as longitudinal columns of neuron cell bodies (many of which extend the entire length of the cord) with a similar function or structure. Table 6-1 presents a summary of the location and function of these nuclei.

The ten laminae designated by Rexed are different in anatomic arrangment from the

Table 6-1. Characteristics of spinal cord nuclei

Nucleus	Location and extent in cord	Function
Substantia gelatinosa (Rolandi)	Occupies apex of dorsal horn throughout entire length	Relay of pain, temperature, and other spinothalamic sensations; probably modification of transmission of sensory input
Nucleus dorsalis (Clarke column)	Ventromedial part of dorsal horn from C8 to L2 or L3	Relay of unconscious proprioceptive data; axons form dorsal spinocerebellar tract
Nucleus proprius	All of dorsal horn except area occupied by substantia gelatinosa and nucleus dorsalis	Relay of ascending and descending data
Visceral afferent nucleus	Base of dorsal horn between T1 and L2 or L3 and between S2 to S4	Relay of visceral afferent data to autonomic preganglionics in cord and brainstem and to autonomic regulatory centers in brain
Medial motoneuron column	Medial portion of ventral horn through entire length of cord	Innervates neck and trunk muscles
Lateral motoneuron column	Lateral portion of ventral horn between C4 and T1 and between L2 and S3	Innervates muscles of extremities
Phrenic nucleus	Medial portion of ventral horn between C3 and C5	Innervates diaphragm
Spinal accessory nucleus	Lateral region of ventral horn between C1 and C6*	Origin of spinal root of cranial nerve XI
Intermediolateral horn	Base of lateral border of ventral horn between T1 and L2 or L3	Origin of sympathetic preganglionic neurons
Sacral parasympathetic nucleus	Base of lateral division of ventral horn between S2 and S4	Origin of parasympathetic preganglionics for pelvic viscera

Adapted from Barr, M.: The human nervous system, New York, 1972, Harper & Row, Publishers.
*Fibers of spinal root of XI also originate from supraspinal nucleus, which is continuation of ventral horn into lower medulla.

Table 6-2. Organization of lamina of spinal cord

Lamina	Characteristics
I	Thin covering of cells that form apex of dorsal horn. There are only a few synapses of dorsal root fibers in this zone.
II	Densely packed with small neurons and corresponds to area of substantia gelatinosa. There are few synapses from dorsal root fibers, which disputes previously accepted beliefs. It is thought that these cells may instead project short intersegmental collaterals, possibly for reflex effects.
III	Band of less densely packed cells that receives greatest number of synapses from entering dorsal root fibers.
IV	Band of cells that receives dorsal root afferents. Included in this layer are most cells of nucleus proprius.
V	Broad zone extending across neck. Lateral portion is reticular nucleus. This nucleus is a continuation of brainstem reticular formation and is particularly prominent in cervical cord. Medial portion, which is most prominent in cervical and lumbosacral areas, receives dorsal root afferents and descending suprasegmental motor fibers (corticospinal and rubrospinal tracts).
VI	Broad band at base of dorsal horn, found only in cervical and lumbar areas. Medial portion receives large number of Ia and Ib afferents. Lateral portion receives fibers from descending pathways. Some axons of these neurons enter fasciculus proprius system.
VII	Occupies most of intermediate zone of gray matter, although its boundaries may vary at different levels. Both nucleus dorsalis (of Clarke) and intermediolateral horns are found here. In cervical and lumbar areas lamina VII projects into ventral horn, which contains many internuncials, Renshaw cells, and gamma motoneurons known to be present. Dorsal root afferents and descending motor and autonomic fibers terminate in lamina VII.
VIII	Although not always sharply distinguished from lamina VII, lamina VIII is found in medial portion of ventral horn. Many descending motor tracts (vestibulospinal, reticulospinal, MLF) terminate here.
IX	Occur as clusters of alpha motoneurons throughout lateral and anterior portion of ventral horn. There is topographic distribution of these alpha motoneurons to different skeletal muscle groups. Fibers from other laminae and descending motor tracts terminate here.
X	Occupies area around central canal and contains mostly neuroglial cells.

nuclei. This is due in part because Rexed used neurons of similar structure (cytoarchitectonics) to delineate the laminae. The advantage of Rexed's laminae is that they eliminate some of the variation in terminology used with the nuclei. The laminae method has become widely used for localizing axonal degeneration studies in the mammalian spinal cord. Table 6-2 describes the laminae of the spinal cord.

The motoneurons of the ventral horn are functionally organized according to their distribution to the body's musculature. The organization of a lower cervical ventral horn is shown in Fig. 6-1. The lumbar ventral horns are arranged similarly in terms of distal-proximal and flexor-extensor motoneurons. Knowledge of this organization can provide useful clues in evaluating traumatic cord lesions.

The spinal cord is extremely organized, like the rest of the CNS. This organization also extends to dorsal root fibers. Just before entering the spinal cord the lightly myelinated and unmyelinated fibers form a lateral division, while the more heavily myelinated fibers form a medial division. The lateral division fibers enter the dorsolateral fasciculus (zone of Lissauer) and give off short ascending and descending collaterals, most of which terminate in laminae II and III. The lateral division fibers are concerned with pain, temperature, crude touch, tickle, itch, and sexual sensation and form the spinothalamic pathways. The medial division fibers enter

the dorsal funiculus directly and ascend to higher centers. These fibers are concerned with discriminative touch, pressure, joint position sense, and vibration. They form the fasciculus gracilis and fasciculus cuneatus tracts (the dorsal column pathway). Some of these fibers terminate in laminae V, VI, and VII for relay to higher centers and lamina IX for spinal reflexes.

Meningeal coverings

The connective tissue coverings of the CNS are known as the meninges. The meninges are continuous throughout the bony confines of the skull and vertebral canal and actually continue as covering around the spinal nerves into the peripheral nervous system (Fig. 6-2). The meninges are organized into a tough outer layer known as the *dura*

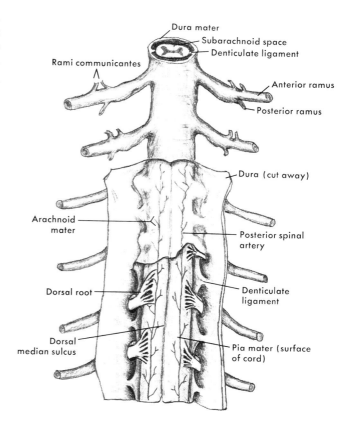

Fig. 6-2. Meningeal coverings and denticulate ligaments of spinal cord. Note that denticulate ligaments come off lateral side of cord and attach to dura. Also note that dura surrounds spinal nerves. (Adapted from Netter, F.: The CIBA collection of medical illustrations. I. The nervous system, Summit, N.J., 1972, CIBA Chemical Co.)

mater, a delicate middle layer called the *arachnoid mater*, and an inner layer adherent to the entire surface of the CNS known as the *pia mater*. In the skull the dura acts as the periosteum during growth, so there is no epidural space. In the vertebral canal there is an epidural space that is filled with fat and veins. Between the dura and arachnoid is the *subdural space*. Between the arachnoid and pia is the *subarachnoid space*. The subdural space is relatively insignificant unless there is hemorrhaging of blood into the area (subdural hematoma). The subarachnoid space is important because it is the location of cerebrospinal fluid. Below the tip of the cord (conus medullaris) around vertebra L-2 is a large area of subarachnoid space containing the cauda equina nerves and CSF. It is here that a lumbar puncture is performed (usually the needle is inserted between the L-4 and L-5 vertebrae).

The composition and specific functions of the meninges are similar for the spinal cord and brain and are discussed in Chapter 7. Unique to the spinal cord are the *filum terminale* and *denticulate ligaments*. The filum terminale is a thread of pia mater from the conus that courses caudally to merge with the dura and attaches to the coccyx. The denticulate ligaments are thick threads of pia that project laterally and attach to the dura. These occur at each segmental level. Both the filum terminale and the denticulate ligaments help to support and anchor the spinal cord within the subarachnoid space.

Vascular supply

The arterial supply to the spinal cord is derived from two sources: the vertebral arteries and the radicular arteries that branch off the aorta (Fig. 6-3).

Before they unite to form the basilar artery, the vertebral arteries give off three descending branches to the cord: an *anterior spinal artery* that travels in the anterior median fissure and two *posterior spinal arteries* that travel medial to the dorsal roots.

The radicular arteries that branch off the aorta travel along the spinal nerves, pass through intervertebral foramina, and then bifurcate into *anterior* and *posterior radicular arteries*. Interconnecting branches between these and the spinal arteries act to form a ring (arterial vasocorona) around the spinal cord.

All cord segments receive blood from the spinal arteries. In general the anterior spinal artery supplies the anterior two thirds of the cord except for peripheral portions of the lateral and anterior funiculi, which are supplied by branches from the arterial corona. Each

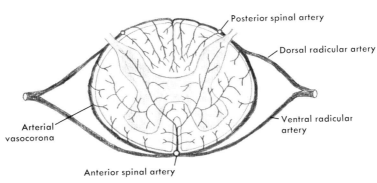

Fig. 6-3. Arterial supply of spinal cord. Note that there are no anastomoses within cord.

posterior spinal artery supplies a posterior funiculus.

The radicular arteries do not uniformly supply all spinal segments, resulting in areas with less collateral circulation for possible shunting of blood in conditions of ischemia. This is complicated by the fact that there are no anastomoses between arteries *within* the spinal cord. The result is that there are areas of the cord that are susceptible to ischemia. These areas are found at C2-C3, T3-T4, and L1-L2.

The venous drainage from the spinal cord follows the spinal arteries traveling along the dorsal and ventral roots to large epidural venous plexuses. These veins drain into the large veins of thoracic, abdominal, and pelvic cavities. These large venous plexuses lie on the posterior surfaces of the vertebral bodies in the spinal canal. They can become troublesome in surgical procedures for a herniated disc.

ALPHA MOTONEURON

Of all of the neurons in the spinal cord, and indeed in the CNS, the alpha motoneuron is probably the best known. It is, after all, the source of voluntary movement. It is large and easy to study and is often used as a baseline neuron with which other neurons are compared.

External features

The alpha motoneuron is among the largest of spinal neurons and may measure up to 100 μm in diameter. There are from 6 to 12 primary dendrite stalks, some of them projecting as much as 1 mm from the soma. These dendrites yield many short branches that act to receive many thousands of synaptic terminals. There is only one axon from the soma. It leaves the spinal cord with other motor axons, forming rootlets that converge to form the ventral root for that segment. One motor axon may extend more than 1 m and supply from 10 to 25 to 2000 muscle fibers, forming a motor unit. A motoneuron and its processes are known as the *final common pathway* or the *lower motoneuron.*

Although some motor axons lack collateral branches, many give off a branch in the spinal cord that travels back toward the soma. These branches are known as the *recurrent collaterals.* They are known to synapse on *Renshaw cells.* These cells synapse on the dendrites of the same neuron and act as an autoinhitor to dampen the output of the motoneuron.

Method of excitation

Like any neuron, the alpha motoneuron responds to the algebraic summation of all of the inhibitory and excitatory synapses converging on the dendrites. The number of synapses coverging on the soma and dendrites may be 10,000 or more. All of these synapses either cause positive ions to flow away from or toward the axon hillock. If enough positive ions leave the axon hillock area, that part of the membrane will depolarize to the critical firing level and an action potential will be produced and propagated dromically toward the motor unit. The axons of alpha motoneurons are heavily myelinated, and conduction is rapid. Remember, the axon hillock has the lowest threshold to excitation.

Of interest, when the axon hillock fires, an AP is not only conducted dromically along the axon but also antidromically to the soma and dendrites. The afterpotentials of the axon hillock following the AP are small. Those on the soma-dendrite membrane are quite long. It is felt that this may be a major factor in regulating the output frequency of a motoneuron.

It is important to realize that the local synaptic potentials (EPSPs or IPSPs) are additive. Assume, for example, that in each excitatory synapse one presynaptic AP causes enough neurotransmitter to be released to depolarize the postsynaptic membrane 10 mV. This depolarization wave is conducted *with* decrement along dendrite and soma membranes to the axon hillock. At this point the amount of depolarization is only 2 mV. In order to produce an AP the axon hillock must be depolarized 20 to −50 mV. Several options are available: the one synapse can be activated 10 times consecutively, or ten different excitatory synapses can be activated, or any combination producing the necessary 20 mV of depolarization can be activated.

Although the numbers are somewhat arbitrary, this example explains the concepts of *spatial* and *temporal summation*. In spatial summation synapses from many sources are activated; in temporal summation a few synapses are activated at a higher frequency. The end result is decrease or increase of the output from the axon hillock. Spatial and temporal summation can be illustrated clinically in a flaccid muscle. To add tone to this muscle spatial summation can be used by recruiting several sources of input such as a tendon tap and the tonic labyrinthine reflex. Temporal summation can be used by vibrating the muscle.

In the discussion of synaptic activity above and in Chapter 5 the implication is that either the synapse is "on" or "off." This is not actually the case, as the motoneuron membrane shows fluctuations in its resting membrane potential (RMP). These fluctuations resemble EPSPs and occur spontaneously in the absence of any overt stimulation. This activity is at least partly the result of (1) presynaptic APs from afferent fibers and internuncials that are spontaneously active and (2) the spontaneous release of a quantal amount of transmitter substance. The functional significance of this activity is unclear. Since it brings the motoneuron closer to firing, this spontaneous activity may play a role in posture and may explain why some individuals have a better aptitude for motor skills.

Methods of inhibition

If the neurotransmitter released at a synapse causes K^+ ions to leave the postsynaptic area a local hyperpolarization results (IPSP). If enough IPSPs are generated in excess of EPSPs there will be a flow of positive ions along the dendrite and soma membranes toward the axon hillock. This hyperpolarization of the axon hillock will depress the generation of APs. A previously excitatory stimulus may have a reduced effect or no effect if inhibitory synapses are active at the same time. By altering the amount of inhibition and excitation on a motoneuron in an appropriate sequence, coordinated patterns of movement can be produced.

The above type of inhibition is known as *postsynaptic inhibition*. It uses neurotransmitters such as glycine and gamma-aminobutyric acid, which are different from excitatory transmitters (ACh, norepinephrine). Postsynaptic inhibition can occur with excitatory synapses on one motoneuron or on a motoneuron that supplies an antagonist muscle. Whenever flexor motoneurons are excited antagonist extensor motoneurons may be inhibited, thus allowing movement. This pattern is known as *reciprocal inhibition (innervation)*. Skills involving rapid alternating movements such as typing require efficient reciprocal inhibition.

Another form of inhibition is known as *primary afferent depolarization (PAD)* or *presynaptic inhibition* of an excitatory synapse. This is somewhat confusing, as it actually involves an excitatory synapse. In presynaptic inhibition an axon synapses with a presynaptic axon or bouton (Fig. 6-4). When activated the axon-axon synapse (which is an excitatory synapse) causes a local depolarization on the presynaptic terminal. This membrane is technically the postsynaptic membrane for the first synapse. This local depolarization reduces the amplitude of an AP traveling down axon number two. If the RMP is raised from −70 to −60 mV the amplitude of an AP is reduced by 10 mV. Recall from Chapter 5 that the amount of transmitter released is directly proportional to the amount of calcium that enters the bouton. With less transmitter substance released the EPSP is smaller and there is a reduced output from the axon hillock.

Although the significance of presynaptic inhibition is not completely understood,

there is at least one advantage to the motoneuron. Presynaptic inhibition reduces excitatory input to a neuron, thereby reducing its output *without* directly inhibiting the soma. In other words, the excitability of the neuron is unaltered and the neuron is free to respond to other pathways.

Presynaptic inhibition has been well documented in the spinal cord involving the

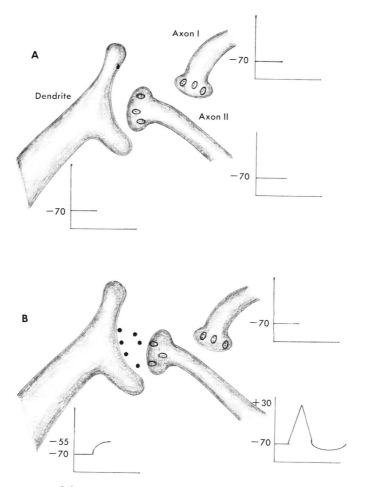

Fig. 6-4. Presynaptic inhibition. **A,** Resting state. RMP of dendrite and both axons is −70 mV. **B,** AP on axon II releases specific amount of transmitter, which depolarizes dendrite to −55 mV (EPSP).

Continued.

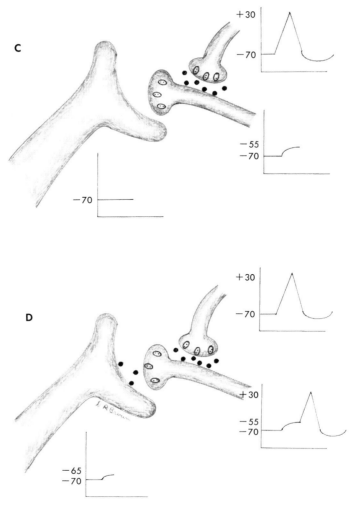

Fig. 6-4, cont'd. C, AP on axon I releases specific amount of transmitter, which depolarizes axon II to −55 mV. **D,** AP on axon II during AP of axon I is smaller in amplitude (−55 mV to +30 mV), causing less transmitter to be released from bouton II and smaller EPSP (−65 mV).

Ia afferents. It is presumed to exist in other areas of the CNS. The existence of presynaptic inhibition of an inhibitory synapse and a presynaptic excitation (involving a hyperpolarization of the bouton) have not as yet been found.

Relationship of motoneuron to muscle fiber

There is an interesting relationship between a motoneuron and the muscle cells it supplies. Basically there are three kinds of muscle cell: slow-fatigue, fast-fatigue, and fatigue-resistant fibers.

Slow-fatigue fibers are the red muscle cells that have a high mitochondrial density and oxidative ability to produce sufficient amounts of ATP for a maintained level of tension. These fibers have long twitch times and thus produce tetanus at low frequencies of stimulation. These fibers function in the tonic maintenance of posture.

Fast-fatigue fibers are the white muscle cells that have large supplies of glycogen to meet sudden requirements for ATP but cannot maintain ATP production. These fibers have a short twitch time and require a high fusion frequency for tetanus. Fast-fatigue fibers are used for brief intermittent phasic contractions.

Fatigue-resistant fibers are between the slow- and fast-fatigue fibers in terms of ATP production and fatigability. These fibers are used in sustained repetitive movements such as walking and running. Of interest, they have the highest capillary supply.

The motoneurons to these types of fibers have different functions related to their muscle fibers. Those to slow-fatigue fibers are smaller and conduct more slowly than the others. The firing frequency is lower, partly because the postspike afterhyperpolarization period, is longer. These motoneurons are powerfully excited by monosynaptic fibers, reciprocally inhibited by Ia afferents (via an internuncial), and tonically activated by the brainstem (extrapyramidal) motor centers.

Those to fast-fatigue fibers are the largest and fastest conducting motoneurons. They produce bursts of high-frequency activity with periods of inactivity. Their afterhyperpolarization period is short. These motoneurons are only slightly influenced by Ia input but are reflexly excited by touch receptors. They are primarily activated by the cortical (pyramidal) motor center.

Those to slow-fatigue fibers are between the other two in size, conduction velocity, firing frequency, and afterhyperpolarization period. They are excited by monosynaptic af-ferents and inhibited by antagonist Ia afferents. These motoneurons are involved with reciprocal (alternating) innervation.

SPINAL CORD INTERNUNCIALS
Characteristics

Any neuron in the spinal cord other than the motoneurons is considered an internuncial. Internuncials are multipolar neurons that vary in size from very small (4 μm) with short axons to those as large as the alpha motoneuron. Internuncials are found throughout the gray matter but are more concentrated in the dorsal horn and intermediate area. They can be classified as follows:

1. Intrasegmental: those located entirely within one segment. These are involved with reflexes, relay of descending motor data to the motoneurons, and autogenic inhibition of the alpha motoneurons.
2. Intersegmental: those that project several segments for involved reflexes.
3. Commissural: those projecting to the opposite side of the segment for crossed reflexes and fiber tracts.
4. Fiber tract: those projecting their axons to brainstem nuclei and the cerebellum.

Electrical activity

Like other neurons, internuncials have RMPs and are excited or inhibited by the same ionic mechanisms. They may use excitatory or inhibitory neurotransmitters and are involved in presynaptic inhibition. Internuncials show two interesting electrical responses to stimulation.

One response involves a prolonged EPSP (even after the stimulation has ceased) that results in a *repetitive discharge* on the axon. Working with this, the APs do not terminate with a long afterhyperpolarization period, thus allowing for higher discharge frequencies. The significance of this repetitive response is that it is a mechanism for amplifying the effect of that internuncial (temporal summation). For example, a pinprick stimulus to the foot of a patient with spinal cord injury may cause flexor withdrawal of the leg that persists after the stimulation, partly because of internuncial amplification.

The other response involves an increased postsynaptic response to a second stimulus that closely follows an initial stimulus. This response is known as *posttetanic stimulation*. It probably involves some mechanism that mobilizes the neurotransmitter in the bouton during an initial volley of presynaptic APs. With a second volley of presynaptic APs larger amounts of neurotransmitter are released. Posttetanic potentiation is a method of amplifying the activity of frequently used synapses, thus making those synapses more efficient.

Renshaw cell

Of all of the internuncials the Renshaw cell is the most intriguing because it receives input from the recurrent collateral of the motoneuron axon. When activated by the recurrent collateral the Renshaw cell acts to postsynaptically inhibit the same motoneuron. This action is known as *autogenic* or *recurrent inhibition* and represents a negative feedback loop (Fig. 6-5). The primary function of the Renshaw cell is to dampen the output of the final common pathway (like the soft pedal on a piano softens the intensity of the note played). This dampening effect is a method of fine tuning alpha motoneurons output. Of interest, the motoneurons to the axial muscles in the medial part of the ventral horn do not have recurrent collaterals.

The Renshaw cell also can project to another internuncial that is inhibitory to the alpha that excited the Renshaw cell, or it can be inhibited directly by another internuncial. The effect is inhibition of an inhibitory neuron, or *disinhibition* (Fig. 6-5). This disinhibition can increase the output of the alpha motoneuron. In either case the Ren-

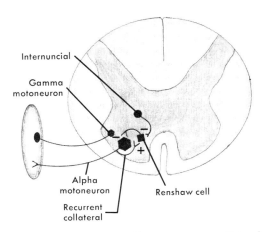

Fig. 6-5. Renshaw cell. Recurrent collateral branch of alpha motor axon excites Renshaw cell, which in turn is inhibitory to homonymous alpha motoneuron. Note internuncial, which is inhibitory to Renshaw cell. This is known as disinhibition.

shaw cell is capable of repetitive discharge, thus prolonging its effect.

Other pathways besides the recurrent collateral excite the Renshaw cell. This may implicate the Renshaw cell as a contributor to spasticity and rigidity in some instances. If a pathway that normally excites Renshaw cells is lesioned, a source of inhibition to the motoneurons is lost. The resulting net increase of excitation on the alpha motoneuron may manifest itself as an increase in tone and an increased resistance to stretch.

SPINAL CORD CIRCUITRY

Within the spinal cord are a variety of circuits usually involving several segments. A neural circuit can be thought of as a chain of neurons that once activated follows a specific set of responses. Neural circuits are found throughout the CNS and are involved in many sensorimotor and cognitive activities. In the spinal cord they are involved with somatic and visceral reflexes.

Somatic reflexes

These spinal circuits involve a series of responses to cutaneous and muscle receptors. The responses evoked are stereotyped patterns of movement (flexion or extension) that are the bases for posture and forward progression. During development these circuits become modified by the brainstem and cerebral motor centers into more purposeful movements. The spinal circuits, however, are retained throughout life and often become prominent in some CNS lesions.

The *myotatic reflex* is discussed in previous chapters (see Fig. 4-10). The adequate stimulus is quick stretch to a muscle. The afferent fibers are large Ia fibers that (1) monosynaptically excite the homonymous alpha motoneuron, (2) excite via internuncials synergistic alpha motoneurons, and (3) inhibit via internuncials antagonistic alpha motoneurons.

The myotatic (stretch) reflex is primarily used for posture and reciprocal movements. It is strongly influenced by gamma biasing and the existing excitability of the alpha motoneuron. Most CNS lesions alter either the sensitivity to stretch or the amount of movement in response to stretch.

The use of tendon taps and vibration to a hypotonic muscle can elicit motion. In fact, vibration is a very powerful stimulus that can produce a tonic contraction for the duration of the stimulus (tonic vibratory reflex). If used alone, however, these stimuli have little long-term benefit, partly because of Renshaw cell inhibition that follows phasic activation.

The *flexor withdrawal reflex* is a protective movement of flexion away from a noxious stimulus. The afferent fibers involved with this reflex are collectively known as flexor reflex afferents (FRA) and are made up of group II, III, and IV fibers from pain, touch, pressure, and temperature receptors. When a strong enough stimulus is applied distally (hand or foot) the FRAs excite internuncials that initiate flexion of the entire limb. This

includes dorsiflexion of the ankle and extension of the toes (physiologic flexors). Components of abduction–external rotation or adduction–internal rotation of the leg may also be present. Extensor motoneurons (ipsilateral) are inhibited via this circuit (Fig. 6-6). A stimulus applied over other parts of the skin such as the thigh do not usually evoke flexion of the whole limb. The movements produced act to remove the area from the noxious stimulus.

The internuncials in the flexor circuit are under the influence of higher motor centers that normally inhibit these internuncials. This inhibition prevents the flexor responses from occurring with any stimulus. In spinal cord injury, however, simple contact of the therapist's hands to the patient's foot may elicit a strong flexor withdrawal. The presence of this higher center inhibition can be seen in someone carrying a very hot and expensive casserole dish from the oven to the table.

The FRA pathway is also involved in activities other than with motoneurons. These activities involve (1) blocking of the perception of pain via tactual stimulation of adjacent areas of skin, and (2) presynaptic inhibition of a weakly activated set of FRAs by a strongly activated set of FRAs. This is a mechanism for filtering extraneous data. Both of these activities may take place in the spinal cord as well as in other areas.

If the stimulus eliciting the flexor withdrawal is strong enough, activity in all four limbs will result. Of interest is the *crossed-extensor reflex* in the leg. If a painful stimulus is applied to one leg, the FRAs will not only excite internuncials for ipsilateral limb flexion but also internuncials that cross in the cord to excite the contralateral limb extensors (Fig. 6-6), including plantar flexion (physiologic extensors). Adduction and internal rotation often accompany the extension. This crossed-extensor reflex can be best seen in patients with spinal cord injury and in in-

A

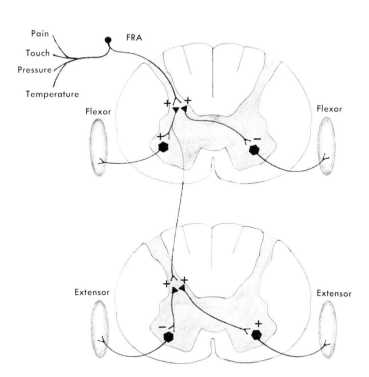

Fig. 6-6. A, FRA pathway. On entering cord, these afferents excite internuncials, which tend to excite ipsilateral flexors for withdrawal and contralateral extensors for support (crossed extensor reflex). Area of skin stimulated and descending influences can alter these responses.

fants in the supine position. A modified response occurs in persons with intact nervous systems and can be seen when one steps on a sharp object or a toe is stabbed. Ths injured limb is held in flexion while the other leg is used for support. The modification involves an abduction and a dorsiflexion component that allows for a functional stance posture. The crossed-extensor circuit is utilized in this modified form for many reciprocal leg activities such as walking and bicycle riding.

The extensor components of the crossed-extensor reflex can be elicited by themselves if pressure is applied to the bottom of the foot (especially the metatarsal area). The input from the pressure receptors enters the cord and excites internuncials, which then excite ipsilateral extensor, adductor, and internal rotator motoneurons and other internuncials that ipsilaterally inhibit flexor motoneurons. There is no contralateral effect (Fig. 6-6).

This reflex response to pressure is known as the *extensor thrust reflex* and can be seen in infants and in patients with spinal cord injury or other CNS disorders. In a mature nervous system the extensor thrust response is modified to allow abduction and dorsiflexion so the lower limb may be more functional

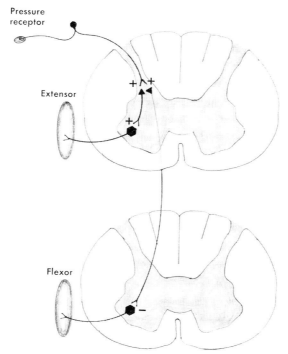

Fig. 6-6, cont'd. B, Extensor thrust pathway. Pressure afferents from bottom of foot excite internuncials, which tend to excite ipsilateral extensors and inhibit ipsilateral flexors. Descending influences may alter this response.

in stance, walking, and jumping. In some instances pounding a patient's heel with the therapist's hand or applying joint compression to the ankle, knee, and hip (in an extended position) can elicit this modified extensor thrust. Palm pounding or compression into the upper extremity joints can facilitate a functional support posture (the all-fours position with wrist extension) of the upper extremity.

All of these somatic reflexes involve segmental and intersegmental pathways that are elicited by somatic receptors. While they can operate independently, they are normally

under the influence of the supraspinal motor centers. With CNS lesions these reflexes may be absent, hyperactive, or visible to varying degrees.

Viscerosomatic reflexes

Visceral afferents entering the spinal cord have been found to excite internuncials that may excite or inhibit alpha motoneurons. The effects of visceral afferent activity include changes in muscle tone and even movements. A common viscerosomatic reflex response is flexor spasm in the legs of a patient with spinal cord injury. This response occurs when a visceral organ such as the bladder becomes distended. Of interest, the visceral afferents have been found to excite internuncials that are also excited by cutaneous afferents.

Somatovisceral reflexes

Segmental pathways have been found to exist between receptors of the skin and the preganglionic neurons of the sympathetic and parasympathetic systems. Although no specific reflexes have been identified by name, the existence of these pathways can explain why high-intensity tactual stimulation (brushing) of the paramedian skin of the back can cause urinary retention and why acupuncture can have such a profound influence on the visceral system.

Pathologic reflexes

Damage to the supraspinal motor centers or their pathways not only alters the sensitivity of the somatic spinal reflexes but can also cause the appearance of the following responses.

Babinski's sign. In the mature nervous system, stroking the lateral border of the sole and across the metatarsals with a pointed but not sharp object will cause plantar flexion of the ankle and toe flexion. In some CNS lesions and in infants the response is dorsiflexion of the ankle and extension of the toes. Babinski's sign is actually part of the flexor reflex.

Clonus. Normally a quick stretch of a muscle results only in a brisk phasic contraction of that muscle. In some CNS lesions the stretch reflex is hyperactive, and a sudden stretch of one muscle can cause an alternating contraction between agonist and antagonist. Clonus is best seen in the ankle, where rapid dorsiflexion stretches the gastrocnemius-soleus muscles, causing plantar flexion, which stretches the anterior tibialis causing dorsiflexion, and so on. Maintaining resistance to plantar flexion can perpetuate the clonus.

Hoffman's sign. This reflex is not really pathologic, since it can occur in normal individuals. However, it is often exaggerated in some CNS lesions. To elicit Hoffman's sign the palm is placed in supination with the wrist slightly flexed. The tips of the fingers are then tapped, briefly extending them. The response is finger and thumb flexion.

Association overflow reactions

Have a friend squeeze a rubber ball as hard as possible with one hand while you watch the other hand. Frequently the other hand will flex as a result of an overflow of impulses presumably from the other side of the cord or brainstem via internuncials. While present in the mature nervous system, these overflow reactions are often more obvious in some patients with CNS disorders. The typical pattern of response is as follows:

1. Resisted upper extremity motion (flexion or extension) on one side causes a similar response (flexion or extension) in the other upper extremity.
2. Resisted flexion (extension) in one lower extremity causes the opposite motion, extension (flexion), in the other lower extremity.
3. Resisted adduction or abduction in one lower extremity causes a similar response in the other lower extremity.
4. The homolateral effect between arm and leg on the same side is to produce similar responses.
5. The diagonal effect between arm and opposite leg is to produce opposite responses.

In many patients with CVA these reactions appear in the involved extremities as the basic synergy patterns. Thus resisted uninvolved side elbow flexion can cause flexion, abduction, and external rotation of the involved upper extremity.

Another overflow response easily demonstrated is the Jendrassik maneuver. Determine the patellar tendon reflex on someone in a relaxed sitting posture. Now have the person place his or her palms together and forcibly push. A test of the patellar tendon reflex during tension will show an increased response. The impulses generated in the cervical cord are conveyed to the lumbar cord.

TEST QUESTIONS

1. An extramedullary tumor compressing the lateral edge of the thoracic spinal cord can initially cause:
   a. Interruption of pain and temperature sensation.
   b. Weakness and incoordination of skilled movements of the distal flexors.
   c. Loss of muscle tone to the proximal postural muscles.
   d. Interruption of discriminative touch sensations.

2. Match the following laminae and nuclei to their appropriate function:
   1. VII
   2. Nucleus gracilis
   3. Nucleus dorsalis
   4. II, III
   5. VI, VII
   6. Substantia gelatinosa
   7. IX

   a. Relay of muscle spindle and GTO data _____
   b. Relay of pain and temperature _____
   c. Motoneurons to muscles of extremities _____

3. The dura mater of the spinal cord:
   a. Is continuous with the perineurium of the peripheral nerves.
   b. Forms the denticulate ligaments.
   c. Terminates at the conus medullaris.
   d. Is adherent to the surface of the spinal cord.

4. A lesion of the anterior spinal artery at C7:
   a. Involves all the ventral horn cells at and below C7.
   b. Causes contralateral loss of pain and temperature below C7.
   c. Interrupts any crossed-fiber tract involving C7.
   d. Causes loss of voluntary motion bilaterally at and below C7.

5. The tetanus bacillus has an ability to block inhibitory neurons in the CNS. Which of the following would be true?
   a. Rapidly alternating movements are difficult.
   b. Increased muscle tone and resistance to stretch are noted.
   c. Fewer positive ions flow toward the axon hillock.
   d. Swallowing and respiration are unimpaired.

6. If an inhibitory internuncial to an alpha motoneuron is presynaptically inhibited while active:
   a. Axon hillock shows reduction in firing frequency.
   b. Postsynaptic IPSP is larger.
   c. RMP of the motoneuron remains the same.
   d. An excitatory internuncial, if activated, will have a greater effect.

7. In excitatory temporal summation:
   a. EPSPs from many synapses are added together to form one larger and longer lasting EPSP.
   b. Motoneuron can respond with a higher firing frequency.
   c. Motoneuron can show posttetanic potentiation during summation.
   d. Amount of neurotransmitter released decreases for each successive AP.

8. Fast-fatigue fibers have motoneurons that:
   a. Have low discharge rates.
   b. Receive strong input from the cerebral cortex.
   c. Are involved with postural control.
   d. Are larger and faster conducting.

9. Internuncials:
   a. Project axons to different cord segments for somatic spinal reflexes.
   b. Project axons that inhibit motoneurons.
   c. Relay descending data from the higher motor centers to the motoneurons.
   d. Interconnect the alpha and gamma motoneurons.

10. The somatic spinal reflexes of a patient with spinal cord injury are often hyperactive. If the left foot is strongly stimulated with a pin:
   a. Right lower extremity should extend, adduct, and internally rotate.
   b. Right ankle will dorsiflex if the heel is pounded at the same time.

c. Response of the left foot will resemble Babinski's sign.

d. Response may outlast the stimulus as a result of repetitive discharge from internuncials.

11. Clonus is commonly seen in CVA and spinal cord injury. Which of the following is true?

a. Clonus can cause knee instability in CVA.

b. Renshaw cells may help extinguish the response.

c. Severing the dorsal roots from the gastrocnemius-soleus muscles has little effect.

d. Eliciting a crossed-extensor reflex *from* the uninvolved leg helps extinguish the response.

12. If the flexors of the left arm of a patient with CVA are hypotonic, movements may be initiated by:

a. Resisted right arm extension.

b. Resisted left leg flexion.

c. Tapping the desired muscle belly.

d. Lying supine and turning the head to the left.

SUGGESTED READINGS

Barr, M.: The human nervous system, New York, 1972, Harper & Row, Publishers.

Everett, N.: Functional neuroanatomy, ed. 6, Philadelphia, 1971, Lea & Febiger.

Netter, F.: The CIBA collection of medical illustrations. I. The nervous system, Summit, N. J., 1972, CIBA Chemical Co.

Noback, C., and Demarest, R.: The human nervous system, ed. 2, New York, 1975, McGraw-Hill Book Co.

Truex, R., and Carpenter, M.: Human neuroanatomy, ed. 6, Baltimore, 1969, The Williams & Wilkins Co.

Willis, W. D., Jr., and Grossman, R. G.: Medical neurobiology: neuroanatomical and neurophysiological principles basic to clinical neuroscience, ed. 2, St. Louis, 1977, The C. V. Mosby Co.

CHAPTER 7

Environmental control systems

In this chapter are presented a concise background of the relationship between the skull and encased brain structures and the systems regulating the internal environment of the cerebrum, brainstem, and cerebellum. These systems are concerned with the connective tissue coverings of the brain structures, the production and circulation of cerebrospinal fluid, the distribution and regulation of cerebral blood flow, and the regulation of the extracellular fluid. Related clinical information concerning hydrocephalus, tumors, anoxia, hypoglycemia, and hypothermia are also presented.

The following areas deserve special attention.

1. Causes of increased intracranial pressure and effects on neural tissue, ventricles, and the cranium
2. Cranial nerves associated with the fossae of the floor of the cranial cavity
3. Mechanisms of neural damage involved in a skull fracture
4. Location of the falx cerebri, tentorium, superior sagittal sinus, and transverse sinus.
5. Components of the blood-brain barrier and its function
6. Production, circulation, composition, and function of cerebrospinal fluid
7. Causes of hydrocephalus and intracranial hemorrhage
8. Structures or area supplied by the three cerebral arteries and the vertebral-basilar arteries
9. Circle of Willis and its function
10. Mechanism that regulates the cerebral blood flow and the composition of the extracellular fluid
11. Effects of anoxia, hypoglycemia, and hypothermia on the nervous system

CRANIAL CAVITY
Cranium and its response to increased intracranial pressure

The cranium, or skullcap, is composed of eight separate bones that fuse together early in childhood to form a solid unit. This unit fits snugly over the entire outer surface of the cerebrum and cerebellum (Fig. 7-1, A). The junctions between the skull bones are known as suture lines. In the newborn the individual bones are separated at the suture lines by 1 to 3 mm of periosteum allowing the skull to be flexible. This can be advantageous to an infant suffering from increased intracranial pressure. The pressure within the cranium can increase from intracranial hemorrhage, tumor, or blockage of the flow of cerebrospinal fluid. The effects of the rise in pressure are compression of neural tissue and dilation or compression of the ventricles (depending on the cause). In the infant the flexibility at the suture lines allows the skull to expand. This relieves some of the pressure, reduces the severity of damage, and allows for a bet-

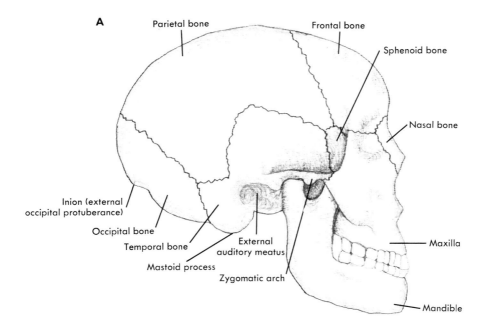

A

Parietal bone

Frontal bone

Sphenoid bone

Nasal bone

Inion (external occipital protuberance)

Occipital bone

Temporal bone

Mastoid process

External auditory meatus

Zygomatic arch

Maxilla

Mandible

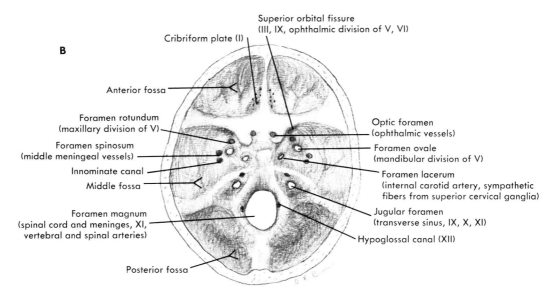

B

Cribriform plate (I)

Superior orbital fissure (III, IX, ophthalmic division of V, VI)

Anterior fossa

Foramen rotundum (maxillary division of V)

Foramen spinosum (middle meningeal vessels)

Innominate canal

Middle fossa

Foramen magnum (spinal cord and meninges, XI, vertebral and spinal arteries)

Posterior fossa

Optic foramen (ophthalmic vessels)

Foramen ovale (mandibular division of V)

Foramen lacerum (internal carotid artery, sympathetic fibers from superior cervical ganglia)

Jugular foramen (transverse sinus, IX, X, XI)

Hypoglossal canal (XII)

Fig. 7-1. A, Lateral view of skull. Note suture lines between bones of skullcap. Keep in mind that there are two parietal, two temporal, and two sphenoid bones. Try palpating zygomatic arch, mastoid processes, and inion on yourself. **B,** Superior view of floor of cranial cavity showing three pairs of fossae and foramina. In real skull, floor has sharper angles and more irregularities than is indicated in this drawing. Compare this view with Fig. 7-4, A and B, to see how base of cerebrum and brainstem are encased by floor of cranial cavity. Internal acoustic meatus (not labeled) is located just above jugular foramen (VII, XIII). (Adapted from Chusid, J.: Correlative neuroanatomy and functional neurology, ed. 16, Los Altos, Calif., 1976, Lange Medical Publications.)

ter prognosis if treated early enough. Many infants suffering from *hydrocephalus* have an enlarged head (Fig. 7-9).

The suture lines begin to fuse at around 2 months and are generally complete at around 18 months. Keep in mind that during this time the flexibility in the infant skull provides only minimal protection against head trauma. The suture lines can still be separated up to the age of 8 years. After this time any increase in pressure within the skull can cause severe and rapidly fatal compression of brain structures within a solid cranium (Fig. 7-2).

Floor of cranial cavity

The floor of the cranial cavity is irregular in contour and is divided into three pairs of *fossae* (depressions) that help support the undersurface of the brain. The frontal lobes lie over the anterior fossae, the anterior portion of the temporal lobe and base of the diencephalon lie over the middle fossae, and the cerebellum lies in the posterior fossae. The posterior portion of the temporal lobes and the occipital lobes lie on a connective tissue shelf, the tentorium, which extends from the middle fossae back to the occipital bones (Figs. 7-3 and 7-4, *A* and *B*).

The fossae contain *foramina*, or openings, to allow the passage of blood vessels, nerves, and the spinal cord. Associated with the anterior fossae are foramina for cranial nerve I. Associated with the middle fossae are foramina for cranial nerves II, III, IV, V, and VI and the internal carotid artery. Associated with the posterior fossae are foramina for cranial nerves VII, VIII, IX, X, XI, and XII and the spinal cord and vertebral artery. The latter two pass through the largest foramen, the foramen magnum. These fossae and foramina can be involved with tumors, especially meningiomas, which may produce cortical, cranial nerve, or cerebellar symptoms depending on the location (Fig. 7-1, *B*).

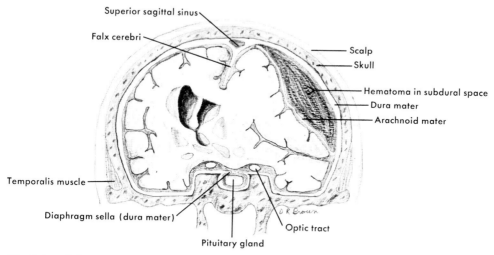

Fig. 7-2. Subdural hematoma with brain compression. Any increase in mass within skull such as from growing tumor, bleeding vessel, or excess amount of cerebrospinal fluid can cause increase in intracranial pressure. This can result in compression and shifting of brain structures. Note shift of ventricles and that both hemispheres are affected. (Adapted from Netter, F.: The CIBA collection of medical illustrations. I. The nervous system, Summit, N.J., 1972, CIBA Chemical Co.)

Protective role

The bones of the cranium provide a formidable source of protection for the enclosed brain structures. This protection is essential because neural tissue has a consistency only slightly firmer than jelly. For the most part the numerous bumps and falls we experience during our lives cause no permanent neural damage. Even when the trauma is severe enough to cause loss of consciousness the skull can still remain intact and there can be complete recovery. Skull fractures can occur, however, especially with impact from a high-velocity object or sudden deceleration impact. The weakest area of the skullcap is the temporal bones above the ears. (For functions of the cortical areas underlying the temporal bones and the kinds of symptoms that might be expected with a skull fracture in this region see Chapters 8 and 9.) When the skull is fractured neural damage can occur for several reasons, including (1) direct penetration from the bone fragments into cortical or subcortical structures and (2) rupture of the meningeal arteries of the dura mater or the cerebral vessels in the subarachnoid space by the bone fragments. This can result in an expanding hematoma, which leads to compression of neural tissues.

Chapter 10 provides further information on the effects of traumatic head injury. It is important to mention here, however, that the extent of neural damage is not directly related to the extent of the skull fracture.

MENINGES
Dura mater

Between the skull and vertebrae and the CNS are three layers of connective tissue known as the meninges. The outermost layer is the *dura* (hard) *mater*, or *pachymeninx*. The dura is composed of two fused layers of tough collagenous fibers. The dura is adherent to the inner surface of the cranium and floor of the cranial cavity and actually functions as a periosteum during the growth of the skull. The dura receives blood from the meningeal arteries and is supplied by sensory afferent fibers from the trigeminal and vagus nerves.

The dura is difficult to visualize in its entirety. It completely surrounds the cerebrum, brainstem, cerebellum, and spinal cord. It even extends out with the cranial and spinal nerves, forming the epineurium. The dura also projects into the medial longitudinal fissure, forming the *falx cerebri*, into the fissure between the cerebellar hemispheres, forming the *falx cerebeli*, and between the occipital lobes and cerebellum, forming the *tentorium*. The tentorium extends around the midbrain and then surrounds the infundibulum, forming the *diaphragm sella*. It then becomes continuous with the dura that covers the anterior fossae. The tentorium and two falces are basically at right angles to each other (Fig. 7-3). By dividing the brain and cerebellum into right-left and top-bottom compartments the tentorium and falces help to anchor the brain structures. This helps to reduce the rotary and linear forces that occur during head trauma.

At the margins and junctions of the falces and tentorium the two layers of dura split to form channels known as the *dural sinuses* (Fig. 7-4, *C*). The top layer of dura remains adherent to the skull, while the inner layer splits to form a sinus. The inner layer then forms a falx or the tentorium. The major dural sinuses are the *superior sagittal sinus*, which runs along the apex of the falx cerebri, the *transverse sinus*, which runs along the lateral margin of the tentorium, *the straight sinus*, which runs in the junction of the falces and tentorium, and the *superior petrosal sinus*, which runs in the anterior margin of the tentorium. The dural sinuses receive venous blood from the brain (conveyed via venous sinuses and veins), venous blood from the scalp and nasal sinuses (conveyed via emissary veins), and cerebrospinal fluid from the subarachnoid space (conveyed via arachnoid

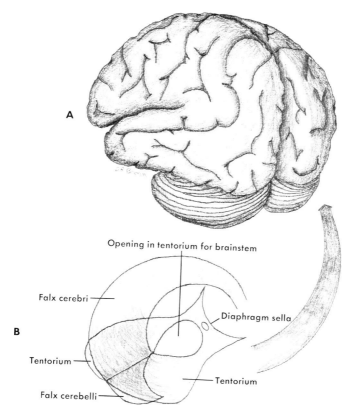

A

B

Opening in tentorium for brainstem

Falx cerebri

Diaphragm sella

Tentorium

Tentorium

Falx cerebelli

Fig. 7-3. Relationship between falces and tentorium and brain. Try to picture how **B** would fit between two hemispheres and between bottom of brain and cerebellum. Note that dural sinuses have been omitted and that pattern of convolutions is somewhat arbitrary for reasons of simplification. (**B** adapted from House, E., and Pansky, B.: A functional approach to neuroanatomy, New York, 1967, McGraw-Hill Book Co.)

villi). This venous blood–CSF mixture in the dural sinus flows dorsally to the transverse sinus. The transverse sinus becomes continuous with the sigmoid sinus, which then becomes continuous with the internal jugular vein. The internal jugular vein then returns the mixture to the general circulation (Fig. 7-4, *A* and *B*).

As mentioned, the dura has its own blood and nerve supply. The major artery is the middle meningeal artery. This artery supplies most of the lateral convexity of the dura. Because the temporal bone is the weakest part of the skullcap, this artery can become ruptured from skull fractures in this area. Arterial pressure is usually high enough to cause a rapidly increasing hematoma, which may be *epidural* or *subdural* in location. These are often rapidly fatal as the increase in pressure from hemorrhaging blood can severely compress the brain.

The dura is richly supplied by branches

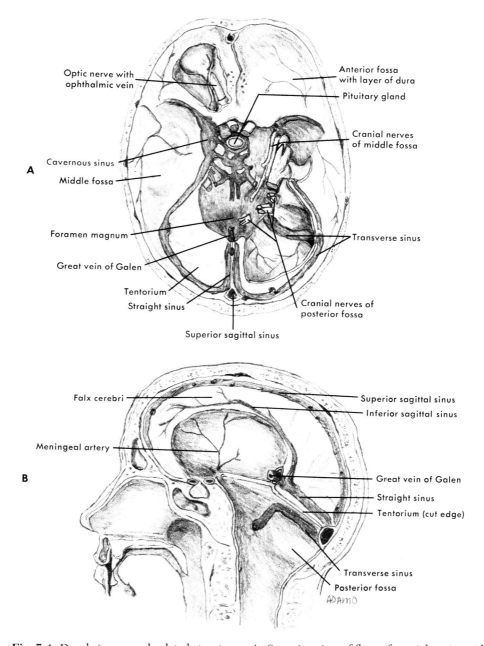

Fig. 7-4. Dural sinuses and related structures. **A,** Superior view of floor of cranial cavity with tentorium, transverse sinus and other sinuses, and related structures. Note that venous blood from front of head and base of brain drains into cavernous sinus and finally into transverse sinus. Other sinuses drain into either transverse sinus or jugular vein. Sigmoid sinus and meningeal artery are not labeled. Note extensive network of venous sinuses in area of pituitary gland. Also note dura has been cut from anterior fossa on left and middle and posterior fossae on right. **B,** Sagittal view showing falx cerebri and superior sagittal sinus. Compare with **A** to note how tentorium has been cut, how great vein of Galen drains into straight sinus and then into transverse sinus, and how transverse sinus exits jugular foramen (it is called sigmoid sinus at this point) to jugular vein.

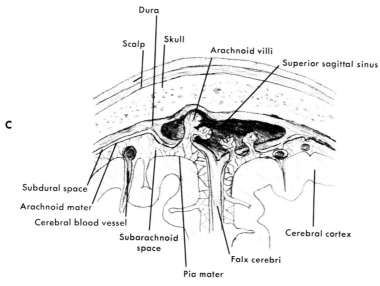

C

Dura

Scalp Skull

Arachnoid villi

Superior sagittal sinus

Subdural space
Arachnoid mater
Cerebral blood vessel
Subarachnoid space

Pia mater

Falx cerebri

Cerebral cortex

Fig. 7-4, cont'd. C, Frontal view through superior sagittal sinus showing relationship of scalp, skull, and meninges with brain. Note that dura splits to form sinus and then falx cerebri. Also note subdural space, which is frequent site of hemorrhage. (Adapted from Netter, F.: The CIBA collection of medical illustrations. I. The nervous system, Summit, N.J., 1972, CIBA Chemical Co.)

from the trigeminal and vagus nerves and by sympathetic fibers that supply the meningeal arteries. The receptors of the dura are very sensitive to changes in intracranial pressure, rubbing, traction, and pressure in the meningeal arteries. Any space-occupying lesion such as a tumor or hematoma, loss of CSF, or excess dilation of the meningeal arteries can irritate these receptors and cause severe headache. Since the brain itself is insensitive to pain, this reaction from the dura can be an important premorbid indicator.

Arachnoid mater

The inner two layers of the meninges are the *arachnoid* (spider) *mater* and the *pia* (tender) *mater*. They are intimately related structurally and together are known as the *leptomeninx*.

When viewed in cross-section (Fig. 7-4, C), the arachnoid appears as a thin layer of cells directly below the dura with many fine threadlike projections that attach to the surface of the brain. Like the dura, the arachnoid covers the entire CNS and even follows the fibers of the cranial and spinal nerves, helping to form the perineurium. Unlike the dura, the arachnoid is very delicate and offers no protection to CNS structures. Its sole purpose is to act as a watertight envelope around the entire CNS. This is essential because between the arachnoid and surface of the CNS is the *subarachnoid space*, which is filled with a clear fluid known as *cerebrospinal fluid* (see discussion under Ventricular System). The cerebrum, brainstem, cerebellum, and spinal cord therefore "float" within this fluid. In some areas such as between the medulla and cerebellum the subarachnoid space is much larger than it is over the hemispheres. These larger spaces are often known as *cisterns* (Fig. 7-5). Although all are named, only two are mentioned here: the cisterna magna, which is sometimes used in hydrocephalus to receive excess cerebrospinal fluid (via a shunt) from a blocked ventrical, and the spinal cistern (not labeled in Fig. 7-5), located below the conus medullaris, which is a frequent site of the lumbar puncture (spinal tap) procedure.

Along the superior sagittal sinus small tufts of arachnoid project into the sinus channel. These tufts are known as *arachnoid villi* or *granulations* (Fig. 7-4, *C*) and function in the reabsorption of CSF into the bloodstream by acting as a one-way valve. When the pressure in the sinus is lower than in the subarachnoid space the meshlike structure of the villi is

open, allowing CSF to flow. When the sinus pressure is greater, as in coughing or lifting something heavy, the villi become compressed and the meshlike openings become occluded. This prevents backflow of fluid into the subarachnoid space.

Like the subdural space, the subarachnoid space is a common site of hemorrhage because of the large number of blood vessels that lie on the surface of the cerebrum. Trauma and ruptured aneurysm are common causes of subarachnoid hemorrhage. As with any intracranial hemorrhage, those of arterial origin are the most destructive.

Pia mater

The pia mater is a layer of cells that is continuous with the arachnoid trabeculae above and is adherent to and follows the contours of

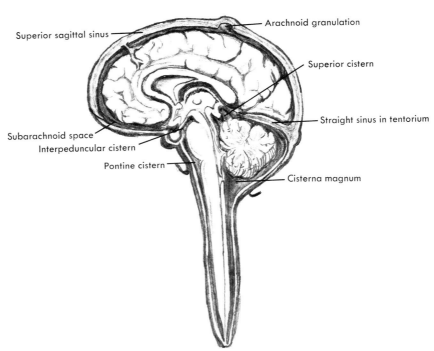

Fig. 7-5. Midsagittal view of CNS showing major subarachnoid cisterns. (Adapted from Noback, C., and Demarest, R.: The human nervous system: basic principles of neurobiology, ed. 2, New York, 1975, McGraw-Hill Book Co.)

the CNS. It even follows the blood vessels that penetrate the surface of the CNS. In the peripheral nervous system it joins with the arachnoid mater to form the perineurium. In the spinal cord it forms the filum terminale and denticulate ligaments.

The pia mater participates in several anatomic and physiologic systems known as the brain barriers. These "barriers" regulate the exchange of nutrients, ions, water, and waste products between the blood plasma, cerebrospinal fluid, and neural cells. These exchanges are regulated by selectively permeable membranes that may prevent, slow, or enhance the transport of these substances between the blood, extracellular fluid (or neural cells), and the cerebrospinal fluid. These barrier systems thus function to stabilize the physical and chemical environ-

ment of the highly sensitive neurons of the CNS. The following barriers involve the pia mater:

1. Blood-brain barrier, composed of the arteriole and capillary wall, which penetrate into the neural tissue, the layer of pia mater and subadjacent neuroglia (astrocytes) membrane, which blend with the vessel wall before it penetrates the substance of the CNS, and end-feet of astrocytes, which cover 85% to 99% of the remaining perivascular surface (Fig. 7-6)
2. Brain-CSF barrier in the subarachnoid space, which involves the pia-glial membranes that form a lining around the outer surface of the CNS
3. Blood-CSF barrier in the choroid plexus, composed of the capillary wall,

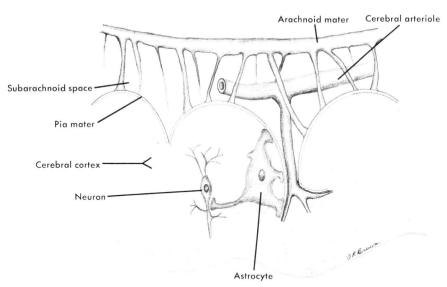

Arachnoid mater

Cerebral arteriole

Subarachnoid space

Pia mater

Cerebral cortex

Neuron

Astrocyte

Fig. 7-6. Blood-brain barrier. Note that layer of pia mater and membrane of astrocyte blend with wall of capillary as it penetrates cortex. Between 85% and 99% of capillary surface area in brain is covered by end-foot processes of astrocytes. Arachnoid trabeculae, which are fine filaments between arachnoid and pia, are not labeled. (Adapted from Noback, C., and Demarest, R.: The human nervous system: basic principles of neurobiology, ed. 2, New York, 1975, McGraw-Hill Book Co.)

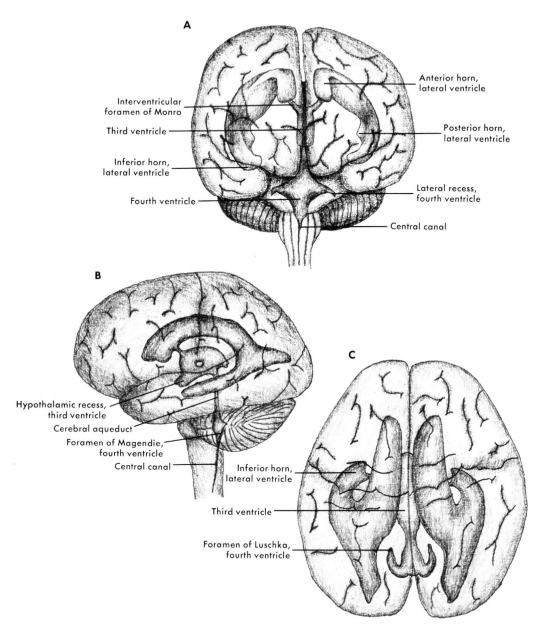

Fig. 7-7. Ventricular system in relation to surrounding brain structures. **A,** Frontal view. **B,** Sagittal view. **C,** Superior view. (**A** and **B** adapted from House, E. L., and Pansky, B.: A functional approach to neuroanatomy, New York, 1967, McGraw-Hill Book Co.)

a layer of pia mater, and the lining of ependymal cells of the ventricle

Another barrier system is the brain-CSF barrier in the ventricles. This barrier is composed of the ependymal lining of the ventricles and the glial membrane lining most of the ventricular surface in the brain.

VENTRICULAR SYSTEM
Structure and components

The ventricles in the cerebrum and brainstem are an irregularly shaped elaboration of the cavity of the original neural tube. The entire system is therefore lined with a layer of ependymal cells. In spite of their irregular contours the ventricles have a fairly standard shape and location in different brains. Therefore they are useful in the diagnosis of some space-occupying lesions (Fig. 7-2). Within each of the four ventricles are vascular structures known as *choroid plexuses*, which produce cerebrospinal fluid.

The individual components of the ventricular system are as follows (Fig. 7-7):

1. There are two *lateral ventricles*, one in each hemisphere. Each lateral ventricle has three "horns" that arise off a central "body." The anterior horn projects into the frontal lobe, the inferior horn projects into the temporal lobe, and the posterior horn projects into the occipital lobe. The body and anterior horn of each lateral ventricle lie under the corpus callosum and are separated by the septum pellucidum and fornix.
2. The *third ventricle* is surrounded by structures of the diencephalon. It therefore has an irregular shape with recesses for the optic chiasm, infudibulum, and pineal gland. There is also a hole in the center to account for the massa intermedia.
3. The *cerebral aqueduct* is a small channel in the midbrain that connects the third ventricle above with the fourth ventricle below.
4. The *fourth ventricle* is located between the pons and upper medulla and the cerebellum. It is somewhat rhomboid in shape with two lateral recesses.
5. The *central canal* is continuous with the fourth ventricle. It extends into the lower medulla and spinal cord.
6. The *interventricular foramina of Monro* connect the two lateral ventricles with the third ventricle.
7. The *foramen of Magendie* is a medial opening from the fourth ventricle into the subarachnoid space below the cerebellum.
8. The *two foramina of Luschka* are openings in the lateral recesses of the fourth ventricle to the subarachnoid space below the cerebellum.

Note in Fig. 7-7 the relationship of the ventricular system with the rest of the CNS. Also note how the ventricles are connected to allow a downward flow of CSF to the fourth ventricle, where the CSF can then enter the subarachnoid space via the foramina of Magendie and Luschka.

Cerebrospinal fluid

In each of the ventricles are capillary-rich structures known as *choroid plexuses*. These plexuses draw off a clear, colorless fluid from the blood known as *cerebrospinal fluid (CSF)*. Precisely how CSF is produced from the blood is beyond the scope of this text. The basic mechanism involves both passive diffusion and active transport systems. The majority of CSF is formed in the ventricles at a continous rate of 0.3 to 0.4 ml/min. Smaller amounts are formed at other sites, probably by vessels in the subarachnoid space. The total amount of CSF in the ventricles and subarachnoid space is around 135 ml, which produces a recordable pressure of from 60 to 180 mm of water in the side-lying position. Of interest, the formation of CSF is relatively independent of the pressure within the ventricles, subarachnoid space, or the blood systolic pressure. This finding has implication in hydrocephalus (see next section).

The composition of CSF is quite consistent. It is normally clear, mostly free of cells, contains little protein, and has specific amounts of CO_2, O_2, sodium, potassium, chloride, glucose, and other organic components. This "normalcy" of CSF composition makes it useful as a diagnostic aid in certain neurologic disorders (see Chapter 10 for a discussion of the lumbar puncture procedure and Table 10-1 for a summary of CSF findings).

CSF is continuously produced at a daily rate of around 300 ml and therefore must flow out of the ventricular system to prevent a buildup of fluid pressure. CSF circulates through the ventricles and exits the fourth ventricle via the foramina of Magendie and Luschka. Once inside the subarachnoid space the fluid slowly circulates around the CNS and up to the arachnoid villi for absorption into the dural sinus system (Fig. 7-8). The circulation of CSF is "powered" by a negative pressure gradient between the dural sinuses and the subarachnoid space. As mentioned, when the pressure in the superior sagittal sinus is low (as it normally is) CSF can flow through the arachnoid villi and reenter the general circulation. Of interest, as the CSF

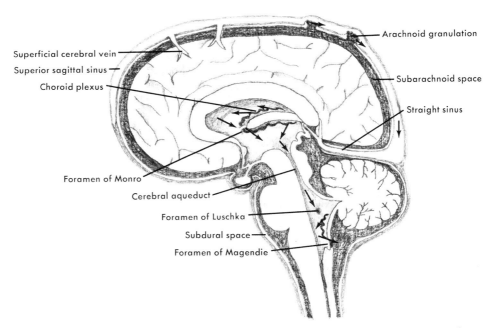

Superficial cerebral vein

Superior sagittal sinus

Choroid plexus

Foramen of Monro

Cerebral aqueduct

Foramen of Luschka

Subdural space

Foramen of Magendie

Arachnoid granulation

Subarachnoid space

Straight sinus

Fig. 7-8. Circulation of cerebrospinal fluid (CSF) produced by choroid plexus in lateral ventricles travels through interventricular foramen of Monro to join with CSF formed in third ventricle. CSF then flows through cerebral aqueduct and mixes with CSF formed in fourth ventricle. CSF leaves fourth ventricle by foramina of Luschka and Magendie. It circulates in subarachnoid space around brainstem, spinal cord, cerebellum, and up and over cerebral hemispheres to arachnoid granulation. Note veins draining into superior sagittal sinus and straight sinus. (Adapted from Netter, F.: The CIBA collection of medical illustrations. I. The nervous system, Summit, N.J., 1972, CIBA Chemical Co.)

circulates, its composition is modified by the exchange of substances such as gases, electrolytes, and sugars with the brain tissue. The pia-ependymal membranes lining the ventricles and the pia-glial membranes lining the surfaces of the CNS have a role in this exchange.

CSF has varied and vital functions in maintaining the integrity of the CNS:

1. Its buoyancy protects the brain, especially during sudden movement. Of interest is the fact that a brain weighing 1400 g only weighs 50 to 100 g while suspended in CSF.
2. It functions as an exchange for nutrients and waste products and helps maintain homeostasis in the chemical environment. This is especially important for the regulation of ion (Na^+ and K^+) fluxes between the blood and extracellular fluid.
3. It has value as a diagnostic aid because of its normal composition.
4. It may serve as a transport medium for certain hypothalamic and pineal hormones.

Hydrocephalus

Because CSF is continuously produced, any obstruction to its circulation or absorption or any increase in its rate of formation can result in an increased volume of CSF. As with space-occupying lesions such as hemorrhage or tumor, the excess CSF will cause an increase in pressure within the skull with resulting compression of neural tissue, (Fig. 7-9). Hydrocephalus (water on the brain) is commonly associated with infants who have a congenital abnormality that blocks the flow of CSF. The cerebral aqueduct and foramina of the fourth ventricle are common sites of obstruction. The flexibility of the infant skull causes the head to enlarge in response to the increase in pressure. Initially, therefore, the compression of neural tissue is moderate. Surgical intervention is usually required, whereby a tube is placed in a ventricle above the blockage. The excess fluid can then be shunted into one of the several areas distal to the block, including the cisterna magna, jugular vein, or auricle of the heart.

Hydrocephalus can also occur in adults because tumors, meningitis, and traumatic hemorrhage can obstruct the flow of CSF. Neural tissue can become rapidly compressed, and immediate surgical procedures are necessary to save the patient's life. Additional information on hydrocephalus is given in Chapter 10.

VASCULAR SUPPLY OF BRAIN
Metabolic demands

Unlike other body tissues, the brain maintains a constant level of activity regardless of whether a person is sleeping, exercising, studying, and so on. Since neurons have almost no metabolic reserves such as glycogen and are unable to undergo anaerobic metabolism, a constant blood flow is necessary to supply oxygen and glucose. About 800 ml of blood flow through the brain each minute (about one third of the output of the left ventricle). So delicate is the margin of physiologic safety that consciousness can be lost if the blood supply is cut off for less than 1 minute, and neural tissue can die in a few minutes of anoxia.

Major arteries

Four arteries supply blood to the brain. The two *common carotid arteries* arise, one off the aorta (left) and the other off the innominate artery (right). The carotid arteries ascend in the neck just medial to the sternocleidomastoid muscles. Just below the jaw the common carotid divides into the *external and internal carotid arteries*. The external carotid gives off branches that supply the neck, face, scalp, and most of the dura mater. The internal carotid enters the skull and ascends to the level of the optic chiasm.

The two *vertebral arteries* arise off the subclavian arteries. They ascend deep in the neck just superficial and lateral to the cervical cord and medulla. At the level of the pon-

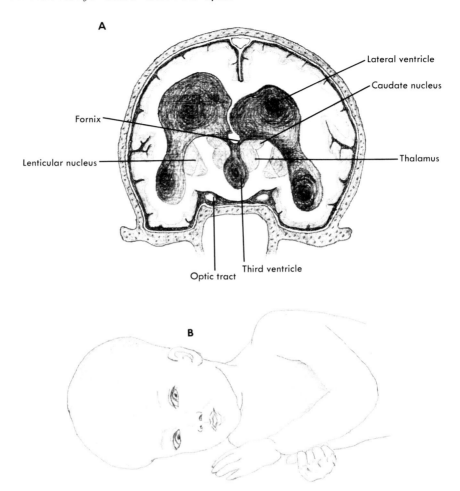

A

Lateral ventricle

Caudate nucleus

Fornix

Lenticular nucleus

Thalamus

Optic tract

Third ventricle

B

Fig. 7-9. **A,** Hydrocephalus in adult. Note dilation of ventricles, compression of cortex against skull, and displacement of subcortical structures. **B,** In infant, head will enlarge in response to buildup of pressure. (Adapted from Netter, F.: The CIBA collection of medical illustrations. I. The nervous system. Summit, N.J., 1972, CIBA Chemical Co.)

tomedullary junction the two vertebral arteries unite to form the *basilar artery*, which ascends on the anterior surface of the pons to the level of the midbrain (Fig. 7-10).

Major branches to CNS structures and their distribution

Vascular disorders are common among patients with CNS disorders. It may be helpful at this time to review the illustrations in Chapter 2 for the location of the major CNS structures, in Chapter 5 for the location of the cranial nerve nuclei, and in Chapter 6 for the location of the major fiber tracts in the brainstem.

The branches of the internal carotid artery are presented below in ascending order and illustrated in Figs. 7-10 and 7-11.

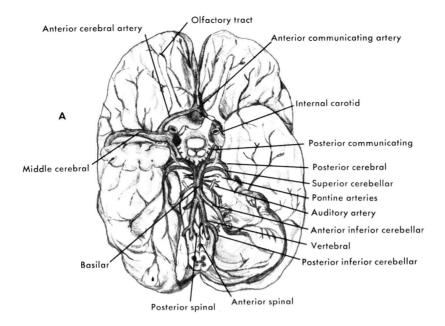

A

Anterior cerebral artery
Olfactory tract
Anterior communicating artery
Internal carotid
Posterior communicating
Posterior cerebral
Superior cerebellar
Pontine arteries
Auditory artery
Anterior inferior cerebellar
Vertebral
Posterior inferior cerebellar
Middle cerebral
Basilar
Posterior spinal
Anterior spinal

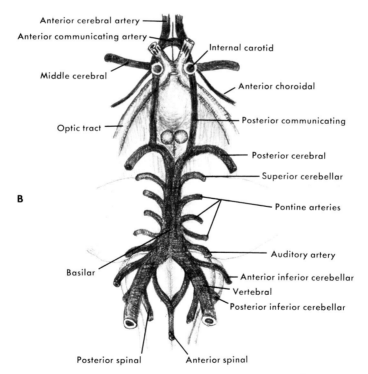

B

Anterior cerebral artery
Anterior communicating artery
Internal carotid
Middle cerebral
Anterior choroidal
Optic tract
Posterior communicating
Posterior cerebral
Superior cerebellar
Pontine arteries
Auditory artery
Anterior inferior cerebellar
Vertebral
Posterior inferior cerebellar
Basilar
Posterior spinal
Anterior spinal

Fig. 7-10. A, Four major arteries supplying brain and their major branches. Arteries are shown in relation to brain, brainstem, and cerebellum. Note that one temporal pole and one cerebellar hemisphere have been removed to show underlying arteries. Also note arteries that form circle around base of brain. **B,** Closeup schema of **A.** (**A** adapted from Everett, N.: Functional neuroanatomy, ed. 6., Philadelphia, 1971, Lea & Febiger; **B** adapted from House, E., and Pansky, B.: A functional approach to neuroanatomy, New York, 1967, McGraw-Hill Book Co.)

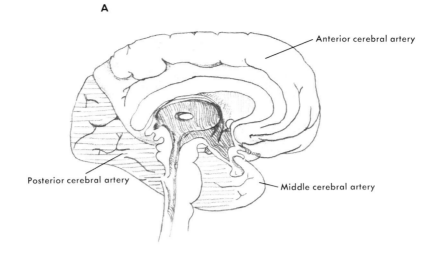

A

Anterior cerebral artery

Posterior cerebral artery

Middle cerebral artery

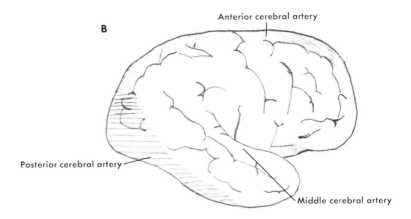

B

Anterior cerebral artery

Posterior cerebral artery

Middle cerebral artery

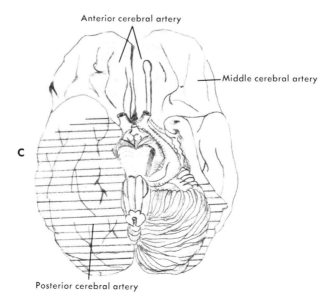

C

Anterior cerebral artery

Middle cerebral artery

Posterior cerebral artery

Fig. 7-11. For legend see opposite page.

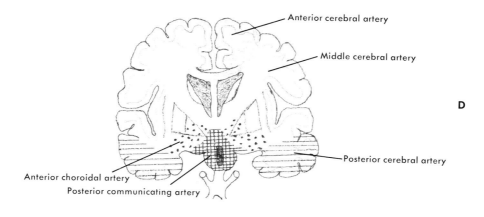

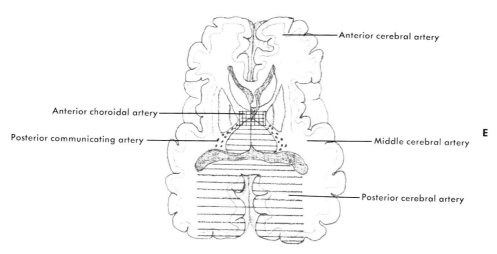

Fig. 7-11. A to **C,** Cortical distribution of three cerebral arteries: midsagittal, lateral, and inferior views, respectively. **D,** Frontal section through cerebrum at level of anterior commissure showing major arterial distribution. **E,** Horizontal section through cerebrum at midthalamus level showing major arterial distribution. (**A** to **C** adapted from House, E., and Pansky, B.: A functional approach to neuroanatomy, New York, 1967, McGraw-Hill Book Co.; **D** and **E** adapted from Chusid, J.: Correlative neuroanatomy and functional neurology, ed. 16, Los Altos, Calif., 1976, Lange Medical Publications.)

The *ophthalmic artery* supplies the optic nerve, retina, eye, and some of the nasal sinuses.

The *posterior communicating artery* connects the interal carotid with the posterior cerebral artery, thus helping to form the circle of Willis. Its branches help supply parts of the hypothalamus, subthalamus, thalamus, internal capsule (genu), and midbrain.

The *anterior choroidal artery* helps supply the choroid plexuses of the lateral ventricles, optic tract, cerebral peduncle, globus pallidus, posterior limb of the internal capsule, and lateral geniculate body.

The *anterior cerebral artery* branches off the internal carotid artery near the olfactory tract. It travels along the corpus callosum in the medial longitudinal fissure almost to the occipital lobe. It supplies the medial surface of the frontal and parietal lobes, including the superior margin along the medial longitudinal fissure, the anterior parts of the caudate, putamen, and internal capsule, and the undersurface of the frontal lobe (rhinencephalon), corpus callosum, and anterior hypothalamus.

The *middle cerebral artery* is the terminal portion of the internal carotid after the above four branches have been given off. This artery travels laterally in the lateral fissure. When it emerges it bends upward and then branches to supply the lateral surface of the frontal, temporal, parietal, and occipital lobes. In the lateral fissure branches are given off to help supply the caudate and lenticular nuclei, the thalamus, and the internal capsule.

The branches of the vertebral-basilar arteries are presented below in ascending order and illustrated in Figs. 7-10 and 7-12.

The *posterior spinal branches* are given off at the midmedulla level. They curve back and descend along the dorsal surface of the lower medulla and spinal cord. They help supply the dorsal region of the medulla, including the nuclei cuneatus and gracilis, and the dorsal portion of the spinal cord.

The *anterior spinal branches* arise at the level of the inferior olive. They unite immediately to form the single anterior spinal artery. It descends the medulla and spinal cord in the anterior median fissure, supplying a medial wedge of the lemniscus and hypoglossal nucleus and the anterior portion of the spinal cord.

The *vertebral branches* to the medulla supply the lateral portion of the lower medulla and an anterolateral wedge of the upper medulla, including the inferior olive and hypoglossal nerve.

The *posterior inferior cerebellar artery* branches off the vertebral artery at the level of the midmedulla. It courses dorsally along the medulla and then curves upward along the inferior surface of the cerebellum. It supplies the dorsolateral part of the medulla, including the inferior cerebellar peduncle, spinothalamic tract, spinal tract and nucleus of V, components of the vagus nerve, and descending central autonomic fibers to the spinal preganglionic sympathetic neurons. It also supplies parts of the inferior vermis, inferior surface and deep nuclei of the cerebellum, and the choroid plexus of the fourth ventricle.

The *anterior inferior cerebellar artery* branches off the basilar artery at the caudal pons level and curves caudally and laterally to the inferior surface of the cerebellum. It helps supply the tegmentum of the caudal pons, parts of the inferior vermis, and inferior surface and deep nuclei of the cerebellum.

The *basilar branches* to the pons include paramedian branches that supply the medial portion of the pons, excluding the tegmentum, and circumferential arteries that curve backward to supply the lateral and dorsal portions of the pons.

The *superior cerebellar artery* branches off the basilar artery at the level of the midbrain just below the basilar bifurcation. It then courses straight back to the superior surface of the cerebellum. It contributes branches to the circumferential arteries that supply central and lateral parts of the crus cerebri and substantia nigra and lateral parts of the midbrain tegmentum. It also supplies the superior vermis, superior surface of the cerebellum, and the rest of the deep nuclei.

The *posterior cerebral artery* branches off the basilar artery at the terminal bifurcation and curves laterally around the midbrain and

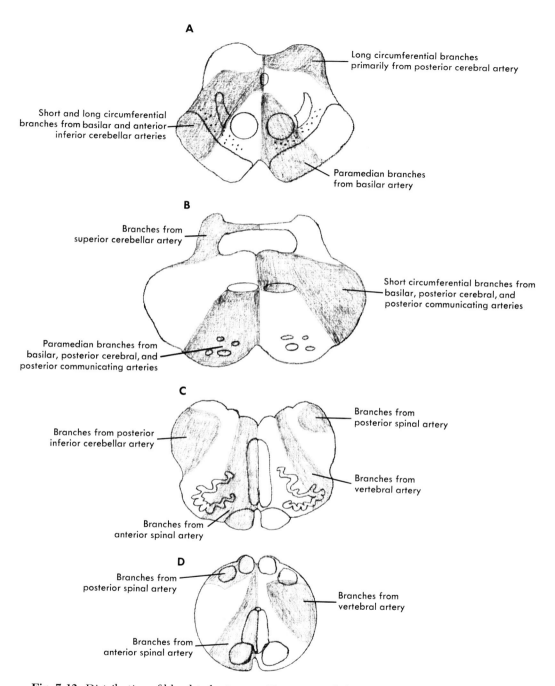

Fig. 7-12. Distribution of blood to brainstem. Keep in mind that variations in distribution do exist and that there can be overlap of adjacent areas. **A,** Midbrain at superior colliculus level. **B,** Midpons. **C,** Upper medulla. **D,** Lower medulla. (Adapted from Truex, R., and Carpenter, M.: Human neuroanatomy, ed. 6, Baltimore, 1969, The Williams & Wilkins Co.)

Labels in figure:

A
- Long circumferential branches primarily from posterior cerebral artery
- Short and long circumferential branches from basilar and anterior inferior cerebellar arteries
- Paramedian branches from basilar artery

B
- Branches from superior cerebellar artery
- Short circumferential branches from basilar, posterior cerebral, and posterior communicating arteries
- Paramedian branches from basilar, posterior cerebral, and posterior communicating arteries

C
- Branches from posterior inferior cerebellar artery
- Branches from posterior spinal artery
- Branches from vertebral artery
- Branches from anterior spinal artery

D
- Branches from posterior spinal artery
- Branches from vertebral artery
- Branches from anterior spinal artery

then dorsally to the occipital lobe. Paramedian branches are given off and join with branches of the posterior communicating artery to supply medial midbrain structures, including the nucleus and cranial nerve of III, the red nucleus, and medial portions of the crus cerebri. Circumferential branches are given off, which with branches of the posterior cerebral artery supply the rest of the midbrain tegmentum and tectum. Posterior choroidal branches help supply the dorsal thalamus, subthalamus, choroid plexus of the third ventricle, and part of the internal capsule. The remaining posterior cerebral artery supplies the medial and inferior surface of the temporal lobe and medial surface and pole of the occipital lobe.

The following summarizes the above in terms of structures:

1. The cerebral cortex is primarily supplied by the three cerebral arteries.
2. The internal capsule is primarily supplied by branches from the anterior and middle cerebral, anterior choroidal, and posterior communicating arteries.
3. The basal nuclei (caudate, putamen, and globus pallidus) are primarily supplied by branches from the middle and anterior cerebral arteries and the anterior choroidal artery.
4. The thalamus is supplied by branches from the posterior communicating artery and the middle and posterior cerebral arteries.
5. The hypothalamus and subthalamus are supplied by branches from the posterior and anterior cerebral and posterior communicating arteries.
6. The cerebellum is supplied by the three cerebellar arteries.
7. The brainstem is supplied by branches from the spinal, cerebellar, vertebral, basilar, and posterior cerebral arteries. The pattern of distribution is roughly mediocentral (paramedian branches) or dorsolateral (circumferential branches).

It is important to keep in mind that there can be variances and overlap in the distribution of these arteries such that the area damaged by occlusion of one artery is often smaller than would be expected. Also many arteries send an anastomosing (connecting) branch to another artery. This branch can help keep an artery filled with blood in the event of occlusion.

Circle of Willis

At the base of the brain, surrounding the infundibulum and optic chiasm, is a circle of interconnected arteries known as the *circle of Willis*. The components of the circle of Willis include parts of the internal carotid, anterior cerebral, and posterior cerebral arteries, which are connected to form a circle by the two posterior communicating and the single anterior communicating arteries (Fig. 7-10).

Under normal conditions the circle of Willis serves no function. Blood does not flow through in a circular pattern, since the arterial pressure in the carotid arteries is similar to that in the basilar arteries. The circle of Willis can function as a safety valve under conditions that produce a differential pressure gradient in one of the supplying arteries. For example, if one of the posterior cerebral arteries becomes occluded where it branches from the basilar artery, the pressure distal to the clot will drop severely. Blood from the internal carotid on that side can flow into the posterior cerebral artery via the posterior communicating artery. In this manner the tissues supplied by this posterior cerebral artery can receive vital oxygen and other nutrients.

Unfortunately a complete circle of Willis is found in only about 20% of the population. The most common deficiency is an absence of one of the communicating arteries. Even in persons with a complete circle any narrowing of the already small lumen in the communicating arteries such as from arteriosclerosis will reduce or prevent the circle from functioning as a safety mechanism.

Venous drainage

The drainage of the structures within the skull can be divided into the superficial and deep veins of the cerebrum, the veins draining the cerebellum and brainstem, and the emissary veins. The general pattern of venous blood flow is as follows.

Superficial cerebral veins drain most of the blood from the cerebral cortex and subcortical white matter into the dural sinus system.

Deep cerebral veins drain blood from internal structures such as the thalamus and basal nuclei into the two *internal cerebral veins*. These veins travel dorsally on the thalamus just above the massa intermedia. At the pineal gland these two veins unite to form the *great vein of Galen*, which continues dorsally to just past the splenium of the corpus callosum, where it merges into the straight sinus (Fig. 7-13). Some venous blood from the base of the brain around the infundibulum collects into *the cavernous sinuses*, a network of venous channels that drain into the transverse sinus.

Cerebellar and brainstem veins have a distribution pattern similar to the arterial supply. Veins draining the cerebellum, midbrain, pons, and upper medulla drain into the dural sinuses at the base of the brain, the internal cerebral veins, or the great vein of Galen; veins from the lower medulla drain into the spinal veins.

Emmisary veins connect veins from the scalp and nasal sinuses to the dural sinus system.

Once inside the dural sinus system the venous blood travels to the transverse and then to the *sigmoid sinuses* (they are continuous with each other) and finally into the *internal jugular vein*.

Regulation of cerebral blood flow

The physiologic boundaries within which neural tissue can survive are extremely narrow. Even slight variations, whether excessive or insufficient, can be damaging. In addition the metabolic machinery of a neuron operates at a very even level of activity throughout the day and night. Therefore the flow of cerebral blood must be kept as constant as possible regardless of the state of activity in the body. The following mechanisms have a role in providing a homeostatic flow of blood.

Pressure receptors in the carotid sinus (at the bifurcation of the common carotid artery) and aortic arch respond to changes in systemic blood pressure. A drop in pressure stimulates cardiac centers in the medulla to increase the heart's output; a rise in pressure decreases cardiac output.

Chemoreceptors in the carotid body (located near the carotid sinus) respond to the level of CO_2 in the blood. Increases in PCO_2 stimulate respiratory centers in the medulla to increase the rate of respiration. Decreases in PCO_2 decrease the respiratory rate. Increases (decreases) in cardiac output are made in response to increases (decreases) in PCO_2 levels.

The cerebral vessels themselves respond to systemic blood pressure. Increases in systemic pressure cause constriction, reducing cerebral blood flow, and decreases in systemic pressure cause dilation, increasing cerebral blood flow. This effect is known as the *Bayliss effect* and seems to be an intrinsic property of the vessel wall.

The cerebral vessels also respond to the concentration of CO_2 in the blood. Increases in PCO_2 produce dilation of cerebral vessels and increased CBF. Decreases in PCO_2 cause the opposite to occur. The precise mechanism of this effect is not known. CO_2 may relax the smooth muscle of the arterial wall.

The first two mechanisms ensure that the heart and lungs provide functional levels of blood and oxygen to the carotid and vertebral arteries. The last two mechanisms maintain a delicate balance of blood pressure and oxygen within the cranium.

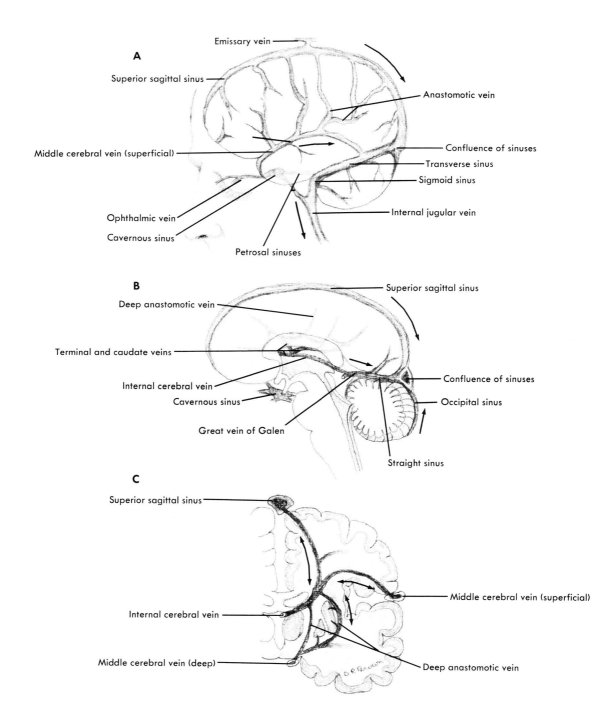

Fig. 7-13. Venous drainage of brain. **A,** Lateral view. **B,** Median view. **C,** Frontal view. Note that venous blood and CSF flow dorsal to where superior sagittal, straight, transverse, and occipital sinuses meet to form area known as confluence of sinuses. (Adapted from Noback, C., and Demarest, R.: The human nervous system: basic principles of neurobiology, ed. 2, New York, 1975, McGraw-Hill Book Co.)

Regulation of extracellular fluid

The composition of the extracellular fluid of the brain must be delicately regulated also, since neuronal functioning depends on the ratio of the concentration of ions across the cell membrane. Four mechanisms regulate and buffer the extracellular fluid against change:

1. The blood acts as a buffering and excretory agent, especially for maintaining a proper balance of positive ions and CO_2.
2. The endothelial lining of cerebral capillaries, the single-cell layer of pia mater, and the processes of the astrocytes (the components of the blood-brain barrier) act to prevent or retard the diffusion of some substances such as large protein molecules and highly enlarged ions.
3. Some neuroglial cells (probably the astrocytes) seem to have the ability to take up K^+ during excess neuronal activity. Remember that excess K^+ on the outside of a neuron can cause the RMP to drop.
4. Central chemoreceptors and osmoreceptors (probably located in the hypothalamic-pituitary and pontomedullary areas) assess the pH and osmotic concentration of the blood. Appropriate changes in the rate and depth of respiration and the diuretic activities of the kidney can then be initiated.

Keep in mind that the interrelationship between the extracellular fluid, CSF, cerebral blood flow, and intracranial pressure are very complex and still poorly understood.

ABNORMAL ENVIRONMENTAL CONDITIONS
Anoxia

There are two vital metabolic requirements for a neuron to function: oxygen and glucose. The effects of lack of oxygen (anoxia) are the result of the neuron's inability to manufacture molecules of ATP. Since neurons have only very minimal stores of ATP, the metabolic machinery quickly fails. The synaptic areas on the dendrites are the most sensitive to anoxia. Transmission starts to fail in 45 to 90 seconds and can result in loss of consciousness. Within minutes the dendrites and soma lose their RMP (remember that the sodium-potassium pump "runs" on ATP). The neuron will quickly perish unless oxygen is brought to the neuron. In general the more specialized the neuron (the more advanced on a developmental scale) the more susceptible it is to anoxia. Cortical neurons therefore will succumb to anoxia in 3 to 5 minutes, while some subcortical structures can survive up to 10 minutes of anoxia.

Anoxia can occur for two reasons: inadequate blood flow or inadequate exchange of gases. Inadequate blood flow can result from (1) a ruptured aneurysm, (2) hemorrhage as from trauma, (3) sudden blockage (embolism), (4) cardiac arrest, or (5) slowly progressive blockage as from a developing thrombus. The first four conditions generally have a sudden onset, and the damage can be extensive. Collateral circulation such as in the circle of Willis can be of benefit in the first three conditions by stabilizing the blood flow to deprived areas. However, many persons (especially older individuals) have little, if any, collateral circulation. In the last condition there may be a history of *transient ischemic attacks* (TIA). These attacks occur during vasospasm or periods of decreased blood pressure in which the already narrow lumen is briefly closed. Usually there is transient hemiparesis-hemisensory deficit, and there may be brief lapses of consciousness. Damage may be minimal but will vary with the severity and frequency of the attacks and the extent of collateral circulation.

An inadequate exchange of CO_2 for O_2 can result from (1) a blocked trachea as from a piece of food, (2) drowning or other incidence of suffocation, (3) spasm of the diaphragm or abdomen as in prolonged siezures, and (4) fluid in the lungs as in severe respiratory infections.

In blocked trachea or drowning the onset is usually sudden and the prognosis is poor unless immediate steps are taken to ventilate the lungs. During the spasm phase of an epileptic seizure the respiratory muscles can be rigid and breathing arrested. This rarely lasts more than 30 seconds, but repeated seizures can severely lower the level of oxygen in the blood. Fluid in the lungs is usually more insidious in onset and is often the cause of death in conditions of prolonged and severe weakness.

Hypoglycemia

Glucose, like oxygen, is vital to a neuron's metabolic production of ATP. Complete absence of glucose in the blood causes necrosis of brain tissue (beginning with the cerebral cortex) within minutes. However, the conditions that usually lower the levels of glucose in the blood (hypoglycemia) do so progressively. Early symptoms are irritability and mental confusion. Later symptoms are convulsions and coma. Irreversible neural loss occurs during the coma stage, and the prognosis is poor. The primary cause of hypoglycemia is too much insulin, which carries the available glucose from the blood into the body cells.

Hypothermia

It is unusual for hypothermia (decreased body temperature) to be listed as a neurologic symptom. It is mentioned with Parkinson's disease and pituitary-hypothalamic lesions involving the endocrine system or temperature regulation center.

Two related effects occur in the nervous system with hypothermia. One is reduction in metabolic activity such that spontaneous action potentials cease. For example, cooling the body by just 4 C to 33 C causes respiration to cease. Cooling to 27 C can cause cardiac arrest. A marked reduction in action po-

tentials is also seen with cooling of the peripheral nerves. The other effect is a reduced need for glucose and oxygen. By cooling the body to 30 C neural tissue can withstand about twice the normal period of ischemia. Additional cooling increases this time even further. This reduced metabolic requirement can be of great value during certain neurosurgical procedures. There is also potential value in the treatment of ischemic or anoxic lesions of sudden onset if cooling can be initiated early enough.

TEST QUESTIONS

For each question select *all* correct choices.

1. Cerebrospinal fluid normally:
   a. Has a pressure of 60 to 180 mm of water in the side-lying position.
   b. Contains large amounts of protein and glucose.
   c. Is pale yellow.
   d. Helps to protect the CNS with its bouyancy.

2. A person suffering from an increase in intracranial pressure often complains of headache, nausea, and vomiting as initial symptoms. Which of the following statements are true?
   a. Increased pressure could be due to a blocked central canal.
   b. There could be a tumor of the choroid plexus causing an excess production of CSF.
   c. There could be a blockage of one of the cerebral arteries.
   d. The increased pressure could cause the ventricles to shift position.

3. A vascular lesion of the genu of the internal capsule could involve:
   a. Middle cerebral artery.
   b. Anterior choroidal artery.
   c. Posterior cerebral artery.
   d. Posterior communicating artery.

4. The basilar artery:
   a. Supplies the descending corticospinal tracts in the pons.
   b. Gives off branches that supply the spinal cord.
   c. Is formed by the two vetebral arteries.
   d. Gives off the middle meningeal artery.

5. A meningioma on one side of the middle fossa could cause:
   a. Symptoms of endocrine and autonomic nervous system involvement.
   b. Double vision.
   c. Anesthesia of one half of the face.
   d. Paralysis of one half of the tongue.
6. In head injury surgery may be required to:
   a. Replace lost CSF.
   b. Drain a hematoma.
   c. Remove bone fragments.
   d. Repair a severed blood vessel.
7. The superior sagittal sinus:
   a. Is located in the apex of the medial longitudinal fissure.
   b. Drains CSF from the subarachnoid space.
   c. Is at right angles at its base to the transverse sinus.
   d. Contains arachnoid granulations.
8. The blood-brain barrier:
   a. Is a system of membranes that acts as an interface between the bloodstream and the extracellular fluid around the neurons.
   b. Blocks the transport of gases and ions from the blood to the delicate neurons.
   c. Is composed of pia mater, the capillary wall, and astrocytes.
   d. Must be considered in establishing dosage levels of some CNS drugs.
9. Which area(s), if obstructed, could cause the fourth ventricle to dilate?
   a. Cerebral aqueduct.
   b. Foramen of Monro.
   c. Arachnoid granulations.
   d. Foramina of Magendie and Luschka.
10. Which of the following is(are) not a component of the circle of Willis?
    a. Anterior cerebral artery.
    b. Posterior cerebral artery.
    c. Anterior choroidal artery.
    d. Posterior communicating artery.
11. The circle of Willis:
    a. Surrounds the midbrain.
    b. Can redistribute blood to areas of the brain receiving insufficient amounts.
    c. Is ineffective if occlusion is distal to the circle.
    d. Normally has blood flowing in a circular direction to balance the arterial pressure.
12. Under conditions of anoxia:

a. Synaptic transmission is affected initially.
b. Vegetative functions can persist for a while even after a person becomes unconscious.
c. Some cortical neurons can use the oxygen in the CSF to survive.
d. Neurons are unable to manufacture ATP.
13. The effects of hypoglycemia can include:
    a. Coma and death.
    b. Changes in personality.
    c. Sharp drop in blood pressure.
    d. Headache, nausea, and vomiting.
14. The composition of the extracellular fluid around the neurons must be maintained within very narrow boundaries for the neurons to survive. Which of the following statements is(are) true about the regulation of this extracellular fluid?
    a. Drop in systemic blood pressure increases cardiac output and causes dilation of the cerebral vessels.
    b. Increased levels of PCO_2 increase cardiac output and cause dilation of the cerebral vessels.
    c. Astrocytes are able to absorb K^+ during excess neuronal activity.
    d. pH of the blood is partially regulated by chemoreceptors in the brainstem that alter the rate and depth of respiration.
15. The venous drainage of the brain involves:
    a. Superficial veins that drain into the great vein of Galen.
    b. Emissary veins that drain the scalp and nasal sinuses.
    c. Dural sinuses that eventually drain into the internal jugular vein.
    d. Arachnoid granulations.
16. The third ventricle:
    a. Is found in each hemisphere.
    b. Is a midline structure.
    c. Receives CSF from the lateral ventricles.
    d. Contains the foramen of Magendie.
17. A subdural hematoma:
    a. Produces blood in the CSF.
    b. Is probably from a ruptured middle meningeal artery.
    c. Is often rapidly fatal.
    d. Causes the ventricles to dilate.
18. The tentorium cerebelli:
    a. Separates the two cerebellar hemispheres.
    b. Is bisected in a midsagittal section.

c. Contains the transverse sinus along its circumference.

d. Is located under the cerebellum.

19. The dura mater:
    a. Is one of the components of the blood-brain barrier.
    b. Surrounds the entire CNS but does not dip into the sulci.
    c. Is continuous with the lining of the ventricular system.
    d. Is a single layer of tough fibrous tissue.

20. In the dominant hemisphere the language function centers for speaking and for understanding speech are located in the area around the junction of the central sulcus and lateral fissure. This area is supplied by:
    a. Anterior cerebral artery.
    b. Posterior cerebral artery.
    c. Middle cerebral artery.

21. In the case of a possible skull fracture there is initial concern about:
    a. Area of the skull that might be fractured.
    b. Increase in intracranial pressure.
    c. Infections traveling to the brain via the emmissary veins.
    d. Padded headboard to protect the skull.

22. In a rupture of the left middle cerebral artery in the area where it emerges from the lateral fissure one would expect:
    a. Blood in the CSF.
    b. Shift of the ventricles to the *right*.
    c. Anoxia of the neurons in the pre- and post-central gyri.
    d. Downward compression of the cerebrum.

SUGGESTED READINGS

Carpenter, M.: Core text of neuroanatomy, ed. 2, Baltimore, 1978, The Williams & Wilkins Co.

Chusid, J.: Correlative neuroanatomy and functional neurology, ed. 16, Los Altos, Calif., 1976, Lange Medical Publications.

House, E., and Pansky, B.: A functional approach to neuroanatomy, New York, 1967, McGraw-Hill Book Co.

Netter, F.: The CIBA collection of medical illustrations. I. The nervous system, Summit, N. J., 1972, CIBA Chemical Co.

Noback, C., and Demarest, R.: The human nervous system, ed. 2, New York, 1975, McGraw-Hill Book Co.

Willis, W. D., Jr., and Grossman, R. G.: Medical neurobiology: neuroanatomical and neurophysiological principles basic to clinical neuroscience, ed. 2, St. Louis, 1977, The C. V. Mosby Co.

CHAPTER 8

Higher centers and their sensory processing systems

In this chapter are discussed the various functions of the sensory system, including its influence on the motor system, the major sensory pathways from the spinal cord to the brainstem and cerebrum, the role of the reticular formation, thalamus, and cerebral cortex in sensory processing, recent concepts of pain, and clinical considerations of lesions of the sensory system. The following areas deserve special attention:

1. Function of the sensory system in the unconscious sensorimotor activities involving the somatic and visceral reflexes
2. Definition of awareness in terms of the perception and interpretation of a stimulus
3. Mechanism by which the CNS determines the quality, intensity, and location of a stimulus
4. Location of the first-, second-, and third-order neuron cell bodies, the location of the first and second synapses, and the site of crossing for the dorsal column, spinothalamic, and trigeminothalamic tracts
5. Function of the unconscious sensory tracts
6. Course traveled by the special somatic, general visceral, and special visceral afferent fibers

7. Input to and output from the reticular formation and its sensory-related functions
8. Role of the thalamus and cerebral cortex in the perception and interpretation of sensory data
9. Location and function of the primary receiving and association areas of the cerebral cortex
10. CNS circuitry involved in the gate theory of pain
11. Symptoms of the Brown-Séquard and thalamic syndromes
12. Definition of and areas involved in paresthesias, radiating pain, and the three agnosias

FUNCTIONS OF SENSORY SYSTEMS
Unconscious activities

The sensory systems have a vital role in the unconscious responses of the body, which include the following:

1. Intra- and intersegmental spinal reflexes such as the myotatic and flexor withdrawal reflexes
2. Suprasegmental brainstem reflexes such as the tonic and righting reflexes
3. Cerebral reflexes such as the equilibrium reactions and the rhythmic automatic responses such as walking
4. Visceral (autonomic system) reflexes

225

such as respiration and blood pressure that primarily involve the spinal cord and brainstem

The purpose of the above reflexes is to provide for (1) protective reactions to noxious stimuli and loss of equilibrium, (2) maintenance of posture, (3) movement patterns such as walking and chewing that do not require constant attention, and (4) homeostasis of the internal environment.

In general these unconscious activities represent a stereotyped pattern of muscular or glandular reactions in response to a volley of action potentials from somatic (skin, muscle, tendon, joint, eye, or ear) receptors or from visceral receptors (internal organs, blood vessels, and mucous membranes). Although much of these afferent data do not reach consciousness, some volitional control over these reflexes is possible. The extreme of this is walking and chewing, which can be completely controlled by conscious thought or delegated to subcortical centers. Under less control is respiration. Both the rate and depth can be altered at will but only for brief periods. Holding one's breath is easy initially, but the chemoreceptors in the carotid body and in the CNS will normally override conscious effort as the arterial P_{CO_2} increases and the pH associated with this decreases. Even some of the spinal reflexes such as withdrawal in response to pain can be altered voluntarily. Of interest is the fact that some people can control heart rate, blood pressure, and other autonomic functions. The procedures utilized are shrouded in the mystique of eastern meditation.

Conscious activities

The conscious component of the sensory systems utilizes parts of the thalamus and cerebral cortex to create an awareness of the environmental energies and forces that influence the body. These environmental data are primarily conveyed to the CNS by the cutaneous and special sense receptors, although some visceral receptors produce a conscious awareness.

The concept of awareness begins with the *perception* of a stimulus. Perception primarily involves an ability to recognize the quality (type) of stimulus, its intensity, and its location in the body. Other modalities such as form, size, direction, and velocity are also important to the perception of a stimulus.

Simple recognition that a stimulus with specific characteristics is invading a portion of the sensory field of the body is insufficient for functional awareness. The *interpretation* of the meaning of the perceived data is necessary to form judgments as to the usefulness of the stimulus, the emotional value of the stimulus, and whether any physical or cognitive response is required. For example, the perception of a door key is that it is (1) a small irregularly shaped object with some rough edges, and (2) it is lightweight but solid. The interpretation of these data is that this object is a key and is used to open doors. Physical movements may be planned to unlock the door. If it is cold and raining the interpretation of the usefulness of this object may bring joy and relief. Of interest clinically is the fact that the CNS areas for perception and interpretation are different.

ORGANIZATION OF SENSORY SYSTEMS

A variety of mechanical, chemical, thermal, and photic energies are capable of exciting the various receptors in the body. These receptors can be generally classified as either somatic or visceral. Both types react in a similar physiologic manner to create sensations.

Physiologic concepts

The conversion of a stimulus energy to action potentials and the transmission of these impulses to levels of consciousness utilize four concepts that influence the perception of sensation. The first concept concerns *receptor specificity*, which dictates that a given receptor will usually respond to only one form of energy and in some cases to only a portion of that form of energy. For example, both a hair follicle ending and a pacinian corpuscle are sensitive to mechanical stimulation. The latter, however, is not sensitive to light stroking because of its *location* in the skin and its *cellular organization*. Somewhat confusing are free nerve endings. They are all basically alike, since they are the bare terminals of an afferent fiber. However, factors including location and cellular organization allow some free nerve endings to be mechanoreceptors (touch, blood pressure), thermoreceptors, and chemoreceptors (pain). Receptor specificity is thus essential in order for the CNS to determine the *quality* or type of stimulation.

The second concept is *intensity coding*. It involves receptors and their ability to transform the strength of the stimulus and the rate of stimulation into varying frequency patterns. For example, placing a hand in a bowl of warm (36 C) water will excite a particular number of receptors that respond within a specific frequency range. Placing the same hand in a bowl of hotter (42 C) water will excite more receptors and cause an increased output from those already firing. A given receptor therefore can tell the CNS both the type of stimulus and the intensity of stimulation.

The third concept concerns the *preprocessing of incoming afferent data* in the CNS relay nuclei that are involved with the ascending fiber tracts. This preprocessing involves internuncials within the relay nuclei and descending fibers from higher centers that terminate in the nuclei. Both systems function to modulate the frequency of incoming sensory data. One form of modulation has been found to occur from presynaptic inhibition of dorsal column data synapsing in the nuclei cuneatus and gracilis. Another example of sensory suppression is seen when one is intently reading a book in a noisy location such as an airport. All of the extraneous auditory data are blocked to allow attention to the visual data.

The fourth concept involves a *point-to-point representation* of receptive fields in discrete areas of the cerebral cortex. This allows for the localization of a stimulus. For example, a hair pulled on the back of the hand will excite a small cluster of neurons in a single location in the postcentral gyrus.

In summary, the sensory data reaching consciousness are coded in terms of strength, focused by filtering extraneous data, localized, and identified as to the type of stimulus.

Somatic sensory pathways

The emphasis in this section centers on the pathways and synapses involved in relaying data from the cutaneous receptors of the body and face to the cerebral cortex. These "conscious" pathways include the dorsal column–medial lemniscus tract, the spinothalamic tract, and the trigeminothalamic tract. The other ascending tracts and the pathways for vision and audition are discussed at the end of this section.

The three conscious pathways are described in terms of the following:

1. Stimulus conveyed
2. Receptor
3. Type of afferent fiber
4. Location of the first-order neuron cell body
5. Course of the central process of the first-order neuron
6. Location of the first synapse (with the second-order neuron)
7. Course of the second-order neuron axon
8. Location of the second synapse (with the third-order neuron)
9. Course of the third-order neuron axon

These nine characteristics are presented in Table 8-1 and diagrammed in Figs. 8-1 and 8-2. For a better perspective of the location of these and the other fiber tracts in the

Text continued on p. 232.

Table 8-1. Characteristics of conscious sensory tracts

	Anterior spinothalamic	Lateral spinothalamic
Stimulus conveyed.	Crude touch, tickle, itch, orgasm.	Pain, temperature.
Receptors.	Free nerve endings, hair follicle ending.	Free nerve endings, Krause's end-bulb.
Type of afferent fiber.	Groups III and IV.	Groups III and IV.
Location of first-order neuron cell body.	Dorsal root ganglia.	Dorsal root ganglia.
Course of axon of first-order neuron.	Enter cord through lateral division of dorsal root. Enter dorsolateral tract of Lissauer and ascend or descend 1 to 3 segments from point of entry. Terminate in laminae I, II, and III.	Enter cord through lateral division of dorsal root. Enter dorsolateral tract of Lissauer and ascend or descend 1 to 3 segments from point of entry. Terminate in laminae I, II, and III.
Location of first synapse (with second-order neuron).	Laminae V, VI, and VII. There are interneurons intercalated between first- and second-order neurons.	Laminae V, VI, and VII. There are interneurons intercalated between first- and second-order neurons.
Course of second-order neuron axons.	Cross in anterior white commissure and form anterior spinothalamic tract. Ascends in anterior portion of cord. Joins with lateral spinothalamic in lower brainstem.	Most cross in anterior white commissure and form lateral spinothalamic tract or enter spinoreticular tract. Uncrossed fibers enter ipsilateral spinoreticular tract. Ascends in lateral portion of spinal cord and brainstem to thalamus.
Location of second synapse (with third-order neuron).	Ventral posterolateral nucleus of thalamus.	Ventral posterolateral, posterior, and intralaminar nuclei of thalamus.
Course of third-order neuron axons.	Enter posterior limb of internal capsule. Terminate in postcentral gyrus (SI) and SII.	Enter posterior limb of internal capsule. Terminate in postcentral gyrus (SI) and SII.

Dorsal column–medial lemniscus	Trigeminothalamic
Discriminative touch, pressure, vibration, kinesthesia.	All cutaneous sensations, kinesthesia.
Merkel's disc, Meissner's corpuscle, pacinian corpuscle, Ruffini ending.	Free nerve ending, hair follicle ending, Krause's end-bulb, Merkel's disc, Meissner's corpuscle, pacinian corpuscle, Ruffini ending.
Groups II, III.	Groups II, III, IV.
Dorsal root ganglia.	Trigeminal ganglia, geniculate ganglia (VII), superior ganglia (IX, X).
Enter cord through medial division of dorsal root. Enter dorsal funiculus. Fibers below T6 form fasciculus gracilis; fibers above T6 form fasciculus cuneatus. Ascend in dorsal columns to medulla.	Fibers conveying pain, temperature, crude touch, tickle, and itch enter midpons and descend to medulla and first two levels of cervical cord. These fibers form spinal tract of V. Fibers conveying discriminative touch, pressure, vibration, and kinesthesia enter midpons. Most ascend to main sensory nucleus of V. Some descend short distance in spinal tract of V.
Nuclei cuneatus and gracilis.	Neurons in spinal tract of V that form nucleus of spinal tract V. Main sensory nucleus of V.
Cross midline forming internal arcuate fibers. Ascend brainstem as medial lemniscus. In pons, shifts dorsolaterally. In midbrain, becomes associated with spinothalamic tract.	Axons from spinal nucleus of V cross in lower brainstem and upper cervical cord to form ventral trigeminothalamic tract. Those from main sensory nucleus either cross at that level and enter ventral trigeminothalamic tract or ascend ipsilaterally as dorsal trigeminothalamic tract. These latter fibers are from mandibular division of V. Both sets of fibers project colaterals to ascending reticular tracts.
Ventral posterolateral nucleus of thalamus.	Ventral posteromedial, posterior, and intralaminar nuclei.
Enter posterior limb of internal capsule. Terminate in postcentral gyrus (SI) and SII.	Enter posterior limb of internal capsule. Terminate in postcentral gyrus (SI) and SII.

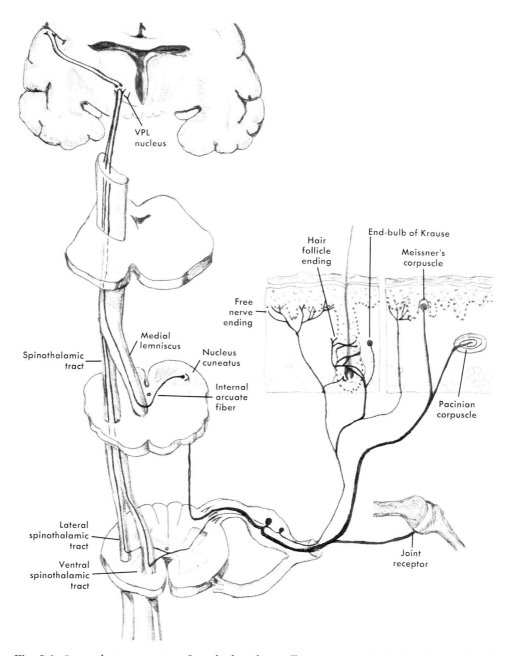

Fig. 8-1. Somesthetic sensations from body. Those afferents conveying highly localized and discriminative sensations are larger (more myelin), faster conducting, and travel in dorsal column–medial lemniscus pathway. Those afferents conveying less defined sensations are smaller, slower conducting, and travel in spinothalamic tracts. Collateral branches for reflexes have been omitted, as have intersegmental branching of incoming pain and temperature afferents. (Adapted from Netter, F.: The CIBA collection of medical illustrations. I. The nervous system, Summit, N.J., 1972, CIBA Chemical Co.)

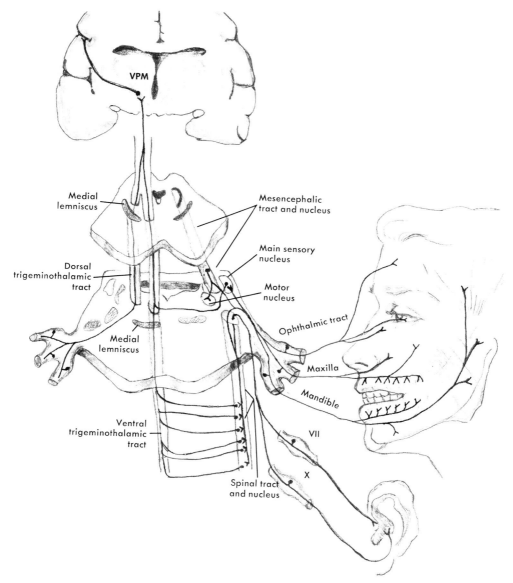

Fig. 8-2. Somesthetic sensations from face. Those afferents conveying highly localized discriminative sensations (epicritic) synapse in main sensory nucleus. Those afferents conveying cruder sensations (protopathic) synapse in spinal nucleus. Second-order neurons cross and form ventral trigeminothalamic tract, which travels in close association with medial lemniscus. Unconscious proprioception afferents have their first-order neuron cell body in mesencephalic nucleus. Muscle spindle afferents from extraocular, facial, and tongue muscles may also have their cell bodies in mesencephalic nucleus. Synapses from unipolar neurons of this nucleus are with trigeminal and other brainstem motor nuclei. Some somesthetic data are relayed to VPM nucleus ipsilaterally via dorsal trigeminothalamic tract.

spinal cord see Fig. 6-1; for those in the brainstem see Figs. 5-12 to 5-19. It is useful to formulate a perspective of the location of the ascending and descending tracts in terms of the effects of a vascular lesion, an intermedullary (in the center) lesion, and a parasagittal (along the outside) lesion of the brainstem and cord. Such lesions can produce varying combinations of sensory and motor symptoms that when evaluated can often indicate the level of involvement.

The other ascending tracts are considered the unconscious pathways. They basically convey somatic information to the cerebellum and brainstem centers such as the reticular formation, inferior olive, and tectum. Although some of this information may reach the thalamus and cerebral cortex, it is generally not considered a major contributor to the perception and interpretation of somatic sensations. These tracts are as follows.

Spinoreticular tract. This pathway arises from cells in the posterior horn and intermediate zone. The fibers ascend ipsilaterally and contralaterally in the anterior and anterolateral white matter of the cord. This tract conveys touch, pain, and probably visceral afferent data to the medial two thirds of the pontomedullary reticular formation. This system is part of the ascending reticular system that projects to the intralaminar nuclei of the thalamus, hypothalamus, and limbic system. Similar sensory data from the head probably enters this pathway (Fig. 8-3). The ascending reticular system is considered phylogenetically old. It is not somatopoietically organized and is multisynaptic. It is concerned with the emotional and visceral reactions to pain such as disagreeable, burning quality, and diffuseness and with gastrointestinal tract activity.

Spinotectal tract. Cells in the intermediate zone give rise to fibers that cross in the anterior white commissure. They ascend between the spinothalamic tracts to the superior colliculus and central gray matter of the midbrain. Although its function is not completely understood, it probably relays noxious stimuli from tactual, thermal, and pain receptors to help initiate head and eye movements for localizing a harmful stimulus (Fig. 8-3).

Posterior spinocerebellar tract. Cells in Clarke's nucleus receive monosynaptic excitation from Ia and Ib fibers as well as collaterals from touch and pressure afferents. These cells send axons that travel in the posterolateral border of the lateral funiculus to the medulla on the same side. Here the fibers become incorporated into the inferior cerebellar peduncle and terminate in the ipsilateral rostral and caudal portions of the vermis. Clarke's nucleus is found only between C8 and L3. Dorsal root fibers for this tract that enter below L3 ascend in the fasciculus gracilis to the upper lumbar segments, where they then terminate in Clarke's nucleus. Those dorsal root fibers that enter above C8 ascend in the ipsilateral dorsal columns to the accessory cuneate nucleus in the dorsolateral part of the lower medulla. Cells from this nucleus form the cuneocerebellar tract, which enters the adjacent inferior cerebellar peduncle to terminate in the ipsilateral vermis. This is considered the upper extremity equivalent of the posterior spinocerebellar tract.

This tract is somatotopically organized and rapidly conducting (up to 120 m/sec). It is believed that the impulses transmitted via this tract are utilized in the fine coordination of posture and movements of individual limb muscles (Fig. 8-4).

Anterior spinocerebellar tract. Cells in the lumbosacral portion of laminae V, VI, and VII receive monosynaptic excitation from Ib afferents of widely separate tendons in the lower extremity and collaterals from flexor reflex afferents (FRAs). Axons of these cells cross midline and ascend along the lateral border of the lateral funiculus just anterior to the posterior spinocerebellar tract.

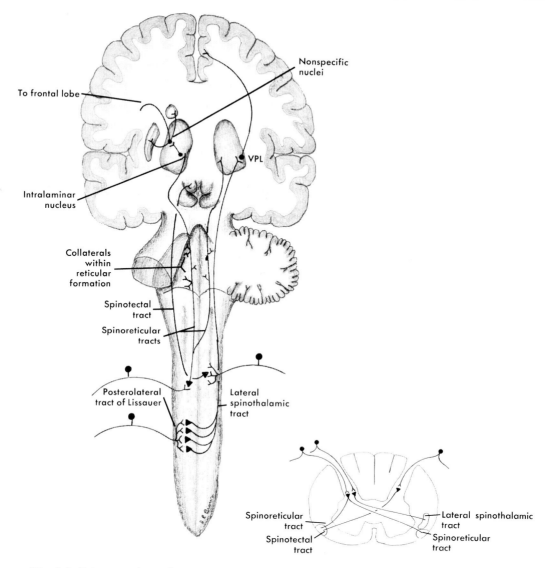

Fig. 8-3. Spinoreticular and spinotectal pathways. Note collaterals given off by spinoreticular tracts and projections from nonspecific thalamic nuclei. Lateral spinothalamic tract is also included to demonstrate intersegmental branching that occurs before synapsing with second-order neuron.

The fibers of the anterior spinocerebellar tract ascend to the lower midbrain, where they enter the superior cerebellar peduncle and terminate contralaterally in the anterior part of the vermis. This tract is involved with coordinated movements and posture of the entire lower limb (Fig. 8-4).

Spino-olivary tract. This tract arises from cells in all levels of the cord. These cells receive input from cutaneous afferents, Ib af-

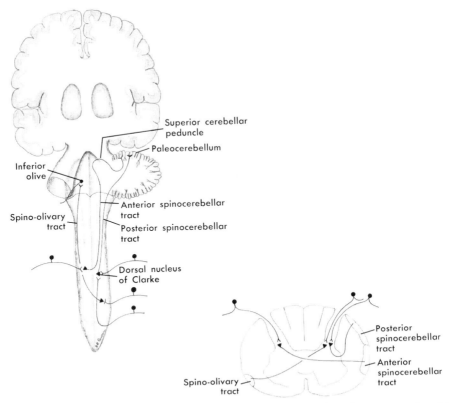

Fig. 8-4. Spinocerebellar and spino-olivary pathways. Note fiber from lower part of cord that ascends in fasciculus gracilis before synapsing in Clarke's nucleus.

ferents, and II and III muscle afferents. Axons from these cells ascend contralaterally in the anterior funiculus and terminate in the dorsal and medial accessory olivary nuclei (part of the inferior olivary complex). Axons from cells of these nuclei cross midline and enter the inferior cerebellar peduncle. They terminate mainly in the vermis. The exact role of this pathway is uncertain (Fig. 8-4).

Other ascending tracts. Direct connections between spinal cord neurons and the vestibular nuclei, pontine nuclei, and cerebral cortex have been established. The functional significance of these tracts remains unclear. Of interest clinically is the *spino-cervical* tract, which conveys touch, pressure, and pain sensations. Afferents from all levels enter the cord and synapse on neurons in laminae IV and V. These axons ascend ipsilaterally to the lateral cervical nucleus in the dorsal horn of the upper cervical cord. The third-order axons cross midline and ascend in the ventral funiculus to the medulla, where they join with the medial lemniscus. They end in the VPL nucleus, where fourth-order neurons relay this data to the postcentral gyrus. The spinocervical tract, which is smaller in humans than in carnivores, lies in the dorsalmost part of the lateral funiculus. This may account for the failure of the anterolateral cordotomy procedure used to relieve intractable pain (see p. 248).

Special somatic pathways. The special somatic pathways include the senses of vision, audition, and equilibrium. In Chapters 4 and 5 impulses from these systems are traced from their origin at the receptors to their entry in the CNS. This section continues discussion of these pathways within the CNS.

VISUAL SYSTEM. The fibers of the optic tract are involved with three pathways. The first is the geniculocalcarine system. Fibers in this system travel in the optic tract and terminate in the lateral geniculate body. In this nucleus of the posterior thalamus are cells that respond to specific color wavelengths. Although not well understood, there is some degree of preprocessing of visual data as corticogeniculate fibers have been identified. The only output from the lateral geniculate body is to area 17 of the occipital lobe (Fig. 9-6) via the geniculocalcarine portion of the internal capsule. The calcarine sulcus is a major fissure of the occipital lobe that divides it into upper and lower portions. The fibers conveying data from the upper hemiretinas are located in the upper half of the optic radiations and terminate in the visual cortex above the calcarine sulcus. Fibers from the lower hemiretinas are located in the lower half of the optic radiations and terminate below the calcarine sulcus. These latter fibers are organized such that those carrying data that "sees" the ipsilateral temporal upper quadrant and contralateral nasal upper quadrant loop into the temporal lobe before turning occipitally. This is known as Meyer's loop. A lesion here results in contralateral upper quadrant anopsia (Fig. 8-5).

The second visual pathway involves those fibers of the optic tract that terminate directly on the cells of the superior collinculus (Fig. 8-5). This pathway forms a component of the tectospinal tract (see Chapter 9) that

reflexly regulates many head and eye movements. The superior colliculus also serves as an integrative center for vision, since a lesion of one superior colliculus can alter the responsiveness of an animal to visual input and cause ipsilateral head rotation. The tectal system is concerned with locating the visual data, while the geniculocalcarine system is more concerned with identifying the visual data. The phylogenetically older tectal system is thus involved with the protective head and body movements that automatically occur when an object suddenly crosses into the peripheral visual field.

The third pathway involves fibers from the optic tract that terminate in the pretectum of the midbrain (just rostral to the superior colliculus), bypassing the lateral geniculate body. Internuncials cross to the opposite side (in the posterior commissure) to excite other internuncials (probably in the reticular nuclei) that terminate on the parasympathetic preganglionics of the Edinger-Westphal nucleus. This pathway is involved with constriction of the pupil to bright light. Of interest, when one eye is exposed to bright light both pupils constrict. The response in the stimulated eye is known as the direct light reflex and in the nonstimulated eye the *consensual light reflex* (Fig. 8-5).

Lesions of the pretectum result in the *Argyll-Robertson pupil*, in which the pupil does not constrict to light but does constrict during accommodation for near vision. The latter utilizes different pathways involving the occipital lobe.

In addition to these responses other internuncials project axons caudally to the sympathetic preganglionics in the upper thoracic cord (via the reticulospinal tract). Axons from these cells ascend in the paravertebral chain ganglia to the superior cervical ganglia, where the postganglionic neuron travels to the dilator muscles of the iris. A mutual antagonism is thus established such that the amount of light entering one or both eyes can

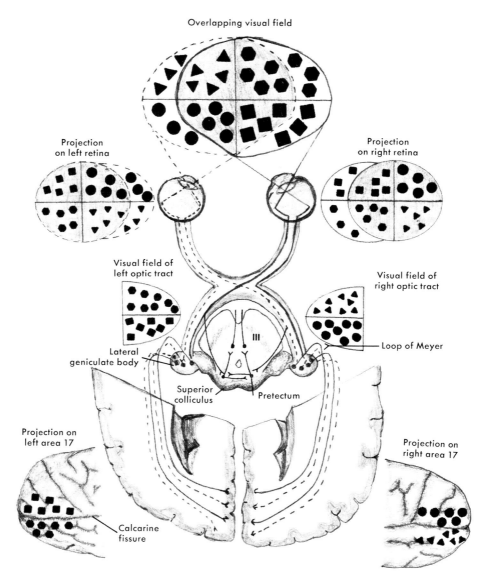

Fig. 8-5. Visual projections. Visual field of each eye can be divided into upper and lower portions and temporal and nasal portions. Each optic tract, however, "sees" only contralateral visual field, which can be divided into upper and lower quadrants. Within geniculocalcarine tract (caudal portion of posterior limb of internal capsule), fibers are arranged so that those originating from lower nasal portion of retina loop anteriorly into temporal lobe and terminate in area 17 above calcarine fissure. Because of this organization it is possible to have contralateral upper or lower quadrantic hemianopia. Note fibers from optic tract terminating in pretectum for light reflex. (Adapted from Netter, F.: The CIBA collection of medical illustrations. I. The nervous system, Summit, N.J., 1972, CIBA Chemical Co.)

reflexly determine the diameter of the pupil.

AUDITORY SYSTEM. The auditory data that reach the dorsal and ventral cochlear nuclei go through several routes to reach the cerebral cortex (Fig. 8-6). Some fibers cross midline in the posterior pons and ascend in the dorsolateral corner as the *lateral lemniscus* to the inferior colliculus. Other fibers synapse in the *superior olivary nucleus* (ventral tegmentum of the lower pons) on that side. Axons from these cells travel in both lateral lemnisci to the inferior colliculi. Those that cross midline form the *trapezoid body*. Still other fibers terminate in the nuclei of the reticular formation, where they are probably relayed to cranial nerves III, IV, and VI.

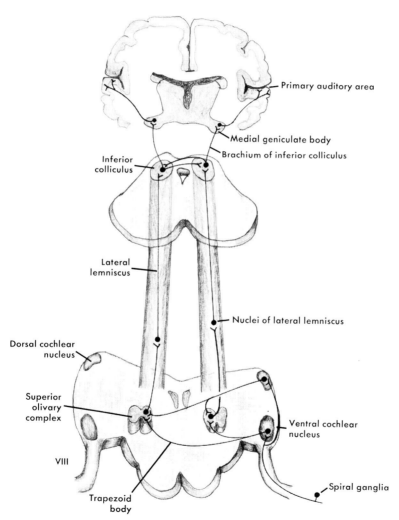

Primary auditory area

Medial geniculate body

Brachium of inferior colliculus

Inferior colliculus

Lateral lemniscus

Nuclei of lateral lemniscus

Dorsal cochlear nucleus

Superior olivary complex

Ventral cochlear nucleus

VIII

Spiral ganglia

Trapezoid body

Fig. 8-6. CNS projections of auditory data. Note bilateral projections to each hemisphere. Reflex connections of III, IV, VI, XI, and spinal cord are not shown.

The neurons of the inferior colliculi project to the medial geniculate body and to the superior colliculi. Neurons of the medial geniculate body project via the internal capsule to the primary auditory cortex (transverse temporal gyrus of Heschl) of the temporal lobe (areas 41 and 42). Although each hemisphere primarily receives input from the contralateral ear, ipsilateral input is present. This bilateral input functions in the localization of sounds.

From the superior colliculi auditory data enter the descending tectospinal pathway for reflex eye, head, and body movements.

In each of the auditory relay nuclei, the lateral lemniscus, and the trapezoid body are internuncials that are involved in the preprocessing of auditory data. These internuncials receive corticofugal (descending) fibers from the temporal lobe. This processing system includes sharpening of ascending data and suppression of unnecessary sounds. The

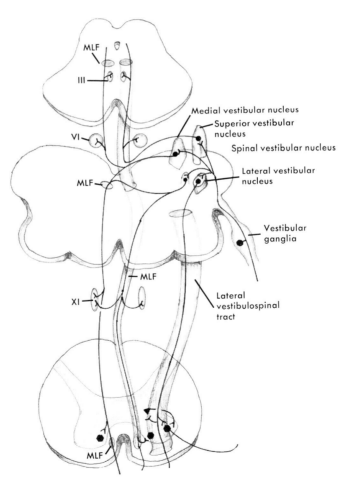

Fig. 8-7. CNS projections of vestibular data. Not shown are connections with cranial nerve IV, cerebellum, reticular formation, and cerebrum.

cochlear nuclei projections to the reticular formation may also serve as a processing station. Some of these descending cortical fibers even travel out to the hair cells of the organ of Corti to regulate the incoming auditory data.

VESTIBULAR SYSTEM. The CNS pathways for the vestibular system arise from the superior, inferior, medial, and lateral (Deiter's) nuclei in the dorsolateral corner of the pontomedullary area. The axons from these nuclei project to the following (Fig. 8-7).

Motor nuclei of cranial nerves III, IV, and VI, lateral gaze center in the pons, and vertical gaze center in the pretectum. These ascending fibers form the rostral portion of the medial longitudinal fasciculus (MLF).

Cervical spinal cord. These descending fibers form the caudal continuation of the MLF. In the cord the MLF is called the medial vestibulospinal tract and acts to increase extensor tone.

Full length of spinal cord. These fibers form the lateral vestibulospinal tract. It arises primarily from Deiter's nucleus, which in addition to labyrinthine input receives direct input from the Purkinje cells of the cerebellar cortex. This spinal tract is concerned with extensor muscle tone (facilitation), postural adjustments, and balance.

Flocculonodular lobe and fastigial nucleus of cerebellum via inferior cerebellar peduncle. This pathway has a role in the coordination of the axial muscles of the neck and vertebral column in posture and balance. These influences are conveyed to the axial motoneurons by the vestibulospinal and reticulospinal tracts.

Nuclei of pontine and medullary reticular formations. Besides the above-mentioned role in the coordination of axial muscles, this pathway may be involved in the nausea, vomiting, dizziness, and sweating (motion sickness) that can result from vestibular stimulation. Components of the hypothalamus and limbic system are probably also involved.

Ascending vestibular fibers. These fibers are thought to project to the medial geniculate body and then to the "head" region of the postcentral gyrus. This pathway adds to the conscious awareness of the position and velocity of the head. Like the auditory system, there are efferent fibers to the vestibular receptors in the labyrinth. Although their role is uncertain, they may act to inhibit output from the labyrinth, suppressing the symptoms of motion sickness. Drugs such as dimenhydrinate (Dramamine) may act to enhance the effects of these efferents.

Visceral sensory pathways

The visceral sensory pathways are concerned with the ascending general visceral afferents from the body and the special visceral afferents from the gustatory and olfactory receptors.

The general visceral afferent system conveys sensations related to pressure, distention, and pain from the heart, lungs, gastrointestinal tract, genitourinary system, blood vessels, and other internal structures. The level of carbon dioxide in the blood is also detected. This afferent system is made up of group II, III, and IV fibers that enter the spinal cord and brainstem via somatic (e.g., sciatic) and visceral (e.g., vagus and splanchnic) nerves. The majority of these afferents are involved in autonomic and somatic reflexes that do not reach consciousness. Some of the afferents ascend in the anteriolateral white matter of the cord and travel near the spinothalamic tract in the brainstem. They terminate in the thalamus, where their data are relayed to the "trunk" area of the postcentral gyrus. This visceral pathway presumably creates an awareness of pain and fullness. Chapters 5 and 6 provide more information about the visceral reflexes and the phenomenon of referred pain.

The CNS pathways conveying the sense of taste are somewhat obscure. On entering the brainstem the taste afferents from VII, IX, and X turn caudally in the solitary tract. They synapse with neurons in this tract (solitary nucleus). These second-order neurons cross midline and ascend in the medial lemniscus to the ventral posteromedial nucleus of the thalamus. Third-order neurons then relay the data to the caudal portion of the postcentral gyrus and possibly to the insula and superior temporal gyrus. While taste normally adds a degree of pleasure to our lives, it may have a role in the preservation of life itself. Many animals seek foods that contain a particular nutrient that may be lacking in their diet.

The CNS pathways for olfaction are discussed in Chapter 5. To summarize, some olfactory tract fibers end in the septal area of the frontal lobe, the anterior perforated substance of the frontal lobe, and the anterior portion of the temporal lobe including amygdala. Note that there are no direct projections to the thalamus. Although olfaction is a relatively minor sense in many ways, there are many complex processing circuits, including corticofugal fibers from the olfactory cortex to the olfactory bulk, that act to code and integrate olfactory data.

ROLE OF RETICULAR FORMATION
General organization

Like the thalamus and cerebral cortex, the reticular formation (RF) is both a sensory and a motor center. The anatomy of the RF is discussed in Chapter 9. For now it is enough to know that the reticular formation is composed of three longitudinal columns of nuclei that are found in the core of the entire brainstem. Within the reticular formation is a tract known as the *central tegmental tract*, which interconnects approximately 24 separate nuclei that make up the reticular formation. The hypothalamus in fact is considered a rostral extension of the RF.

Afferent input

The reticular formation is a true "melting pot" of neural activity. Impulses from many sources converge on reticular neurons. These sources include the following:

1. Ascending somatic and visceral sensory data via the spinoreticular, spinotectal, and spinal thalamic tracts
2. Somatic and visceral afferent data from the cranial nerve nuclei
3. Motor control data from the cerebral cortex, basal nuclei, and cerebellum
4. Autonomic control data from the hypothalamus and limbic lobe

Processing and projections

The extent of interaction of input to the RF is almost beyond comprehension. Each *neuron* in the RF can receive input from over 4000 other neurons and in turn project to over 25,000 other neurons. The result of this interaction is to either suppress or enhance the excitability of many neurons, thus modifying the transmission of neural information to other CNS sensory and motor centers both directly and indirectly.

The output from the RF is as extensive as the input and includes the following:

1. Ascending projections to the hypothalamus, nonspecific nuclei of the thalamus, and the limbic lobe and cerebral cortex (via the thalamus)
2. Lateral projections to the cerebellum and cranial nerve nuclei
3. Descending projections to spinal motoneurons resulting in facilitation of the physiologic extensors

Sensory-related functions

The motor influences are discussed in Chapter 9. The influences on the sensory system involve the following:

1. Role in the sleep-wake cycle. A sleep center may exist in the pontomedullary RF. This center could enduce the drowsiness-sleep state by the suppression of ascending data. In fact, some brainstem lesions result in permanent coma as a result of interruption of ascending sensory data.

2. Cortical arousal and ability to attend to a stimulus.
3. Behavioral characteristics such as motivation and drive.
4. Habituation to a monotonous or repeated stimulus. This results in inattention and decreased sensitivity to the stimulus such as in a boring lecture.
5. Filtering of nonnecessary input.
6. Altering the threshold of excitability to stimuli such as pain.

ROLE OF THALAMUS
General organization

The thalamus is composed of 26 pairs of nuclei that have specific anatomic and functional "boundaries." These nuclei have four basic functions involving the sensory system:

1. Relay and processing of all sensory pathways except for olfaction to primary areas of the cerebral cortex for conscious awareness
2. Perception of the crude aspects of pain, temperature, and touch sensations
3. Adding an effect such as agreeable or disagreeable to sensation, thus influencing emotions
4. Maintaining a background of cortical activity (as recorded in an electroencephalogram) that influences arousal, attention, and the sleep-wake cycle

The thalamus also functions as a relay integrator of data from the cerebellum and globus pallidus to the motor cortex (see Chapter 9).

The nuclei that perform these four functions can be grouped into two major areas: the specific and nonspecific nuclei.

Specific nuclei and their functions

Specific nuclei and their functions are discussed below and shown in Figs. 2-21 and 9-6.

Anterior nuclear group. The anterior nuclei relay data from the mamillary bodies to the cingulate gyrus. This circuit is involved in emotions such as anger and the autonomic responses such as changes in blood pressure that occur with the emotion.

Lateral nuclear group (ventral tier). The ventral lateral nucleus relays data from the cerebellum, red nucleus, globus pallidus, and substantia nigra to the precentral gyrus (area 4 of the motor cortex).

The ventral anterior nucleus relays data from the globus pallidus, substantia nigra, brainstem, and premotor cortex (area 6) to the premotor cortex. Both of these nuclei function to regulate and synchronize the output from the motor cortex (see Chapter 9). A portion of the ventral anterior nucleus is part of the nonspecific system.

There are two ventral posterior nuclei. The ventral posterolateral (VPL) nucleus relays dorsal column–medial lemniscus and spinothalamic tract data to the postcentral gyrus (areas 3, 1, and 2), which is known as the *primary somesthetic area*. The ventral posteromedial (VPM) nucleus relays data from the trigeminothalamic tracts to the primary somesthetic area.

Lateral nuclear group (dorsal tier). The lateral dorsal nucleus is a caudal extension of the anterior nuclear group. Although the input is uncertain, direct connections with the cingulate gyrus are present.

The lateral posterior nucleus receives input from the ventral posterior nuclei and has reciprocal connections with the somesthetic association area (5 and 7) of the parietal lobe. These connections imply a role in the processing of somatic sensory information.

The pulvinar receives input primarily from other thalamic nuclei concerning somatic, visual, and auditory data. Projections are sent to the cortical association areas in the parietal, occipital, and temporal lobes. Although the largest of the thalamic nuclei, lesions do not produce any observable changes in the functioning of the sensory system.

Metathalamus. The metathalamus is composed of the lateral and medial geniculate bodies. The lateral geniculate body relays data from the optic tract to the primary visual cortex (area 17) of the occipital lobe.

The medial geniculate body relays data from the auditory and vestibular pathways to the primary auditory cortex (areas 41 and 42) of the temporal lobe and the "face" area of the postcentral gyrus. These latter projections presumably provide a sense of orientation of the head in space, including static head position and acceleration movements.

Posterior thalamic region. This region (caudal to the ventral posterior nuclei) receives bilateral input from the spinothalamic and reticulothalamic tracts concerning painful and noxious stimuli and from other ascending pathways concerning visual, auditory, tactual, and vibratory data. This information is relayed to the secondary somatic sensory area (SII) of the parietal lobe. This region also functions in the perception of noxious stimuli.

Medial nuclear group. This group (dorsomedial nucleus) receives extensive input from other thalamic nuclei, especially those receiving input from the reticular formation, hypothalamus, amygdala, olfactory tract, and neocortex. The primary projections are to the hypothalamus and prefrontal cortex (rostral to area 6). This nucleus is concerned not with the relay of a specific sensation but with the affective responses of an individual. These can include feelings of euphoria, mild depression, well-being, ill-being, pleasant, and unpleasant. Also included are the autonomic responses that occur with a particular mood.

Point-to-point representation

An interesting property of the somatic sensory relay nuclei (VPL, VPM, geniculate bodies) that project to the primary cortical areas in their point-to-point representation of receptive fields. Any cutaneous, visual, or auditory receptive field that is activated will excite a small cluster of neurons in these thalamic nuclei, which in turn excite a small cluster of neurons in a primary cortical area. Through the mechanism of preprocessing, the activated clusters (which may only be a few neurons) are surrounded by a zone of inhibition, and any nonessential discharging clusters are blocked. This is known as *surround inhibition* and functions to filter and focus incoming sensory data for easier reception in the thalamus and primary cortical areas. These same preprocessing activities occur at the other relay nuclei in the brainstem and spinal cord. The cerebral cortex and the reticular formation are sources of preprocessing input to these areas.

The relationship between receptor field and neural cluster allows for the mapping of receptor fields. This has been best performed on the postcentral gyrus and is known as somesthetic homunculus (Fig. 9-12, *A*). Each receptive field in the skin is represented by a cluster of neurons in the postcentral gyrus. The distortion of the body results because areas such as the lips and hands have a greater density of receptive fields. The homunculi of the VPL and VPM nuclei are different. The visual and auditory nuclei and primary cortical areas do not have homunculi, but they do have point-to-point representation.

Nonspecific nuclei and their functions

Nonspecific nuclei and their functions are described below and shown in Figs. 2-21 and 9-6.

Intralaminar nuclear group. These nuclei receive input from the ascending reticular system, spinothalamic tracts, cerebellum, cerebral cortex (including the motor cortex), and globus pallidus. Projections are made to the putamen, caudate, the nonspecific portion of the ventral anterior nucleus, and the cerebral cortex.

Thalamic reticular nucleus. Input to this nucleus is from the reticular formation and cerebral cortex. Output is to other thalamic nuclei and the midbrain RF. This nucleus seems to have a vital role in the integration of the activity in the other thalamic nuclei. The synchronization of the electrical activity of the cortex (EEG) into the low-frequency, high-amplitude waves seen in light sleep has been attributed to this nucleus.

Midline nuclear group. The primary connections to this nucleus are with hypothalamus and are probably concerned with visceral functions.

In general the nonspecific nuclei can be thought of as a rostral extension of the reticular formation. Although their cortical projections are limited to the basal portion of the frontal lobe, their influence here and on the other thalamic and subcortical nuclei can alter the activity of diffuse areas of the cortex. This diffuse influence on the cerebral cortex is quite powerful. It can be thought of as an energizing force of background neural activity that elevates specific cortical areas to a functional level of excitation. This background activity can be recorded as the electroencephalogram (EEG). Changes in the EEG can be seen in sleep, altered states of arousal, and many CNS lesions.

Summary of functional considerations

The thalamus is an amazingly complex structure. Its connections and sensory-related roles are summarized below. Its anatomy and motor functions are presented in Chapter 9.

Each half of the thalamus receives bilateral ascending sensory input and has reciprocal connections with the ipsilateral cerebral cortex, hypothalamus, and basal nuclei. There are extensive intranuclear connections within each half of the thalamus as well as between the two halves. Despite the ipsilateral cortical projections, the thalamus as a unit has a profound influence on both cerebral hemispheres. These influences include the following:

1. Relay of modified sensory data to specific areas for the purpose of perception of the type, strength, and location of the stimulus (This modification provides for a focusing of important incoming signals.)
2. Applying an effect or emotion to a stimulus
3. Assisting the cortex in the crude perception of pain, temperature, and touch
4. Background of synchronized electrical rhythms that elevate the cortex to a functional level of arousal

ROLE OF CEREBRAL CORTEX

This section focuses on the outcome of the specific thalamocortical projections that travel in the posterior limb of the internal capsule, including the auditory and optic radiations. The course of the nonspecific fibers is to the basal portion of the frontal lobe, although the nonspecific nuclei indirectly influence all areas of the cortex. The course of olfactory data has already been discussed. The anatomy of the cerebral cortex is presented in Chapter 2 and discussed further in Chapter 9 along with the motor system.

Primary sensory areas

The essence of the specific thalamocortical projections concerns the point-to-point representation discussed earlier. Most neurons of a specific thalamic nucleus can be thought of as terminating on a small group of neurons in one of three primary sensory areas. This locus of neurons represents a discrete receptive field of sensation and can only be excited by that receptive field. This enables the primary areas to determine the type of stimulus by the receptors being activated, the intensity of stimulation, and the location of the stimulus. These abilities are the basic requirements for the perception of a stimulus. The three primary sensory areas of the cortex are given below (Fig. 8-8).

Primary somesthetic area. The postcentral gyrus (areas 3, 1, and 2) is responsible for perceiving discriminative touch, pressure,

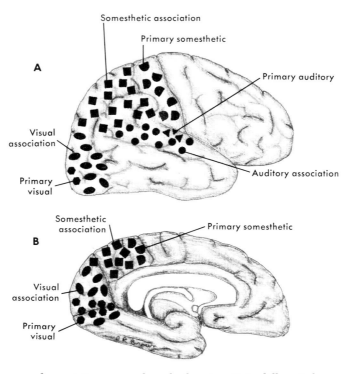

Fig. 8-8. Primary and association areas of cerebral cortex. Note different shapes used to mark each area. **A,** Lateral view. **B,** Medial view.

vibration, kinesthesia, temperature, and pain from the opposite side of the body. This area (sometimes called SI) is the only place for perception of dorsal column sensations. In addition to the type, intensity, and location of the stimulus the postcentral gyrus can also perceive textures, differences in weight, and differences in shapes. At the base of the postcentral gyrus along the upper margin of the lateral fissure is a second primary somesthetic area known as SII. Less is known about SII, but it seems to be somatotopically organized, receives bilateral input primarily from cutaneous receptors, and may play a more important role in the perception of pain than SI. The organization of SI into the sensory homunculus has been discussed.

Primary auditory area. This area is known as the transverse temporal gyrus of Heschl (areas 41 and 42) and is located in the superior temporal gyrus, deep in the lateral fissure and posterior to the central sulcus. This area is tonotopically organized, with the lower frequencies localized laterally and the higher frequencies medially. The neuron clusters in the primary auditory cortex also seem to detect the temporal pattern of sounds, which is used in the comprehension of speech and the localization of sounds.

Primary visual area. This area (17) is found in the tip of the occipital lobe and in areas adjacent to the calcarine fissure. The neural clusters of the visual cortex are organized into three types of circuits: (1) simple circuits that respond to light intensity, (2) complex circuits that respond to a slit of light oriented in a specific direction (length of the slit is not important), and (3) hypercomplex circuits that respond to a slit of light oriented in a specific direction and of a specific length, demonstrating an amazing point-to-point retinotopic representation. That is, impulses generated by a small receptive field of the retina are directed to a small group of cells in the lateral geniculate body, which in turn influence a discrete column of cells in area 17. In fact, one foveal cone may influence 100 or more cortical neurons. Most of area 17 responds to activation of the cones of the macula and is involved with binocular vision (the neurons of area 17 activated by a receptive field from one macula can also be activated by the corresponding receptive field in the other macula). In other words, most of area 17 in each hemisphere "sees" through both eyes. The remaining portion of area 17 (the most rostral area) is devoted to monocular input from the retinal cells that see the periphery.

An interesting reflex activity of the primary visual cortex is the *accommodation reflex* for close vision. In this reflex retinal information that is "out of focus" is sent to area 17, where it is processed and relayed to the motor cortex. From here motor fibers descend in the internal capsule to the superior colliculus, where the signals are relayed to the Edinger-Westphal nucleus and the oculomotor nucleus. The effect is to cause thickening of the lens, pupil constriction, and convergence of the eyes. Look up and focus on something across the room and then look down at this paragraph. The speed of the accommodation reflex is amazing. Realize that the constriction of the pupil to light and for near vision requires different pathways.

Primary vestibular area. This area is located either in the lower part of the postcentral gyrus just behind the general sensory area for the head or just in front of the primary auditory area. Electrical stimulation of the former produces sensations of rotation or bodily displacement. Stimulation of the latter can cause a feeling of vertigo. In either case this information probably contributes to higher motor regulation and conscious orientation in space.

Primary gustatory area. This area is located either at the base of the pre- and postcentral gyri just in front of the general sensory area for the tongue and pharynx or in the insula.

Primary visceral area. General visceral afferents have been traced to the trunk area of SI and to SII. It is presumed that they assist in the sensations of pain-distention and fullness or satiety (e.g., the pressure of a full stomach after a big meal). It should be mentioned, however, that most visceral afferents do not reach levels of consciousness, and those that do are poorly localized, nondiscriminative sensations.

Association areas

Association areas represent a higher level of integrative function than simple perception. In essence, the primary somesthetic, auditory, and visual areas project data related to the type, intensity, and location of the stimulus to their respective association areas for an interpretation of the meaning of the stimulus. In addition the lateral and medial nuclear groups of the thalamus project directly to these three association areas. Besides determining the meaning of a stimulus, these association areas make judgments as to the value of the stimulus, the need for participation of the motor system, and the importance of committing the stimulus to memory and are probably involved in the higher thought processes such as creativity.

The three association areas are discussed below.

Somesthetic association area. Located in the parietal lobe in back of the postcentral gyrus, this area (5, 7, and 10) is concerned with identifying an object by touch and determining how to use that object; with the meaning of differences in textures, temperatures, and pressures; and with the awareness, postural relations, and location of body parts. In other words, concepts such as big versus small, right versus left, and heavy versus light are developed in this association area and used to make judgments about a course of action. In addition this association center integrates data from the visual and auditory association areas to create the communication skills of reading, writing, and calculating. These skills are found only in the dominant hemisphere at the junction of the three association areas (area 39).

Visual association area. This area (18, 19) is located ventral to the primary visual area. This association area determines the meaning and value of what is "seen" by the retina, the need for involvement of the motor system, and the importance of retaining the visual data in a memory circuit. The ability to read, create an art form, and dream are functions of this association area. Even a simple task such as walking to a chair and sitting down requires areas 18 and 19 to determine that the object across the room is used to support the body's weight.

Auditory association area. This area (22) is located in the temporal lobe below and in back of the primary auditory area and is often called Wernicke's area. This area is very important to humans as it comprehends not only the meaning of sounds such as a ringing telephone but also the meaning of the spoken word. As with the other association areas, involvement with the motor system and with memory is also a function.

Psychic center

The lateral and inferior cortex near the temporal pole appears to be involved in the recall of auditory, visual, or tactual experiences. Stimulation of this area may evoke musical or other hallucinations or recall of recent or long forgotten experiences. Such recall is often a clear reenactment of past events, and feelings of joy or fear may be evoked. Entire conversations of over a year past have been recalled during such stimulation. Considering the large amount of white matter in each hemisphere, which is composed of association fibers interconnecting the lobes of one hemisphere and the commissure fibers interconnecting the two hemispheres, it is easy to see how experiences of the different association areas can migrate to this temporal lobe psychic area.

RECENT CONCEPTS ABOUT PAIN

Of all the somatic and visceral senses that reach consciousness, the sense of pain is probably the most significant clinically. For both neurologically impaired patients and healthy persons pain is often an important symptom and frequently is a major block to functional recovery.

Pain receptors and pathways

The sense of pain is experienced in a variety of forms such as a dull ache, sharp jabs, and burning. Unlike other senses, pain has a greater variety of thresholds and sensitivities. Pain can be specifically localized to a small area such as a fingernail or diffusely localized to large areas such as the upper extremity. Pain can also be referred from one area of the body to another.

These characteristics of the sense of pain have led to speculation about the existence of definite pain receptors and pain pathways in the CNS. Recent evidence from the University of North Carolina School of Medicine has demonstrated the existence of free nerve endings in the deep layers of skin, internal organs, periosteum, cornea, and pulp of the teeth that are activated only if the stimulus is strong enough to be painful. The normal method of excitation of the pain receptors seems to involve the release of chemicals such as histamine and bradykinin that occurs when the cells are injured. These chemicals are associated with inflammation and also act to produce volleys of action potentials along the pain afferents. Of interest, one of the effects of aspirin is to inhibit the liberation of bradykinin, thus reducing the pain signal.

The pain afferents enter the CNS primarily with the spinal and trigeminal nerves. The lateral spinothalamic, spinoreticular, and trigeminothalamic tracts convey pain data to the thalamus. Projections to the primary somesthetic area and to the limbic system combine with the thalamus to perceive pain and apply an affective response to the pain.

It is surprising that brain tissue is insensitive to pain even though it perceives pain. Brain surgery has been performed on patients who received only a local anesthetic to cut through the skin and dura.

Gate theory and transcutaneous electrical nerve stimulation

In the mid-1960s a theory proposed by Melzack and Wall evolved to explain the transmission of pain to the higher CNS centers. This theory can be summarized as follows and is diagrammed in Fig. 8-9:

1. The perception of pain is not through specific receptors but is signaled by an enhanced input over pathways whose activity is normally interpreted as touch or pressure.

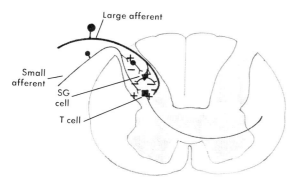

Fig. 8-9. "Gate" theory. Input over large afferents effectively reduces output of T cells by exciting SG cells, which presynaptically inhibit *both* large and small afferent inputs.

2. Cutaneous pain can be blocked by nonnoxious stimulation of adjacent areas of skin. This is the basis for the use of counterirritants and for rubbing a painful area.

3. Information interpreted as pain is carried to the CNS on small afferents (C fibers), while counterirritant input (touch pressure) is carried over larger afferents (A fibers).

4. Large fibers entering the dorsal root synapse on and excite two types of cells: transmission cells (T) and substantia gelatinosa cells (SG). The effect of SG excitation is to cause presynaptic inhibition of both large and small afferents. Thus input over the large fibers produces a modest response of the T cells but interferes with further input.

5. The small afferents also synapse on both cells; however, they inhibits the SG cells. Thus input over the small afferents excites the T cells and also enhances the effect of the small afferents on the T cells by inhibiting the SG cells. Pain results when T cell discharge exceeds some level monitored by higher centers.

6. The effect of counterirritants is to cause the large afferents to fire, which then reduces the discharge of the T cells.

In essence, C fiber stimulation "opens the gates" for pain transmission, while A fiber stimulation "closes the gates."

Some speculation about the "gate" theory has arisen because of an inability to locate its circuits in the dorsal horn. It is probable, however, that the inhibition of pain by large afferent input could take place elsewhere in the CNS such as in the thalamus.

That some form of a gate theory is probable may be supported by a recent noninvasive, nonpharmacologic treatment for chronic pain. This method is known as *transcutaneous electrical nerve stimulation (TENS)*. It involves the stimulation of a peripheral nerve, trigger zone, or acupuncture sites with low-intensity current. This current selectively excites the larger A fibers. Significant clinical results have been reported in terms of a marked temporary relief from chronic pain. TENS units are small and battery powered, allowing the patient to administer self-treatment at any time. Although not a cure-all for chronic pain nor 100% effective in all cases, TENS is a potential method of pain reduction with little or no apparent side effects.

CNS opiates

Perhaps the most exciting development in pain research has been the discovery of "opiate" receptors in the CNS that are highly concentrated in areas such as the thalamus that perceive pain. In addition these areas manufacture amino acid chains that resemble the natural opiates derived from poppy seeds. Three brain opiates have been identified—enkephalin, beta-endorphin, and neurotensin—and seem to function as natural sources of inhibition of pain sensation. Researchers began manufacturing enkephalin because it was the least complex in structure and thus easier to stabilize and transport across the blood-brain barrier. Early results were not encouraging, however, as rats given enkephalin soon developed symptoms that are the same as seen in addiction with morphine and meperidine hydrochloride (Demerol). Other researchers then developed a drug, D-phenylalanine (DPA), that inhibits the enzyme that normally causes a rapid breakdown of the brain opiates. Although still being refined, early reports show excellent potential. The elicitation of the brain's natural analgesics for the treatment of pain has obvious advantages over the use of heavy medication and the current surgical procedures, which include (1) cordotomy (sectioning of the anterolateral white matter of the spinal cord), (2) dorsal rhizotomy (sectioning of the dorsal roots), (3) neurectomy (cutting of a peripheral nerve or the trigeminal nerve), and (4) sympathectomy (sectioning of the paravertebral chain ganglia).

Acupuncture

The link between this ancient Chinese form of treatment and Western medicine may be found in the brain's opiates. In a recent study normal subjects were subjected to experimental pain by electrical stimulation of their teeth. Application of acupuncture effectively raised the threshold of pain. After administering a drug that blocks the action of enkephalin the acupuncture treatment had a significantly lower analgesic effect. This finding suggests that acupuncture may trigger the release of enkephalin, thereby raising the pain threshold.

A person's belief in the ability of another person, a drug, a needle, or own conscious will to control pain can be quite powerful. This belief may be related to the brain's opiate levels. In a study done on the effects of placebo on the pain from a tooth extraction, about one third of the subjects reported a significant reduction in pain. When these persons were subsequently given a drug that blocked the action of beta-endorphin the pain increased almost to the original level.

CLINICAL APPLICATION OF SENSORY SYSTEM LESIONS
Abnormal sensations

There are a variety of alterations in sensory perception that are not anatomically related to one area. These include the following:

1. Anesthesia: complete loss of sensation, usually involving peripheral nerve fields or dermatomes
2. Dissociated anesthesia: loss of some sensations such as pain and temperature while others such as touch are retained
3. Paresthesia: numbness, tingling, and formication (crawling) sensations
4. Hyperesthesia: increased sensitivity to sensations

5. Hypesthesia: decreased sensitivity to sensations
6. Analgesia: loss of pain

Brown-Séquard syndrome

Brown-Séquard syndrome occurs with a hemisection of the spinal cord (a unilateral transverse lesion). The symptoms involve both sides of the body and the motor and sensory systems. The symptoms are listed below and diagrammed in Fig. 8-10:

1. Ipsilateral loss of dorsal column sensations below the level of the lesion
2. Contralateral loss of pain and temperature one to two levels below the lesion
3. Little or no loss of tactual sensibility

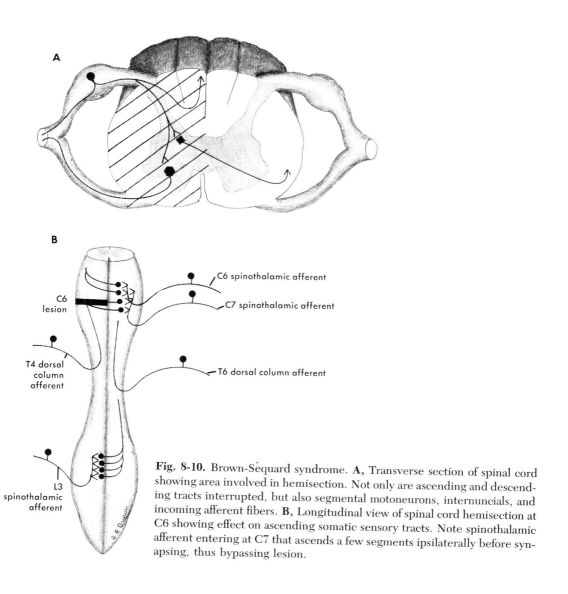

Fig. 8-10. Brown-Séquard syndrome. **A,** Transverse section of spinal cord showing area involved in hemisection. Not only are ascending and descending tracts interrupted, but also segmental motoneurons, internuncials, and incoming afferent fibers. **B,** Longitudinal view of spinal cord hemisection at C6 showing effect on ascending somatic sensory tracts. Note spinothalamic afferent entering at C7 that ascends a few segments ipsilaterally before synapsing, thus bypassing lesion.

4. Ipsilateral spastic paralysis below the level of involvement (upper motoneuron)
5. Ipsilateral Horner's syndrome if the lesion is above T3
6. Ipsilateral flaccid paralysis of the motor units innervated by the lesioned ventral roots and horns (lower motoneuron)
7. Ipsilateral anesthesia in the dermatomes supplied by the lesioned dorsal roots or horn (Overlap from intact dorsal roots may reduce the area of anesthesia.)

Somatotopic organization of ascending tracts

The primary relayers of conscious cutaneous sensations are organized into fibers representing different body parts. The lateral spinothalamic tract is arranged with the fibers from the lower extremity in the posterolateral portion. Fibers from successively higher segments are added on the medial aspect of the tract. The fibers in the dorsal columns are arranged with the lower cord segments placed medially and the upper segments laterally. As the axons from the nuclei cuneatus and gracilis form the medial lemniscus, the upper extremity fibers are ventrally placed, while those from the lower cord segments are placed dorsally. While the spinothalamic tract retains its organization to the thalamus, the medial lemniscus rotates so the upper extremity becomes located medially and the lower extremity laterally.

While this information may seem tedious, knowledge of this somatotopic organization can be useful in identifying clinical levels of involvement. For example, a tumor compressing the lateral funiculus at C5 may show a contralateral loss of pain and temperature in the foot as an early symptom.

Thalamic lesions

Lesions of the intralaminar nuclei or the dorsomedial nucleus of the medial nuclear group or a prefrontal lobotomy alter an individual's response to pain. Pain can still be perceived, but the fear and anxiety about it are gone. Surgical lesions of these areas have been performed to alleviate the intractable pain found with some types of cancer.

Some lesions of the posterior choroidal or posterior cerebral arteries can produce the *thalamic syndrome.* Initially there may be a contralateral hemianalgesia. Later the threshold for pain, temperature, and touch is raised, but nonnoxious stimuli (applied to the contralateral side) may produce painful or disagreeable sensations. Simple touch, a pinprick, heat, cold, or even music can evoke uncomfortable sensations including burning pain in the affected side. Spontaneous (paroxysmal) pain can occur in the absence of any stimulus. Other disagreeable sensations include swelling, drawing, pulling, and tension.

The cause of the thalamic syndrome is uncertain but may involve an interruption of descending corticothalamic inhibitory fibers to the affective thalamic nuclei.

Lesions of sensory cortex

Lesions of the parietal, temporal, and occipital lobes produce varying symptoms, depending on the extent of involvement of the primary and association areas. The symptoms of lesions of the various sensory areas of the cortex can be outlined as follows.

Primary somesthetic area. With lesions of this area pain, temperature, and tactile senses are retained, but the ability to localize and detect small gradients of change is impaired. The patient is unable to determine the shape of an object by touch (astereognosis), its texture, or differences in weight. Position sense is impaired, causing difficulty in walking in the dark if the medial portion of this area is impaired. These symptoms are seen contralaterally.

Somesthetic association area. There are two basic symptoms seen with lesions of area

40. One is tactual *agnosia*. Agnosia is the inability to recognize the nature, meaning, or value of a stimulus even though the stimulus can be consciously perceived. In tactual agnosia a patient (blindfolded) is able to feel and describe the shape of a house key (postcentral gyrus) but cannot name it or describe how to use it. Agnosia is usually seen in lesions of the dominant hemisphere. The other symptom is failure to be aware of or recognize the contralateral side of the body (hemiattention). Patients have been known to wash and dress only the "uninvolved" arm and leg or shave only half of the face. This symptom is more common in right hemisphere lesions.

Primary visual area. The primary symptom of a lesion in this area is contralateral homonymous hemianopia. This creates difficulty with reading and walking (objects on the involved side are often bumped into).

Visual association area. The primary symptom of a lesion in this area is visual agnosia. The patient can see an object in front of him, such as a chair, and can walk around it but cannot name it or know that it is to be sat in. This symptom is most prominent when the lesion is in the dominant hemisphere, along with alexia or inability to read. If the nondominant visual association area is involved there is difficulty with spatial relations, copying figures, and artistic creativity.

Primary auditory area. The primary symptom regardless of the hemisphere is inability to localize sounds. Because of bilateral input, neither ear is deaf. Hearing in both ears may be diminished.

Auditory association area. A lesion in the nondominant hemisphere will probably cause decreased appreciation for music. In the dominant hemisphere there will be auditory agnosia. A patient with auditory agnosia can localize the ringing of a telephone but does not know what the sound means. Inability to comprehend speech (receptive aphasia) is also present.

Headache

A complete discussion of headache is beyond the scope of this text. Since it does involve the sensory system, a brief summary is presented.

Headaches are common to everyone. They vary from mild to severe and are difficult to objectively diagnose. The causes of headaches are varied and can range in seriousness from unimportant to life threatening. The causes of headaches include the following:

1. Trauma to the head or neck (whiplash) or tearing or traction on the dura
2. Inflammations such as from sinusitis or meningitis
3. Tumor in which there is increased intracranial pressure
4. Vascular, in which there is hemorrhage or excess vasodilation, both of which increase intracranial pressure
5. Metabolic, such as from hypothyroidism, excess alcohol consumption, anemia
6. Emotions such as anxiety or tension
7. Miscellaneous, such as from eye strain, menstrual period, lumbar puncture, or to avoid social engagements

In neurologically impaired patients headaches can be a major clue to increased intracranial pressure. It is wise never to assume that a patient complaining of a headache is merely trying to avoid treatment.

TEST QUESTIONS

1. Those unconscious reflexes involving only the spinal cord and brainstem:
   a. Assist in the maintenance of posture.
   b. Provide for pupil constriction in the accommodation reflex.
   c. Provide for protection from noxious stimuli.
   d. Are often hyperactive in CNS lesions.
2. The perception of a stimulus:
   a. Involves point-to-point representation.
   b. Allows for the identification of the type of stimulus.

c. Determines the value or meaning of the stimulus.

d. Requires receptor specificity.

3. A hemisection of the left upper medulla would involve:

a. Contralateral loss of vibration, pressure, discriminative touch, and kinesthesia.

b. Bilateral loss of pain and temperature of the face.

c. Bilateral coordination symptoms in the extremities.

d. Ipsilateral Horner's syndrome.

4. A lesion of one optic tract:

a. Abolishes the light reflex in both eyes.

b. Causes wallerian degeneration of the visual fibers in the internal capsule.

c. Causes contralateral homonymous hemianopia.

d. Results in visual agnosia.

5. Shining a bright light in one eye:

a. Causes constriction of both pupils.

b. Causes the lens to thicken.

c. Activates cells in the superior colliculus.

d. Inhibits sympathetic preganglionics at T1 to T4.

6. The destination of the auditory data in the eighth cranial nerve can include:

a. Lateral lemniscus.

b. Cervical motoneurons via the tectospinal tract.

c. Cranial nerves III, IV, and VI.

d. Medial geniculate body via the inferior colliculus.

7. The destination of the vestibular data in the eighth cranial nerve can include:

a. Precentral gyrus via the internal capsule.

b. Primary visual area of the cortex to coordinate eye movements.

c. Inferior colliculus to coordinate head movements.

d. Extensor motoneurons throughout the spinal cord.

8. The reticular formation is concerned with:

a. Cortical arousal.

b. Sensory filtering.

c. Perception of dorsal column sensations.

d. Response to pain.

9. The specific nuclei of the thalamus:

a. Include the VPL and the VPM nuclei that project to the postcentral gyrus.

b. Include the pulvinar, which projects to the cortical association areas.

c. Are concerned with the affective qualities of sensation.

d. Are the primary source of the background electrical activity of the cortex.

10. A lesion of the entire primary somesthetic area:

a. Interferes with balance at night.

b. Causes anesthesia on the contralateral side of the body.

c. Interferes with conscious skilled activities such as dressing.

d. Causes the thalamic syndrome.

11. A loss of the primary and association auditory cortex results in:

a. Deafness in the contralateral ear.

b. Interference with verbal communications if the dominant hemisphere is involved.

c. Difficulty in localizing sounds.

d. Lack of appreciation of music if the dominant hemisphere is involved.

12. A conscious awareness of pain is probably influenced by:

a. Level of CNS opiates.

b. Level of cortical arousal.

c. Conduction velocity of the fibers in the lateral spinothalamic tract.

d. Ratio of large afferent to small afferent fiber input.

13. A hemisection of the spinal cord results in:

a. Ipsilateral loss of pain and temperature 1 to 2 segments below the level of injury.

b. Dissociated sensory loss.

c. Ipsilateral Horner's syndrome if T1 to T4 segments are involved.

d. Contralateral paresthesia.

14. The thalamic syndrome can include:

a. Contralateral anesthesia.

b. Increased threshold to stimulation contralaterally.

c. Disagreeable affect to nonnoxious stimulation.

d. Spontaneous pain.

15. Headache can result from:

a. Pressure on the dura mater.

b. Dilation of the cerebral or dural vessels.

c. Thrombus of a cerebral artery.

d. Space-occupying lesion.

SUGGESTED READINGS

Everett, N. B.: Functional neuroanatomy, ed. 6, Philadelphia, 1971, Lea & Febiger.

House, E., and Pansky, B.: A functional approach to neuroanatomy, New York, 1967, McGraw-Hill Book Co.

Minckler, J.: Introduction to neuroscience, St. Louis, 1972, The C. V. Mosby Co.

Netter, F.: The CIBA collection of medical illustrations. I. The nervous system, Summit, N.J., 1972, CIBA Chemical Co.

Noback, C., and Demarest, R.: The human nervous system, ed. 2, New York, 1975, McGraw-Hill Book Co.

Tucker, J.: An end to pain, Omni, February 1979.

Willis, W. D., Jr., and Grossman, R. G.: Medical neurobiology: neuroanatomical and neurophysiological principles basic to clinical neuroscience, ed. 2, St. Louis, 1977, The C. V. Mosby Co.

CHAPTER 9

Higher centers and their motor control systems

In this chapter are discussed the structures of the CNS responsible for producing the great variety of programs of movements that we possess and how these structures interrelate. Also presented is the way in which the functional characteristics of these structures manifest themselves when a lesion of one of the structures creates an imbalance. Special attention should be given to the following areas:

1. Overall effects of the five suprasegmental motor centers on the neural circuits and motoneurons of the spinal cord
2. Structure, location, neural connections, and role in the production of movement of each of the motor centers
3. Clinical symptoms commonly associated with a lesion of each center or part of each center
4. Role of the thalamus in the activation of cortical neurons and production of the EEG
5. Function and location of cortical areas 1 to 8, 17 to 19, 41, and 42
6. Blood supply to the motor and sensory homunculi
7. Role of arousal, attention, perception, reinforcement, repetition, interference, and motivation in the learning process

8. Roles of the dominant and nondominant hemispheres
9. Components and projections of the limbic system
10. Function of the limbic system in behavior, emotion, and memory
11. Determinants of muscle tone and methods of clinical assessment
12. Influences on movement patterns derived from synergies, reflexes, centrally programmed circuits, and association patterns
13. Effect on the motoneurons of lesions of each of the suprasegmental motor centers or their descending tracts
14. Innervation of the intraocular and extraocular muscles, facial muscles, and tongue
15. Conditions that produce ptosis, strabismus, facial paralysis, and tongue deviations

SUPRASEGMENTAL STRUCTURES AND THEIR FUNCTIONS

The motor systems of the CNS can be divided into two functional groups: the segmental and suprasegmental systems. The *segmental system* is composed of the numerous neural circuits and motoneurons within the spinal cord. The *suprasegmental system* is composed of the motor centers rostral to the spinal cord (i.e., the reticular formation and

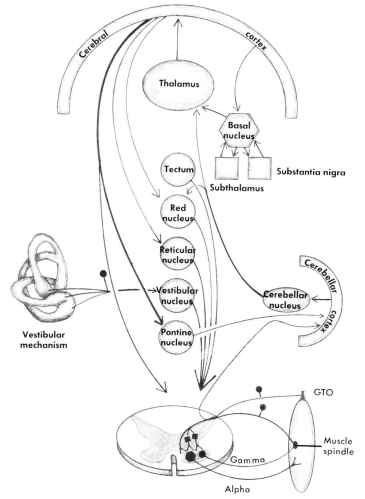

Fig. 9-1. Major suprasegmental motor centers. Influence on these higher centers can be either directly on motoneurons or on internuncials. Effect of higher center influence may be (1) direct depolarization or hyperpolarization of alpha motoneurons, (2) modification of output from muscle spindle (gamma biasing), or (3) enhancement or reduction of influence of spinal reflexes on motoneurons. Note circuits between cerebral cortex and basal nuclei and between cerebral cortex and cerebellum.

brainstem nuclei, the basal ganglia, the cerebellum, the thalamus, and the cerebral cortex). These "higher" centers function as modulators of the segmental circuits and neurons and are often referred to as *upper motoneurons* (Fig. 9-1).

Segmental activities

The spinal cord is capable of independent motor activity. This is very obvious in any patient with a complete spinal cord transection. These patients demonstrate the myotatic, flexor withdrawal, crossed extensor, and extensor thrust reflexes found in all of us. When the foot of such patients is jabbed with a pin the pain afferents enter the dorsal horn and excite a neural circuit composed of thousands of neurons. Axons from these neurons spread to many cord segments and even to the contralateral side. The result is flexion of the jabbed leg and extension of the other. The purpose of these spinal circuits is to provide very crude mechanisms for posture and protection. Because they are pure reflex responses, however, they lack the flexibility to be useful efficient patterns of movement.

The purpose of the higher motor centers is to alter the excitability of these circuits, produce coordinated movements on a background of tonic postural activity, and incorporate these reflex movements into more functional movements.

Influence of suprasegmental centers

The modulating effect of the "higher centers" on the spinal cord is quite dramatic when the activities of a normal person are compared with those of a patient with spinal cord injury. First, the reflex responses of such patients are usually hyperactive (a greater than normal amount of excursion is seen) and hypersensitive (less stimulus is needed to elicit a response). It is often difficult to exercise these patients' legs because any contact of the hands may elicit a withdrawal reflex. In a normal person the higher centers modify this reflex by continuously inhibiting the neural circuit. The result is that there is withdrawal only to noxious stimuli and the amount of movement is much less (unless there is a very noxious stimulus). The other spinal reflexes such as the myotatic are also regulated by the higher centers, generally by inhibiting the appropriate neurons.

The second effect is the lack of goal-directed or voluntary movements in patients with spinal cord injury. The higher centers are responsible for initiating movement patterns and for providing a background for postural stability (tone) on which these move-

ments can occur. The production of postural stability and willed movements is a result of the regulation of excitatory and inhibitory synapses on the alpha and gamma motoneurons.

For posture there is a maintained level of excitation on the tonic motoneurons of the slow-fatigue muscle fibers. This results in a pattern of *cocontraction innervation* in these fibers of supporting joints, creating *stability*. In willed movement there are bursts of excitatory activity on phasic motoneurons to fast-fatigue and fatigue-resistant muscle fibers. This results in a pattern of *reciprocal innervation* in these fibers, creating *mobility* between agonist and antagonist.

Finally, the higher centers can modify the output of these circuits. For example, the extensor thrust reflex is utilized for stance, walking, and in jumping. The ankle, however, is in neutral (functional) rather than plantar flexed (reflex, nonfunctional).

In summary, the higher centers project their axons to alpha and gamma motoneurons or to internuncials projecting to them. They modify the sensitivity and output of inherent neural circuits for the production of useful patterns of movement. By utilizing existing spinal pathways greater efficiency of movement is achieved. Keep in mind that the spinal circuits remain intact throughout life and manifest themselves if higher influences are removed.

Reticular formation and brainstem nuclei

The brainstem from the medulla to the midbrain contains motor centers that strongly influence muscle tone, postural adjustments, and protective reactions. These centers include the reticular formation, the vestibular nuclei, the red nucleus, and the tectum. The substantia nigra, subthalamus, and cerebellum are discussed in later sections. Not included are the cranial nerve motor nuclei discussed in Chapter 5.

The location of the vestibular and red nuclei and the tectum are discussed in Chapter

2. The *reticular formation*, however, is more complex. It is composed of a highly intricate column of neurons that extends throughout the core of the brainstem (including the diencephalon) and the length of the spinal cord. These neurons have considerable collateral branching as well as long ascending and descending axons.

In the spinal cord the reticular formation is located in lamina VII between the dorsal and ventral horns.

In the brainstem below the diencephalon the reticular nuclei are organized into three longitudinal columns (Fig. 9-2):

1. Raphe and paramedian nuclei located adjacent to midline. The dorsal and

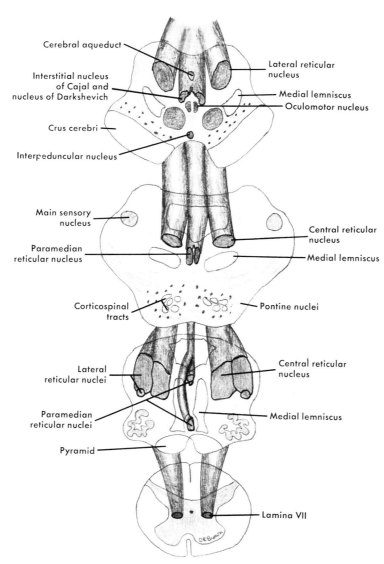

Fig. 9-2. Brainstem components of reticular formation. Individual reticular nuclei for pons and medulla are not labeled. Effect is to show that RF is long column of cells in core of neuraxis.

ventral tegmental nuclei and nucleus linearis of the midbrain are included.

2. Central reticular nuclear group located lateral to the raphe and paramedian nuclei. Included here are the nuclei reticularis centralis and gigantocellularis of the medulla; the nuclei reticular pontis caudalis and oralis and reticulotegmental nucleus of the pons; and the nucleus of Darkschewitsch, interstitial nucleus of Cajal, red nucleus, and interpeduncular nucleus of the midbrain (the latter two are specialized nuclei that are often not included in the reticular nuclei).

3. Lateral reticular group, which includes the lateral reticularis parvicellularis and lateral reticular nucleus of the medulla, no nuclei in the pons, and the cuneiform and pedunculopontine nuclei of the midbrain.

In the diencephalon the reticular core extends primarily into the ventral and medial portions of the thalamus and into the hypothalamus. Some authors include the subthalamus, globus pallidus, and septal nuclei in the reticular formation.

Functionally the reticular formation can be divided into two areas. A suppressor area is located in the medial part of the medulla and pons and is sometimes called the *reticular inhibitory system (RIS)*. A facilatory area is located in the midbrain and lateral part of the pons and upper medulla. This is sometimes known as the *reticular activating system (RAS)*. The RIS functions to inhibit spinal motoneurons and neural circuits, to assist in cortical attenuation by filtering unnecessary sensory data, and as a contributor to the state of sleep. The RAS functions in cortical arousal, consciousness, emotions, the electroencephalogram, and in facilitating spinal motoneurons and the neural circuit. Both systems function together to assist in the regulation of vasomotor, respiratory, gastrointestinal, and other autonomic activities.

Afferent neural connections to the reticular formation are from the following:

1. Suppressor and facilitory areas of the cerebral cortex for voluntary control of proximal and trunk muscles.

2. Cerebellum for the regulation of muscle tone and postural adjustments needed for synergistic movements.

3. Spinal and cranial nerves for the integration of environmental sensory data. This includes cortical arousal and sensory filtering activities.

Efferents from the reticular formation are primarily to the following:

1. Thalamus, where the data are relayed to the cortex. This information is concerned with cortical arousal, sleep-wakefulness, emotions, and electrocortical activity (EEG).

2. Spinal cord for modulation of tone, synergy patterns, complex postural reflexes, and spinal reflexes. These actions are accomplished by direct and indirect excitation and inhibition of the spinal motoneurons and related circuits. The main effect is the stimulation of extensor muscles and inhibition of flexors.

3. Autonomic functions of the reticular formation including respiration, circulation, vomiting, and blood pressure.

The *vestibular nuclei* are an important motor center for the regulation of muscle tone (especially extensors) and balance reactions. The primary afferent input is from the labyrinthine mechanism and the cerebellum. The primary efferent output is to the following:

1. Spinal cord internuncials, which in turn facilitate extensor motoneurons.

2. Cerebellum to form a vital vestibular-cerebellar-vestibular-spinal cord circuit. This circuit is an important part of protective extension reflexes.

3. Cranial nerves III, IV, VI, and XI to cause the head and eyes to turn in the direction of a fall.

4. Thalamus and reticular formation for the production of sensations of motion sickness and dizziness.

The afferent input to the *red nucleus* is primarily from the cerebellum, although fibers have been traced from the cerebral cortex and globus pallidus. The primary efferent output is to spinal cord internuncials, where the effect is facilitation of flexors and inhibition of extensors. In lower animals, in which the corticospinal system is less dominant than in humans, the rubrospinal pathway is a powerful activator of flexor muscles. In humans, however, the red nucleus does not seem to play a significant role in the organization of motor behavior. Electrical stimulation of the red nucleus in monkeys produces contralateral excitation of flexor alpha motoneurons and inhibition of extensor motoneurons.

The final brainstem center is the *tectum*. Auditory data from the inferior colliculus and visual data from the retina and visual cortex converge on neurons of the superior colliculus. These neurons project to cranial nerves III, IV, VI, and XI as well as to the upper spinal cord. The effect is to produce reflex eye, head, and arm movements in response to sudden visual or auditory stimuli. Elicitation of this reflex is not only useful as a protective reaction but is also a valuable source of arousal.

It is difficult to determine the effect of isolated lesions on these individual motor centers. Most diseases or injuries of the brainstem involve other surrounding tissues. Also many brainstem lesions are not compatible with life because of the involvement of the vital respiratory and cardiovascular centers. However, if only the red nucleus is involved there is little noticeable effect. Some reports have found transient tremor and ataxia with midbrain tegmentum lesions. These symptoms were attributed more to destruction of cerebellar afferents close to the red nucleus than to the red nucleus itself. Destruction of the tectum reduces the protective reflex responses to visual and auditory stimuli. Other pathways such as the corticospinal and reticulospinal tracts can be used, but there will be

increased latency in responding. There will also be deafness in both ears. The effects of unilateral tectal lesions are unclear. Isolated lesions of the vestibular nuclei and cranial nerve VIII occur in acoustic neuroma. If the lesion is unilateral *nystagmus* (a rhythmic to and fro oscillation of the eyes), *vertigo* (dizziness), and a tendency to fall to one side occur as a result of the unbalanced labyrinthine input. Bilateral lesions relieve this imbalance, and these symptoms do not appear. It is surprising that patients with bilateral lesions have only mild problems with balance. As long as there are two body orientation systems (visual and general proprioceptive), loss of the third system (the labyrinth) does not seriously affect balance. Auditory symptoms of deafness and *tinnitus* (ringing in the ear) can occur if the eighth nerve is involved.

Any lesion of the reticular formation, in the medulla or pons usually results in death because of involvement of the vital respiratory and circulatory centers. Numerous experimental studies have been performed on animals where the midbrain has been transected at the intercollicular level. The remaining portion of the intact reticular formation dominates the other motor centers caudally. The result is a powerful facilitation of the extensor muscles of the extremities, neck, and back. This condition is known as *decerebrate rigidity* and can be seen in some patients with severe head injury or cerebral palsy. If the cerebellum is removed in the intercollicular transection the extensor facilitation is even more marked. This results in *opisthotonus*, in which the neck and back are strongly arched.

Basal nuclei

As mentioned, the basal nuclei are composed of all of the telencephalic nuclei, that is, the caudate, lenticular, amygdaloid, and claustrum. On a functional level, however, the latter two are excluded and the substantia nigra is added. Some authors also include the subthalamic nuclei, as I do. The location of these structures can be best seen in a frontal section at the midthalamus level. The caudate and lenticular nuclei, subthalamus, and substantia nigra surround the thalamus (see Fig. 2-32).

The basal nuclei are organized into the *neostriatum* composed of the caudate and putamen and the *paleostriatum* composed of the globus pallidus (pallidum). Together they form the *corpus striatum.*

The neural connections of these nuclei are complex. Not only do each of the components interconnect, but there are also pathways between the cerebral cortex, thalamus, red nucleus, and reticular formation. While all of these neural circuits play a role in the output from the basal nuclei, only those of major importance are discussed here.

The main afferent input is from the motor and premotor areas of the cerebral cortex. This input goes to the caudate and putamen, where it is relayed to the globus pallidus. The main efferent output is from the pallidum to the ventral anterior nucleus of the thalamus. The fibers in this pathway are known as the *ansa lenticularis* (Fig. 9-3). The pallidum also projects to the subthalamus, substantia nigra, and reticular formation. Since the ventral anterior nucleus of the thalamus projects its output to the motor and premotor cortex, an important circuit is established between the motor cortex, basal nuclei, thalamus, and motor cortex again. This circuit is probably an important regulator of the autonomic aspects associated with movement. Any activity such as walking that we do not have to concentrate on or any posturing movements such as those used by the

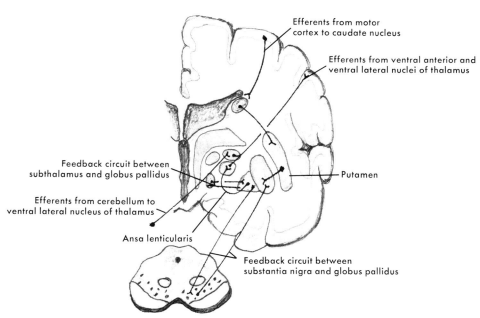

Fig. 9-3. Basic circuitry of basal nuclei. Data from motor cortices are transmitted to corpus striatum (caudate and lenticular nuclei), where they are modulated by substantia nigra and subthalamus. Data are then relayed to thalamus and then sent back to motor cortices along with cerebellar data. (Adapted from Noback, C., and Demarest, R.: The human nervous system: basic principles of neurobiology, ed. 2, New York, 1975, McGraw-Hill Book Co.)

shoulder in painting a picture are probably regulated by the basal nuclei and this circuit.

Two important modifiers to this circuit are the subthalamus and substantia nigra. Both have reciprocal connections with the globus pallidus. When the subthalamus is lesioned violent flinging of the contralateral extremity occurs. This condition occurs in humans and is known as *hemiballismus*. When the substantia nigra is lesioned as in Parkinson's disease (paralysis agitans) two symptoms appear. One is a *tremor* of 4 to 8 oscillations per second that occurs when the patient is awake but *at rest*. This tremor is generally found in the contralateral hand and resembles the motion of *rolling a pill* between the thumb and first two fingers with pronation-supination of the forearm. The tremor can occur at other areas as well, such as the head. A drug known as levodopa (L-dopa) has been developed to treat Parkinson's disease. L-dopa is a precursor of the transmitter substance known as dopamine used by the substantia nigra neurons. L-dopa can pass through the blood-brain barrier, whereas ingested or injected dopamine cannot. It is thought that once inside the CNS L-dopa is converted to dopamine and can replace that lost from the lesioned substantia nigra cells. Thus it seems that the substantia nigra acts to synchronize the cortex–globus pallidus–thalamus-cortex circuit to allow a smooth outflow of motor programming. When this synchronizing influence is interrupted the inherent oscillations of the thalamus that produce the electroencephalogram spill over into the motor system, producing the resting tremor. A similar synchronizing function can be applied to the subthalamus. In fact, the symptom of ballismus can be relieved with an isolated lesion of the globus pallidus. Because much of the ingested L-dopa is broken down in the gastrointestinal tract, some patients experience adverse reactions. The addition of carbidopa to L-dopa prevents this breakdown, allowing more to

reach the brain and the dosage to be reduced. The combination of the two drugs is known as Sinemet.

The other symptom is generalized muscle *hypertonus (rigidity)*. The effect of this state is *bradykinesia (poverty of movement)* that (1) makes it difficult for the patient to initiate or rapidly alternate movement, (2) reduces or eliminates associated movements such as the arm swing during gait, and (3) produces a masklike face (i.e., no facial expressions).

An interesting feature about the rigidity of Parkinson's disease is that there generally is not the constant resistance to movement associated with rigidity *(lead pipe rigidity)*. Parkinson's rigidity is often manifested as a series of catches during the resistance. As the muscle is stretched the resistance is periodically released during the movement (similar to the catches noted when winding a wristwatch). This type of rigidity is referred to as *cogwheel rigidity* and can be thought of as a combination of rigidity and tremor.

The components of the corpus striatum have been studied through ablation and electrical implant procedures to determine specific functions.

Electrical stimulation of the neostriatum causes an inhibitory effect on the globus pallidus, resulting in inhibition of spontaneous activity, of a motion in progress, or of deep-tendon reflexes. Lesions (especially bilateral) result in hyperkinesia. An animal with bilateral ablation of the putamen will wander about continuously with no regard for its environment (it will walk into a wall or off a table). Clinically the symptoms of *athetosis and chorea* are attributed in part to neostriatal lesions.

Electrical stimulation of the globus pallidus increases muscle tone and can produce a tremor on the contralateral side. Lesions of the pallidum cause hypokinesia, hypotonus, loss of associated movements, and even somnolence. Such an afflicted animal seldom moves even if placed in a bizarre posture. It has been mentioned that the globus pallidus acts to activate the gamma motoneurons during alpha-gamma coactivation.

In summary, it is difficult to relate isolated functions of the basal nuclei to an overall function. As a unit the basal nuclei have a role in (1) associated movements such as raising the arms when yawning, (2) synchronizing the output from the thalamus and motor cortex, (3) muscle tone, (4) regulating automatic activities such as walking or chewing, (5) regulating postural adjustments during a skill movement such as typing, and (6) relating the environment to the required movement.

Cerebellum

The total size of the cerebellum is roughly that of two golf balls, yet the intricate, rapid, and graceful movements it regulates are amazing. The hands of a concert violinist or the fluid body of an Olympic ice skater demonstrate what the cerebellum does.

The overall role of the cerebellum is that of an error-correcting device for cortically induced goal-directed movements. For example, when you are reaching to pick up a cup of coffee the cerebral cortex sends out a volley of impulses. Some go to the basal ganglia and reticular formation for postural adjustments and proximal joint stability, some to the cerebellum (via the pontine nuclei) to inform it that a program of movement is about to begin, and some go to the appropriate spinal motoneurons so that the arm and hand can reach for the cup. As the cup is raised the muscle spindles and GTOs communicate to the cerebellum the amount of movement, tension (weight of the cup), and velocity of movement. The cerebellum can then integrate this data and inform the cerebral cortex whether its original signal was too weak (the cup is too heavy), too strong (the cup is moving too fast toward the face), or just right. The cerebral cortex can then make the appropriate adjustments in its output signal.

Each hemisphere of the cerebellum, has two main components: an outer cortex, which is the afferent receiving station, and the deep nuclei, which are the efferent output stations back into the brainstem. Between the hemispheres is the vermis, which has a cortex but no nuclei. The cortical convolutions are called folia.

The input to the cerebellar cortex is complex but may be summarized as follows:

1. Motor cortex data are relayed in the pontine nuclei and projected to the cerebellar cortex via the middle cerebellar peduncle.
2. Data concerning muscle length, tension, and velocity are sent via the posterior spinocerebellar tract and the inferior olive (inferior cerebellar peduncle) and via the anterior spinocerebellar tract (superior cerebellar peduncle).
3. General afferent data from the body are sent from the reticular nuclei via the inferior cerebellar peduncle.
4. General afferent data from the head are sent from the trigeminal nuclei.
5. Data concerning head position and movement are sent from the vestibular nuclei via the inferior cerebellar peduncle.
6. Visual and auditory data are sent from the tectum via the superior cerebellar peduncle.

Afferent data reaching the cerebellar cortex travel over two kinds of fibers: *climbing fibers and mossy fibers.* Climbing fibers originate primarily from the inferior olive. Mossy fibers are those axons from pontine, vestibular, reticular, and trigeminal nuclei and the second-order neurons of the spinocerebellar tracts.

The circuitry within the cerebellar cortex is complex but can be simplified as follows. The cortex is composed of three layers of five kinds of cells (Fig. 9-4). The *Purkinje cells* are the final common pathway out of the cor-

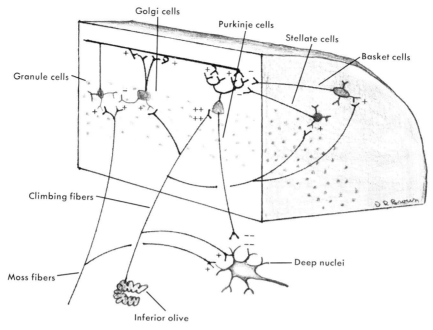

Golgi cells

Purkinje cells

Stellate cells

Basket cells

Granule cells

Climbing fibers

Deep nuclei

Moss fibers

Inferior olive

Fig. 9-4. Basic neural circuitry within folium of cerebellar cortex. Input to cortical neurons is from climbing fibers that originate from neurons of inferior olive and mossy fibers. Output from cortex is via Purkinje cells. Their axons project to deep nuclei within cerebellar white matter. Neurons of deep nuclei are tonically excitatory as result of collateral faciliation from climbing and mossy fibers and tonically active "pacemaker" neurons within nuclei. Purkinje cells are powerfully excited by climbing fibers and also from granule cells, which are excited by mossy fibers. This excitation is modulated by IPSPs from stellate and basket cells, which are activated by climbing fibers and by granule cells (not shown). Golgi cells are excited by granule cells and climbing fibers. They act to inhibit granular cell excitation of Purkinje cells. Overall effect on Purkinje cells is very precise, fine tuning or focusing of inhibitory output to deep nuclei. Three cortical layers—molecular (outer), Purkinje (middle), and granular (inner)—are not labeled. (Adapted from Noback, C., and Demarest, R.: The human nervous system: basic principles of neurobiology, ed. 2, New York, 1975, McGraw-Hill Book Co.)

tex to the deep nuclei. The Purkinje cells can be excited directly via climbing fibers or indirectly via mossy fibers and granule cells. Purkinje cells can also be inhibited by mossy fibers through a longer circuit involving basket, stellate, and Golgi cells. This latent inhibition is similar to the Renshaw inhibition on alpha motoneurons in that it acts to finely tune the output of the Purkinje cells.

The Purkinje cells project inhibitory impulses to the four *deep nuclei:* dentate, globus, emboliform, and fastigial. The myelinated Purkinje axons make up a large portion of the white matter found between these nuclei and the cortex. The deep nuclei have a constant output of excitatory discharge, as they receive excitatory collaterals from the climbing and mossy fibers. The Purkinje effect is to regulate the tonic discharge of the deep nuclei.

Although the cerebellum is a primary source of coordinated (synergistic) movement, there is no direct cerebellospinal pathway. The output from the deep nuclei is to two sources: the ventral lateral nucleus of the thalamus and the brainstem nuclei.

The ventral lateral nucleus of the thalamus projects to the motor and premotor cortices, completing an important circuit involving the motor cortex–pontine nuclei–cerebellar cortex–deep nuclei–thalamus–motor cortex. It is this circuit that acts as an error-correcting device for cortically willed movement.

The brainstem nuclei (reticular, vestibular, and red) are involved in the coordinated alterations of posture necessary during any skilled movement (Fig. 9-5).

Functionally the cerebellum can be organized into three divisions: the *archicerebellum*, *paleocerebellum*, and *neocerebellum*.

The archicerebellum is composed of the paired flocculi of each hemisphere and the nodulus of the inferior vermis. This is the flocculonodular lobe. It receives input from the vestibular nerve and nuclei and is concerned with muscle tone, equilibrium, and posture, primarily of the trunk muscles. The flocculonodular efferents project to the vestibular nuclei and reticular formation via the inferior cerebellar peduncle.

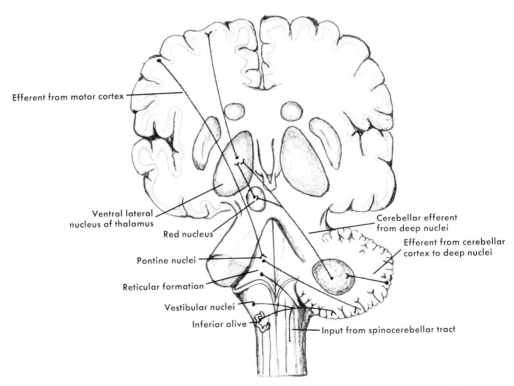

Fig. 9-5. Neural connections of neocerebellum. Note feedback between cerebral cortex–pontine nuclei–neocerebellum–deep nuclei–thalamus and cerebral cortex. Other efferent connections of cerebellum have been omitted. (Adapted from Noback, C., and Demarest, R.: The human nervous system: basic principles of neurobiology, ed. 2, New York, 1975, McGraw-Hill Book Co.)

The paleocerebellum consists of most of the vermis and the anterior lobe. This portion of the cerebellum receives general proprioceptive and exteroceptive data. The paleocerebellum plays an important role in muscle tone. Its output is presumably to the vestibular nuclei and reticular formation via the superior cerebellar peduncle.

The neocerebellum consists of the bulk of each hemisphere and parts of the vermis. This portion is primarily concerned with coordination of the extremities. It receives input primarily from the motor cortex (via pontine nuclei) and inferior olive.

The cerebellum is susceptible to trauma, vascular insufficiency, disease, and tumors. Different symptoms result depending on the area involved. If the archicerebellum is involved uncoordination (ataxia) of the trunk may result. Patients tend to fall backward and sway from side to side, or they may not be able to maintain an upright posture. If the paleocerebellum is involved, especially the anterior lobe, an increase in the facilitation of ipsilateral alpha motoneurons occurs, resulting in muscle rigidity (lead pipe). When the neocerebellum is involved a variety of ipsilateral symptoms occur, including the following:

1. Hypotonia, which results in hyporeflexia.
2. Pendular knee jerk in which the knee swings freely back and forth (related to rebound phenomenon).
3. Scanning (staccato) speech. Speech is hesitating, slurred, and explosive in quality with pauses in the wrong places (related to dysarthria).
4. Asynergy of movements. Voluntary motions are jerky and puppetlike. The term ataxia is used synonymously with asynergy but is usually used to describe incoordination of the legs as in ataxic gait.
5. Dysmetria, or inability to judge distances, resulting in over- or undershooting a desired object (past pointing).
6. Adiadochokinesia, or inability to perform rapidly alternating movements such as pronation-supination of the forearm with equal excursion.
7. Intention tremor, or oscillations of the distal extremity, especially at the end of the movement. Intention tremors are absent or diminished during rest.
8. Rebound phenomenon, or inability to regulate reciprocal innervation circuits. This effect is best seen when the elbow flexors are actively contracting against resistance and the hand is only a few inches from the face. The sudden release of resistance will cause the hand to hit the face. A normal cerebellum would detect the sudden flexion and break it by inhibiting the elbow flexors and stimulating the triceps.
9. Ataxic gait, which is halting, lurching, and with a broad base of support.

Thalamus

The thalamus proper is composed of two egg-shaped clusters of nuclei completely surrounded by other structures. Because of its centralized location the thalamus is ideally suited to act as a coordinator and regulator of cerebral activity.

The nuclei of the thalamus are divided into the *anterior*, *lateral*, and *medial nuclear groups* by a plate of neural tissue called the *internal medullary lamina*. Within this lamina is the *intralaminal nuclear group*. On the lateral surface adjacent to the internal capsule is the thalamic *reticular nucleus*. On the medial surface adjacent to the third ventricle is the *midline nuclear group*.

These nuclei are often classified into the *nonspecific* and the *specific nuclei*. The former include:

1. Intralaminar nuclear group
2. Thalamic reticular nucleus
3. Midline nuclear group

The latter include:

1. Specific cortical relay nuclei
   a. Anterior nuclear group

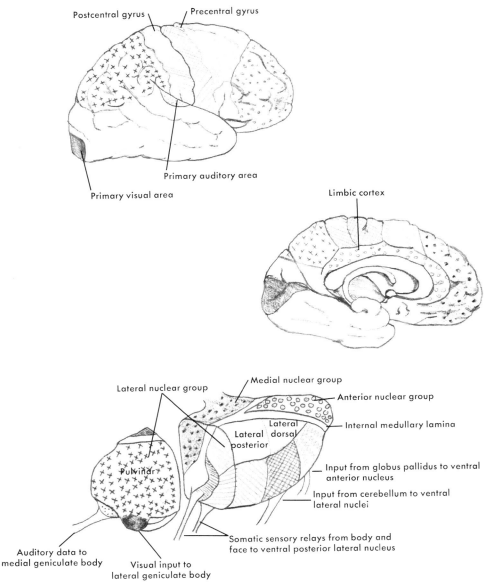

Fig. 9-6. Thalamocortical projections. Thalamus relays data to all parts of cerebral cortex. Note relays from geniculate bodies, ventral tier of lateral nuclear group (VA, VL, VPL), and anterior nuclear group. These are specific relays to primary cortical centers. Medial nuclei and dorsal tier of lateral nuclear group (including pulvinar) are specific relays to association areas. Not shown are intralaminar, midline, and reticular nuclei, which are nonspecific relays to diffuse areas of cortex. Refer also to Figs. 2-18, 9-10, and 9-11 to further understand these projections. (Adapted from Netter, F.: The CIBA collection of medical illustrations. I. The nervous system, Summit, N.J., 1972, CIBA Chemical Co.)

b. Lateral nuclear group (ventral tier)
   (1) Ventral anterior nucleus
   (2) Ventral lateral nucleus
   (3) Ventral posterior lateral nucleus
   (4) Ventral posterior medial nucleus
   (5) Metathalamus (geniculate bodies)
2. The specific association nuclei
   a. Lateral nuclear group (dorsal tier)
   b. Medial nuclear group

The massive efferent projections of the thalamus are primarily to the cerebral cortex via the posterior and anterior limbs of the internal capsule (Fig. 9-6). These projections are from the specific and nonspecific systems. The specific projections are direct connections from certain thalamic nuclei to specific cortical areas. For example, the ventral anterior and ventral lateral nuclei project strictly to the motor and premotor gyri from the globus pallidus and cerebellum. Other nuclei such as the ventral posterolateral and ventral posteromedial are even more specific. These nuclei are somatotopically organized. This means that the neurons of these nuclei respond to only one modality from one area of the body and project to a specific set of cortical neurons (in this case, in the postcentral gyrus). Other specific thalamic projections are to the primary visual and primary auditory cortices via the geniculate bodies, to the limbic cortex via the anterior nuclear group, and to the association (integrative) areas of the cortex via the medial and lateral nuclear groups (especially the pulvinar).

The nonspecific system includes the thalamic nuclei that have widespread, multineuronal connections with the cerebral cortex. These nuclei receive their major input from the reticular formation and also from the cerebral cortex. These nonspecific nuclei are the primary source of the background electrical activity that is responsible for arousal and consciousness. This electrical activity is present all of the time, even during sleep, and can be easily recorded. Such recordings are known as *electroencephalograms* (EEG) and are made by placing 16 sensitive electrodes on specific predetermined spots on the scalp. The EEG of different cortical regions can be recorded by comparing the electrical variations between the appropriate electrodes. The cause of this electrical activity is partly the result of closed self-reexciting chains of neurons. These chains (Fig. 9-7) produce an oscillation in the electrical potential of the cortical neurons by exerting pre- and post-synaptic influences on the cortically projecting thalamic fibers. The result is described in terms of oscillating cycles per second. Typical frequencies of an awake adult are 8 to 12 Hz (alpha pattern) and 15 to 30 Hz (beta pattern).

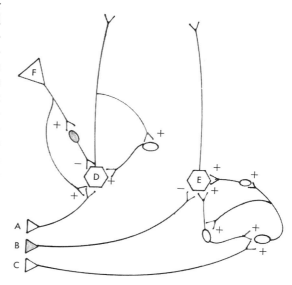

Fig. 9-7. Schema of complex neural circuitry in thalamus. Input from sources *A* and *C* can cause maintained output on thalamic neurons *D* and *E* to cerebral cortex. This is accomplished by reverberating circuit of neurons. Sources *B* and *F* can dampen this output through direct or presynaptic inhibition. Effect on *D* and *E* could be oscillation in output similar to EEG pattern.

The primary driving force for these thalamocortical circuits is the afferent input to the thalamus and reticular formation. An isolated piece of cortex exhibits little or no spontaneous electrical activity. Of interest, if the brainstem is lesioned just caudal to the thalamus, leaving these circuits intact, a rhythmic synchronous alpha pattern can be recorded. When certain areas of the reticular formation are stimulated the EEG becomes desynchronized into fast irregular low-voltage beta patterns. Alpha patterns are typical of an awake but resting adult, while beta patterns characterize an alert (thinking) brain.

During sleep, when the RIS is active, there is an interesting EEG. In deep sleep there is no dreaming or rapid eye movements (non-REM sleep). The thalamocortical circuits are very rhythmic and synchronized at 2 to 4 Hz. At intervals of approximately 90 minutes we shift into light sleep with dreams and rapid eye movements (REM sleep). The EEG of REM sleep resembles that of an alert person. Thus it seems that the thalamus acts as a regulator for the rhythmic depolarization-hyperpolarization of cortical neurons producing a state of nonarousal (deep sleep) to sharp alertness (responding to stimuli and thinking) (Fig. 9-8).

To summarize, the specific nuclei act as conveyors and processors of sensorimotor data from subcortical centers to the cerebral cortex. The nonspecific nuclei serve as energizers for large areas of the cerebral cortex.

Functionally the thalamus has the following roles:

1. Sensory relay of all sensations to specific primary and associative receiving areas except for olfaction.
2. Relay of data from the cerebellum and corpus striatum to the motor cortex in a synchronized pattern. These data are critically important in the production of somatic movement. The extent of preprocessing of data by the thalamus is uncertain.
3. Involvement in cortical arousal, consciousness, the EEG, and sleep. These activities are expressions of the nonspecific nuclei.
4. Role in emotion and behavior through its connections with the limbic system. Although still unclear, the thalamus seems to be able to add affects (feelings) such as agreeable or unpleasant to sensations.
5. Perception of pain, temperature, and crude touch. Localization and intensity discrimination, however, are poor.

Because of these varied functions thalamic lesions can create a variety of symptoms. Clinically the most common thalamic lesion is from vascular insufficiency resulting in the *thalamic syndrome*. In this syndrome the threshold for pain, temperature, and tactual sensations is usually raised. However, nonnoxious stimuli may produce exaggerated pain or unpleasant sensations on the opposite side of the body. The application of heat or ice or the pressure of clothes may be enough to evoke this reaction. Tumors also affect the thalamus, resulting in a variety of symptoms including unconsciousness, emotional disorders, hypothalamic disturbances, and possibly epilepsy. Isolated damage to the ventral lateral and ventral anterior nuclei does not normally occur. However, surgical lesions of these nuclei in patients with Parkinson's tremor and cerebellar asynergy have produced amelioration (though not always permanent) of symptoms. After surgery a decrease in muscle tone of these patients is noticed, which suggests that these nuclei may act to modulate the higher center output to the gamma motoneurons. Perhaps the error-correcting data from the cerebellum and automatic movement data from the pallidum are used by these nuclei to inform the cortex of the appropriate amount and proper timing of gamma activation needed. It is interesting that when a thalamotomy is performed on a patient with Parkinson's disease transient

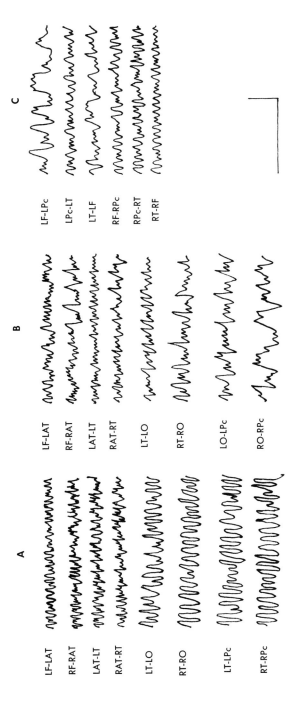

Fig. 9-8. **A**, EEG of healthy adult. Note alpha and beta rhythms typical of awake state. **B**, Epilepsy in 6-year-old boy. Two of three convulsions started in left extremities. Note theta and delta waves in right hemisphere. **C**, Left frontal brain tumor in 58-year-old man with 2-year history of seizures, right hemiparesis, and aphasia. Calibration: vertical, 50 µV; horizontal, 1 sec. *R*, right; *F*, frontal; *P*, parietal; *AT*, anterior temporal; *L*, left; *O*, occipital; *Pc*, precentral; *T*, temporal. (Adapted from Chusid, J. G.: Correlative neuroanatomy and functional neurology, Los Altos, Calif., 1976, Lange Medical Publications.)

cerebellar signs appear. Their disappearance with time suggests the existence of compensating pathways.

Cerebral cortex

If the cortex were peeled off, it would weigh only about 600 g (1 lb), cover a flat area of just more than 0.75 m² (2.5 ft²), and vary in thickness from 1.5 mm to 4.5 mm. Yet in spite of its smallness it contains 10 to 15 billion neurons and 50 billion glial cells. In the neocortex (see below), which forms over 90% of the cortex, these neurons are arranged in six layers. Each layer is distinguished by the type, density, and arrangement of the five types of cortical cells. The complexity of the circuitry between these neurons is beyond comprehension. One only need think of the immense variety of motor, behavioral, and creative functions that occur in each of us to appreciate this complexity (Fig. 9-9).

The cerebral cortex (pallidum) can be divided on a developmental and functional basis into three areas. The *archicortex* contains the cortical components of the limbic system and is located in the medial part of the frontal, parietal, and temporal lobes. The *paleocortex* includes the medial inferior part of the frontal lobe and the medial superior part of the temporal lobe, which are associated with the olfactory system (Fig. 9-10). The *neocortex* is all of the rest of the cortex (about 90% in bulk).

The entire cortex has been mapped according to function by several neuroanato-

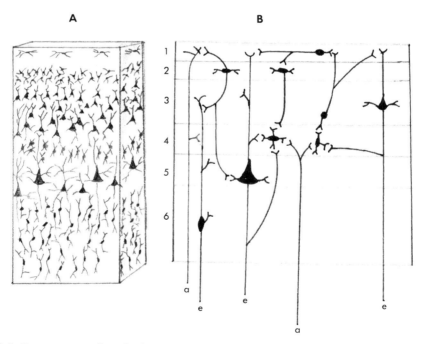

Fig. 9-9. Cross-section of cerebral cortex. **A,** Diagram of cells in six layers as they would appear with Golgi staining technique. **B,** Simplified drawing of intracortical circuitry. *a,* Afferent input from thalamus and other cortical areas; *e,* efferent output to other cortical areas and CNS centers. (Adapted from Barr, M.: The human nervous system, New York, 1972, Harper & Row, Publishers.)

mists. Through the use of electrical stimulation, surgical ablation, and retrograde degeneration studies specific cortical areas have been designated. The most widely used cytoarchitectural mapping was devised by Brodmann. His 47 cortical areas are presented in Fig. 9-10. Fortunately it is not necessary to know the functions of all 47 areas, only the more important ones. The archicortex and paleocortex are composed of areas 24 through 36 and are discussed in the next section.

The remaining numerical areas make up the neocortex, and most of them can be divided into three areas: (1) afferent (sensory areas), (2) integrative (association) areas, and (3) efferent (motor) areas. Comparing Figs. 9-10 and 9-11 will be useful in this next section.

Since the functions and effects of lesions of the individual sensory and association areas are presented in previous chapters, only their relation to Brodmann's areas is presented at this time.

Primary sensory areas. The primary sensory areas are concerned with perceiving sensations in terms of quality, intensity, and localization. These areas are as follows.

SOMATIC SENSORY AREA I (SI). This area is the postcentral gyrus (3, 1, and 2) and is for

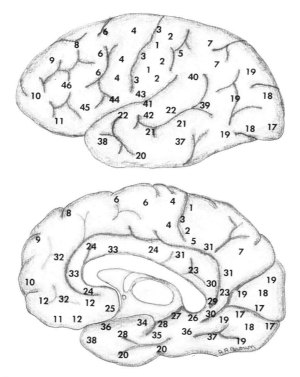

Fig. 9-10. Cortical mapping of Brodmann. Numbers represent general location of 47 cytoarchitectural areas designated by Brodmann. The paleocortex (areas 25, 28, 34, 36, and 38) and archicortex (areas 23, 24, 26, 27, 29, 30, 33, 35, and 37) compose limbic lobe. Remaining areas compose neocortex. Refer also to Fig. 9-11 to compare function with cortical area. (Adapted from Noback, C., and Demarest, R.: The human nervous system; basic principles of neurobiology, ed. 2, New York, 1975, McGraw-Hill Book Co.)

the perception of somatic sensations. Thus data received by the VPL and VPM nuclei of the thalamus (contralateral dorsal column, spinothalamic and trigeminal sensations) are transmitted to the ipsilateral postcentral gyrus. It is important to mention again that the postcentral gyrus is primarily concerned with spatial relations (kinesthesia), intensity discrimination, and stereognosis. The sensations of pain, temperature, and crude touch are perceived by both the thalamus and postcentral gyrus.

AREAS 41 AND 42 FOR THE RECOGNITION OF SOUNDS. These adjacent areas are located in the temporal lobe on the lower border of the Sylvian fissure, just in back of the central sulcus. Each hemisphere receives input from both medial geniculate bodies.

AREA 17 FOR THE RECOGNITION OF VISUAL DATA. This area is the dorsal tip of the occipital lobe (occipital pole). Each hemisphere receives data from the ipsilateral lateral geniculate body (and contralateral visual field of each eye).

AREA 43 FOR THE RECOGNITION OF VESTIBULAR SENSATIONS (vertigo, nausea). This area is located in the parietal lobe at the corner of the central sulcus and Sylvian fissure.

SOMATIC SENSORY AREA II (SII). This area (not numbered by Brodmann) is located just in back of area 43. It receives input from the somatic sensory pathways from both sides of the body. The somatotopic representation is less well defined and its function less clear than for SI. Ablation of SII produces no obvious deficit. Some investigators have suggested a role in pain perception. It is sometimes called the secondary somatic sensory area.

Association areas. The association areas are concerned with interpreting and making judgments about the meaning of the data received from the primary area. The association areas project to other cortical areas, many subcortical nuclei, and to the descending motor tracts. The association areas include the following.

AREAS, 5, 7, AND 40. These are the somatic sensory association areas and are located in the parietal lobe dorsal to the postcentral gyrus.

AREA 22 (WERNICKE'S AREA). This is the auditory association area and is located in the temporal lobe around the primary auditory area.

AREAS 18 AND 19. These are the visual association areas and are located in the occipital lobe in front of area 17.

Motor areas. Although many areas of the cerebral cortex contribute to motor activity, the principal motor areas are in the frontal lobes. These areas receive extensive direct stimulation from the ventral anterior and ventral lateral nuclei of the thalamus, the thalamic reticular nuclei, and all lobes of the ipsilateral and contralateral hemispheres. The major motor areas are as follows.

AREA 4 (PRECENTRAL GYRUS). All somatic muscles are represented in this area in a very specific distribution known as a homunculus (Fig. 9-12; next section). However, area 4 is primarily concerned with the facilitation of the contralateral distal muscles (especially flexors). This facilitation is involved with voluntary skilled movements. Electrical stimulation can produce isolated movements of one muscle, but generally groups of muscles respond in a precise pattern of movement. The primary input to area 4 is from the ventral lateral nucleus of the thalamus (cerebellar data).

AREA 6 (PREMOTOR AREA). This area is located in front of area 4. The primary subcortical input to area 6 is from the ventral anterior nucleus of the thalamus (corpus striatum). Stimulation of area 6 causes generalized movements such as turning of the head, twisting of the trunk, and flexion or extension of the limbs.

PART OF AREA 8 IN FRONT OF THE PREMOTOR AREA. This area functions in voluntary eye movements. Electrical stimulation of this area produces eye deviation to the opposite side. Area 8 is often called the frontal eye field.

AREAS 44 AND 45 (OF BROCA). These areas are located in the frontal lobe at the junction of the central sulcus and Sylvian fissure. They are responsible for the movement patterns for the production of speech (i.e., the muscles of the tongue, lips, and larynx). Although these areas in both hemispheres innervate the speech muscles, it is from a person's dominant hemisphere that complicated speech patterns are produced.

SUPPLEMENTARY MOTOR AREA. This area (not numbered by Brodmann) is located in the medial surface of the frontal lobe rostral to the primary motor area (between areas 4 and 6). Movements obtained from electrical stimulation can be seen bilaterally and consist of assuming postures such as raising the arm, bilateral synergy contractions of the trunk and leg, and tonic contractions of postural muscles.

Effects of cerebrocortical lesions. Lesions of the motor cortex produce some interesting symptoms that are partly the result of loss of a specific motor area and partly caused by loss of cortical suppressor areas (Fig. 9-11). These suppressor areas act as sources of cortical inhibition to other cortical and subcortical motor centers. A good example of this can be seen in a monkey after removal of area 4 only. The monkey will show initial severe paralysis, especially in the distal muscles, hypotonia, and decreased deep-tendon reflexes on the contralateral side. In time the sypmtoms improve and include only impairment of skilled movements of the hands and feet. If the suppressor zone in front of area 4 is included in the lesion a degree of spasticity and hypertonicity is also evident.

Isolated lesions of area 6 produce only transient grasp reflexes (an object placed in the palm causes flexion of the fingers and thumb).

Lesions of areas 4 and 6 including the suppressor area may initially produce flaccid paralysis. This usually gives way to spastic paralysis. The spasticity is predominantly in the

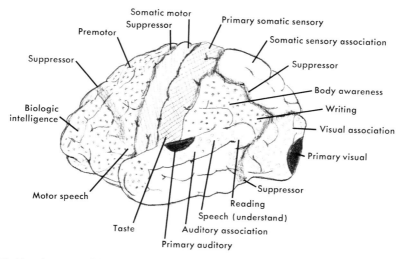

Fig. 9-11. Localization of function of lateral surface of dominant hemisphere. Nondominant hemisphere is functionally organized in similar pattern except for communication areas.

antigravity muscles. This results in the typical hemiplegic posturing (dystonia): adducted and internally rotated shoulder, flexed elbow, and flexed writs and adducted and internally rotated hip, extended knee, and plantar flexed ankle. Other symptoms include clonus, increased deep-tendon reflexes, positive Babinksi sign, and absent superficial reflexes (see Chapter 10).

If the supplementary motor area is included in lesions of areas 4 and 6 the sypmtoms are even more severe. Isolated lesions of the supplmentary motor area produce only minimal effects contralaterally. These include slowness of movement and mild hypertonia of the shoulder muscles.

Lesions of area 8 can cause transient conjugate deviation of the eyes to the side of the lesion.

Lesions of area 44 in the dominant hemisphere produce difficulty with verbal expression, known as expressive asphasia. It involves inability to produce the appropriate

commands to the speech muscles. The person often knows what he or she wants to say but is unable to do so. Expressive asphasia is different from dysarthria, in which one or more of the speech muscles are paralyzed or lacking coordination.

An interesting feature of primary motor and sensory areas in terms of function is the representation of the body muscles and receptors. Extensive mapping of these gyri have produced the *motor and sensory homunculi* presented in Fig. 9-12. The distorted appearance of the parts of the body indicate the relative importance of that part in terms of movements or sensory input. For example, the hands and face are very important for exploring and manipulating our environment, whereas the elbows and knees are not.

Knowledge of the homunculi and location of the different motor, sensory, and association areas is very useful when related to the distribution of the three cerebral arteries (see

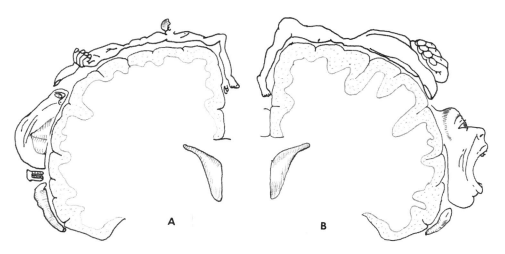

A   B

Fig. 9-12. Sensory (**A**) and motor (**B**) homunculi. Distortion in representation of body indicates relative importance of these areas. Thus hands and face have more motor and sensory cortical neurons than elbow or knee. Lower lateral areas are for throat and are involved with speech and swallowing. Note in **A** separation of face from head, teeth from face, and genitalia from pelvis. (Adapted from Penfield, W., and Rasmussen, T.: The cerebral cortex of man, New York, 1957. Copyright 1950 by The Macmillan Co.)

cortical distribution of these arteries in Chapter 7). This information is valuable in determining the various symptoms associated with CVA, head injury, and brain tumor. It is important to mention that not all brain-lesioned patients have a cluster of symptoms. For example, typically in CVA there is involvement of the middle cerebral artery and both motor and sensory symptoms and perhaps speech difficulties. However, if only an isolated branch of the middle cerebral artery is involved, such as to part of area 4, there is paresis of a specific area such as the shoulder. While this example is not common, it does happen and can also occur with tumors or trauma to the brain.

Integrative functions. Although discussed as individual entities, the components of the motor, sensory, and association cortices do not function in isolation. For example, the sensory and association areas of the parietal lobe evaluate the somatic sensory data from your finger when you reach in your pocket for change. When the appropriate coins are found (based on auditory or visual data that indicated you need 30¢ for an ice cream cone), the motor system acts to pull the coins out.

The above is a very simple example of the integrative abilities of the neocortex. Evaluating a patient's symptoms and planning an appropriate treatment program is a higher level cortical function requiring judgment and creativity. Although not directly involved with the motor systems, the following integrative activities influence our daily functions and are important in the treatment of any patient.

PERCEPTION. The ability to perceive a stimulus first of all requires that the person be aroused to an alert state of consciousness. Many patients with brain lesions are in states of lethargy or semiconsciousness initially after the lesion. They are slow to respond to commands to move and have little carryover (learning) to the next day. Simple instructions repeated slowly and often are helpful. The second requirement is the ability to pay attention to the stimulus. Many patients with CNS involvement, especially of the frontal lobes, have difficulty concentrating. This can be very frustrating to the therapist who is trying to accomplish a treatment goal. Placing the patient in a quiet area away from extraneous sights and sounds, using simple commands, and working on only one or two activities per session are helpful. The process of perception not only assumes recognition of the stimulus in terms of quality, intensity, and position in space, it also involves comparing the stimulus with previously learned stimuli, placing the stimulus in a particular category, and making judgments as to its usefulness. Thus a chair presented upside down is still a chair and, depending on your previous experiences, may be used to sit on, to stand in, or to throw clothes on. Patients with lesions of the association areas (agnosia) lack the ability to apply meaning to the objects they come in contact with. The world is very strange to these patients, and being brought into a busy therapy department can be too confusing for them. A quiet room with few objects can be beneficial.

LEARNING. There are many types of learning, ranging from conditioned reflexes where the response to one stimulus is trained to occur with a new stimulus (Pavlov's dogs) to discovery learning where the person makes individual associations in achieving the desired response. The most common type of learning involving patients is skill learning. Activities such as ambulation and dressing are skills that some patients must learn again, for the first time, or in an altered manner using an assistive device. Skills are best learned by a process of explanation, demonstration, practice with guidance, and then independence. The role of the therapist is to provide a suitable learning environment. He or she must (1) determine whether the appropriate level of arousal and attention span are present, (2) provide simple but useful directions, (3) demonstrate the skill broken down into its components, (4) provide practice with guidance and verbal reinforcement, and (5) provide encouragement and motivation.

The importance of practice can be explained on a neurophysiologic basis. Any sensorimotor skill such as walking requires the development of long and complicated neural circuits. The efficiency of a skill is basically a function of the efficiency of the synapses in the circuits. Repetitive stimulation of a synapse can improve this efficiency by altering the structure of the pre- or postsynaptic membranes, or the production or release of the transmitter substance. Outside interference such as a noisy room and insufficient repetitions requiring the use of the circuit retard or prevent these structural-chemical changes. It is important to keep in mind that many patients, and especially older persons, may require more practice than others would to learn even simple skills. The elderly not only have slower conduction velocity and fewer neurons but are also less capable of forming motor skill circuits.

The importance of motivation is that it alone is the driving force to want to learn. An unmotivated patient is difficult to work with, and unfortunately there is no simple recipe for increasing motivation. Methods that work with one patient will not work with all patients.

COMMUNICATIONS. The ability to communicate is one of our most valued integrative skills. Communication involves the auditory, visual, and tactile systems for receiving data and the motor systems for transmitting data through the use of speech, writing, and gestures. Loss or impairment of any of the receiving or transmitting components seriously affects our ability to communicate and thus our whole personality. Any lesion to cranial nerve V, VII, or XII, the cervical nerves or cord, cerebellum, descending motor projections, dorsal thalamus, internal capsule, or many parts of the cerebral cortex will interfere with communications. In other words, many patients with neurologic disorders have a communication disorder. The frustration they experience is often acute enough that it interferes with treatment. These persons should be treated with respect and maturity appropriate for their age. Many persons talk to such patients in childlike language even though their intelligence is intact. Proper eye contact and touching are very important for their self-esteem, especially when they are attempting to communicate. Extra time should be provided for the patient to attempt to communicate. This is especially true in speech disorders such as asphasia and dysarthria.

INTELLIGENCE. Our intelligence is the highest of all integrative functions. It is partly based on our genetic makeup and partly on the kind of environment we have been exposed to. The acuity of our perceptual mechanism, the speed and skill of our motor acts, the rate and types of things we learn, our ability to remember, calculate, and verbalize, and our ability to synthesize data into creative products are all related to intelligence. It is important to remember that intelligence can manifest itself in different areas. For example, a highly articulate person appears very intelligent, but he or she may have difficulty learning how to ambulate using an assistive device or tying a shoe with one hand. At the same time a person with a modest vocabulary may have no trouble with such skill learning. It is beneficial to interact with all patients at their own level and at the same time respect each patient for the kind of intelligence he or she possesses.

HEMISPHERIC DOMINANCE. In the course of evolution certain cortical functions, because of their complexity, have been specialized into one or the other hemisphere. In humans the left hemisphere is the dominant hemisphere for handedness and language (reading, speech, and writing) in 90% of the population. The remaining 10% are left-handed (except for a few ambidextrous individuals). Although the motor cortex is dominant in the right hemisphere in left-handed persons, more than half of them have speech and reading centers in the left hemisphere. Experiments have been performed in which the short-acting anesthetic amobarbital (Amytal) has been injected into one of the internal carotids. Most of one hemisphere is temporarily knocked out, while the other remains functional. Other experiments have been performed in which the corpus callosum and anterior commissures have been severed, leaving two intact but separately functioning hemispheres. The results from these experiments are as follows. The left (dominant) hemisphere is responsible for comprehending the spoken and written word, speaking, writing (if right-handed), calculating, and analyzing concrete data. The right hemisphere can perceive tactile, visual, and auditory stimuli, can communicate through gestures, fidgeting, or blushing, can appreciate spatial relations, and has an important creative role in art, music, and poetry.

Some patients with severe epilepsy have had commissurectomy performed with good results. Careful testing of these patients has shown them to be alert, curious, and outwardly normal. Each hemisphere is completely independent with respect to learning, memory, perception, and ideation. An object such as a key placed in the left hand (person blindfolded) cannot be named because the right hemisphere is mute and the sensory data cannot cross to the left hemisphere. Although able to write, the person will not be able to draw because the left hemisphere cannot provide guidance on spatial relations.

Although several gross anatomic differences between the hemispheres have been reported, it is still uncertain whether this has any bearing on dominance. What is interesting is the ontogeny of hemisphere dominance. It is generally believed that up until the age of 2 years neither hemisphere is dominant, although handedness and speech centers are forming. This is evident by the fact that many children (even older than 2 years) who sustain injury to the left hemisphere can still acquire language skills and become left-handed.

Limbic system

The limbic system is composed of many cortical and subcortical centers that are often referred to as the visceral brain. Although not directly related to the motor systems (there are no direct pathways to the motoneurons), the limbic system is involved with a variety of emotional and behavioral functions that affect somatic and autonomic motor responses. The components of the limbic system are as follows:

1. Olfactory pathway and cortex
2. Amygdala
3. Hippocampal formation, which includes the dentate, subiculum, and hippocampal gyri (The hippocampal gyrus is often referred to as the parahippocampal gyrus or piriform cortex [area 28]. The rostral tip of this gyrus is the uncus [Area 34].)
4. Cingulate gyrus
5. Septal region, which includes the septal nuclei in the septum pellucidum and the medial surface of the frontal lobe under the genu and rostrum of the corpus callosum
6. Nuclei of the hypothalamus such as the mamillary bodies
7. Dorsal medial and ventral anterior nuclei of the anterior nuclear group of the thalamus
8. Habenular nuclei
9. Parts of the mesencephalic reticular formation (MRF)

With the exception of the MRF the components of the limbic system occupy the medial surface of the telencephalon and diencephalon. The hippocampal formation, cingulate gyrus, septal region, and olfactory cortex are sometimes called the limbic cortex or limbic lobe.

The neural connections between the limbic components and with other CNS structures is quite complex. Many have yet to be fully documented. It is almost safe to say that each limbic component interconnects with each other limbic component and that many of them send relays to the neocortex, pitu-

itary gland, and autonomic preganglionics.

The major pathways of the limbic system are as follows:

1. Stria terminalis (parallel to the arch of the fornix), from the amygdala to the septal region and hypothalamus
2. Amygdalofugal fibers, from the amygdala to the limbic lobe and neocortex
3. Fornix, from the hippocampal formation to the septal nuclei, and mamillary bodies (Fig. 9-13)

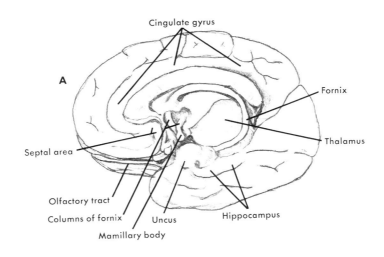

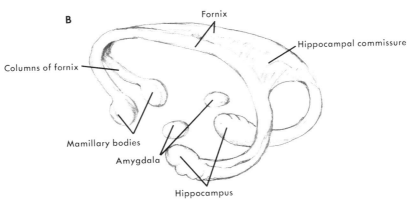

Fig. 9-13. Major components of limbic system. Cortical components form ring around diencephalon and corpus callosum. Note in **B** curve of fornix and close proximity between hippocampus and amygdala. (Adapted from Netter, F.: The CIBA collection of medical illustrations. I. The nervous system, Summit, N.J., 1972, CIBA Chemical Co.)

4. Internal capsule, from the cingulate gyrus to the corpus striatum
5. Stria medullaris thalami, from the septal and preoptic region of the hypothalamus to the habenula (This tract travels along the superior medial border of the thalamus.)
6. Medial forebrain bundle, from many hypothalamic nuclei to centers in the midbrain
7. Mamilothalamic fasciculus, a continuation of fornix data from the mamillary bodies to the anterior nuclear group of the thalamus

The full functional roles of the limbic system are not completely known at this time. The main functions revolve around the expressions of emotion and behavior and the integration of recent memory. The expression of emotions and behavior include the following:

1. Visceral responses to moods, such as changes in heart rate, blood pressure, gastrointestinal tract secretions, and the level of many hormones in the blood
2. Adding a quality of pleasure or disagreeable to sensations, especially olfactory sensations
3. Grooming activities
4. Goal-directed activities involving defense (snarling, clawing) and the acquisition of food (sniffing, licking, chewing, and swallowing)

Many of these responses have been seen in animals and in persons who have undergone experimental electrical stimulation on different limbic structures. For example, electrical stimulation of the cingulate or hippocampal gyri produce generalized arousal and autonomic changes in the respiratory, cardiovascular, and digestive systems. Stimulation of the amygdaloid nucleus produces either aggression, including fits of rage, or friendly behavior. Electrical stimulation to other limbic areas may evoke feelings of well-being,

gaiety, and freeness or feelings of fear or terror, depending on the area stimulated. It should be pointed out, however, that because of the extensive interconnections between the limbic components it is difficult to ascribe one function to only one limbic structure.

The clinical signs of limbic system lesions manifest themselves in a variety of emotional and behavioral reactions. These reactions can be seen after certain surgical procedures are performed on patients with socially unacceptable behavior patterns such as overaggression or anxiety reactions. These procedures include the following:

1. Frontal lobotomy, in which the frontal lobes are severed from the rest of the brain just in front of the lateral ventricles
2. Frontal leukotomy, in which only the fiber tracts of the frontal lobe (just anterior to the lateral ventricle) are severed (this interrupts the thalamofrontal projections)
3. Topectomy, in which only the cortex of the frontal lobe anterior to the lateral ventricles is removed (this leaves the motor and premotor areas)
4. Anterior temporal lobotomy, in which the amygdala, uncus, and anterior part of the parahippocampal gyrus are removed
5. Destruction of the fiber tract from the cingulate gyrus to the neocortex

The effects of these procedures are best seen when the surgery is performed bilaterally. In general there is a decrease in emotional responsiveness, but results vary somewhat with the different procedures.

Frontal lobotomy and leukotomy and topectomy additionally produce (1) decreased attention span, (2) unconcern about certain socially accepted patterns of behavior such as personal appearance, (3) less emotionally driven motor activity, and (4) less inhibition about emotional feelings.

The anterior temporal lobotomy produces a syndrome described by Klüver and Bucy that includes (1) visual agnosia (the patient may not recognize the faces of friends or places, (2) decreased emotional responsiveness such as anger or fear, (3) increased sexual activity, (4) increased manipulation of objects, especially orally, and (5) decreased motivation.

Destruction of the fiber tract primarily reduces agitated behavior and incapacitating anxiety.

Of interest is the fact that many of the tranquilizers and mood elevating drugs have similar effects on emotional responses as the psychosurgical procedures. These drugs act as antagonists of the neurotransmitters (the catecholamines such as dopamine, norepinepherine, and serotonin) used by the limbic system.

The other primary function of the limbic system concerns the integration of recent memory into retrievable memory circuits (engrams). This function seems to be a major function of the hippocampal formation. Patients with extensive damage to the hippocampal area have difficulty remembering recent events, especially emotionally charged events. Also, changes in recorded electrical activity form the hippocampus ahve been reported in learning situations in animals.

It is obvious that the limbic system plays an important role in our daily lives. The limbic system maintains a homeostasis of moods and behavior, assists in the integration of what is to be remembered, and provides emotions of pleasure, anger, and fear. As with the other CNS systems the limbic system performs its duties through complex reciprocal and feedback circuits. Any interruption of these circuits can produce noticeable changes in a person's overall behavior. It is importnat to remember that many patients with brain lesions may have temper tantrums, euphoric moods, hostility, lethargy, poor social behavior, problems remember-ing, or other limbic responses. The therapist must be prepared for these responses, as they can occur suddenly and may change over a period of time.

MUSCULAR ACTIVITIES

Most patients seen by the therapist have some problem involving movement. This includes patients with orthopedic and medical-surgical conditions as well as those with neurologic involvement. For this reason and because motion disorders are particularly acute in neurologicaly impaired patients, a solid knowledge of the mechanisms of motor behavior is essential.

Background activity

The somatic muscles of the body are seldom in a state of complete rest. Even during sleep there is an asynchronous discharge of some of the alpha motoneurons, resulting in what is known as tonus. This discharge is the result of all of the segmental (muscle spindle) and suprasegmental (reticular formation) influences converging on the alpha motoneuron.

Muscle tone is vital to any contraction, as it acts to keep the contractile elements ready (warmed up) to produce tension. When the dorsal roots are cut the muscles become very hypotonic. Movements can occur, but they require great conscious effort.

A muscle tone represents a summation of the EPSPs and IPSPs on the alpha motoneurons and thus vary depending on the following:

1. Degree of alertness of the person
2. Amount of stretch on the muscle
3. Descending output from the suprasegmental motor centers, especially the vestibular nuclei and reticular formation, on alpha and gamma motoneurons
4. Muscle function (postural muscles will have more tone)
5. Influences from the tonic and righting reflexes

The tone of a muscle is best assessed by palpation. A normal muscle at rest has a certain amount of firmness when gently squeezed. It is only with practice that hypo- and hypertonia can be distinguished from normal tone. It is helpful in assessing tone to use a standard position for all patients (e.g., sitting, no head or trunk rotation, and the muscle as relaxed as possible at a standard length).

Synergistic movements

A person with an intact nervous system is able to produce isolated movements of individual muscles. However, during many activities a pattern of movement occurs in which individual muscles contract at just the right moment, usually in a sequential order. Usually the whole limb is involved, and components of flexion-extension, abduction-adduction, and external-internal rotation are incorporated into the appropriate joint. Thus when throwing a ball the shoulder extends, adducts, and internally rotates, the elbow extends, the forearm pronates, the wrist flexes, and the fingers extend. The basic components of synergies are laid down during development as neural circuits in the spinal cord and brainstem. Poking a baby's foot with a pin will excite the circuit for a flexion synergy. This produces flexion, abduction, and external rotation of the hip, knee flexion, dorsiflexion, and toe extension. Pressure on the bottom of the foot will excite the extensor synergy circuit, producing the opposite movements. In the upper limb the flexor synergy is flexion, abduction, external rotation of the shoulder, elbow flexion, supination, wrist flexion, and finger-thumb flexion. The extension synergy again is the reverse. These synergies are very common after CVA, in which the damage to the motor cortex or its efferent pathways releases lower centers from cortical control. Of interest, many patients with spinal cord injury demonstrate adduction rather than abduction with the flexor synergy. This is especially obvious during the withdrawal reflex. Another deviation from the basic synergies is the protective movements that occur with loss of balance. These involve extension and abduction.

Variations in the basic synergies are important for functional movements. During development in an infant components from the basic synergies are combined to produce useful movements. Thus we walk on an extended and abducted hip and we throw with an extended arm and a flexed wrist. Lesions of the neocerebellum or its neural connections interrupt the timing of these components, resulting in awkward, jerky, and clumsy movements as the motor cortices (areas 4 and 6 and the supplemental motor area) attempt to produce coordinated movement patterns on their own. The influences of the neocerebellum are transmitted to the motoneurons via the motor cortices. However, lesions of the motor cortices do not produce cerebellar symptoms, since the primary source of voluntary movement is also interrupted. Thus when a patient with CVA begins to move, the remaining suprasegmental centers succeed only in exciting the basic synergy circuits. Since the patient generally has more tone in the antigravity muscles, the flexor synergy in the upper extremity and extensor synergy in the lower extremity tend to be more prevalent.

Reflex movements

There are two basic types of movements: those that are generated within the CNS and can be altered and those that occur as a result of afferent input and are stereotyped. The former are voluntary movements, and the latter are reflexes. Movements such as walking and association movements are combinations of willed and reflex movement and are discussed later, as are the pathologic reflexes such as Babinski's sign.

Different kinds of reflexes can be found at different levels of the CNS.

282 *Neurosciences for allied health therapies*

Spinal reflexes. The spinal reflexes, which include the myotatic, withdrawal, crossed-extensor, and extensor thrust, are discussed in Chapter 6.

Lower brainstem reflexes. These reflexes are known as the *tonic reflexes*. The receptors for these reflexes are in the labyrinth, cervical spine, neck muscles, and thoracic spine and muscles. These afferents excite neurons in the medulla that descend in the spinal cord to excite varying combinations of the spinal circuits (Fig. 9-14).

TONIC LABYRINTHINE REFLEX. The position of the otoliths in relation to their hair cells alters the amount of tone in the extensors. Maximum extensor tone is seen be-

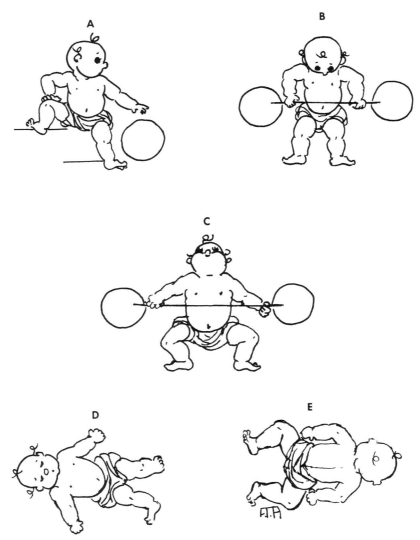

Fig. 9-14. Demonstration of tonic reflexes. **A**, Asymmetric tonic neck. **B** and **C**, Symmetric tonic neck. **D** and **E**, Tonic labyrinthine. Tonic lumbar reflex is not shown.

tween the inverted position and the head angled down 45° below horizontal. Minimum extensor tone is seen between the upright position and the head angled up 45° above horizontal.

SYMMETRIC TONIC NECK REFLEX. Extension of the head and neck facilitates the upper extremity extensors and the lower extremity flexors (like a cat watching a bird). Flexion of the head and neck facilitates upper extremity flexion and lower extremity extension (like a cat ready to jump down from a table).

ASYMMETRIC TONIC NECK REFLEX. Rotation of the head facilitates the extensors in the limbs on the jaw side and facilitates the flexors in the limbs on the occipital side (as in a fencing position).

TONIC LUMBAR REFLEX. Rotation of the trunk to one side facilitates ipsilateral upper extremity flexion and lower extremity extension. The contralateral effects are the opposite. (Take a few steps and stop in midstance on one foot. Notice the position of your arms and legs in relation to the trunk. This simulates the tonic lumbar reflex.)

The tonic reflex circuits are established during the development of the neural tube. They are often present in babies up to about 3 months of age, although it is not unusual to see them weakly present in adults. During the development of the infant nervous system these reflexes normally become integrated into more functional movement patterns.

These basic circuits are retained, however, and some therapists have found them to be useful sources of tone for flaccid muscles resulting from a brain lesion. It is important to mention that when testing for the presence of the tonic reflexes the strongest effect is seen when the person initiates head or trunk movement, as opposed to passive movement by the therapist. Also of importance is that these reflexes may not manifest themselves in actual movement but merely alter the

amount of EPSPs on the appropriate alpha motoneurons. Thus having a patient who has suffered a CVA look toward the involved side not only adds important visual cues but also adds EPSPs to the triceps. This raises the resting potential of the alpha motoneurons closer to the critical firing level. Now a conscious effort from the lesioned hemisphere may be enough to produce some movement.

Upper brainstem reflexes. These reflexes are involved with orienting the body to an upright position and with displacement reactions. The receptors are in the neck, cutaneous receptors in the body, and the labyrinth. Afferents from these receptors excite circuits in the diencephalon, midbrain, and pons, which descend to produce the following reflexes.

NECK RIGHTING REFLEX. Quick rotation of the head causes the body to rotate in the same direction in an attempt to maintain alignment. (It is this reflex in steers that cowboys use to throw them to the ground.)

BODY RIGHTING REFLEX. Asymmetric contact of the body (side lying) facilitates turning of the head and trunk to bring a person to a prone or all-fours position.

LABYRINTHINE RIGHTING REFLEX. Regardless of the position of the body, the otoliths act to maintain the head in a vertical position with the eyes facing the horizon.

DISPLACEMENT REACTIONS. Displacement reactions involve an ability to maintain an erect posture when pushed off balance. The earliest displacement responses are seen with the inclusion of the subthalamus. These responses include crude corrections to lateral displacement.

The purpose of these righting reflexes is to bring the infant to an all-fours position with the head extended. They appear progressively during the first 8 months and reach their peak strength between 10 and 12 months. At this age it is almost impossible for a healthy infant to remain lying down. The appearance of these righting reflexes is an important developmental step, as they indicate the integration of the tonic reflexes into more useful movements and are the key to forward progression.

The purpose of the displacement reactions is to make forward progression more functional by adding a component of equilibrium. The basal nuclei provide a more precise response to displacement.

Cerebral cortex reflexes. These reflexes are involved with optical righting and equilibrium reactions. The ability to keep the head and body righted is primarily derived from the visual system. The vestibular systems and dorsal columns provide additional useful data, but loss of optical input is more serious than loss of either of the two. At about 7 months of age the infant is able to integrate optical data in the visual cortex, which in turn produce postural adjustments to keep the eyes horizontal.

Technically the equilibrium reactions are not a function of the cerebral cortex. They are instead regulated by the vestibular system and cerebellum with help from the basal nuclei. Afferents from the vestibular nuclei and joint receptors project to the primary sensory cortex, but there are no cortical equilibrium centers. The rationale for including equilibrium reactions as a cortical function is that when the motor cortices are lesioned vestibular and cerebellar data cannot by themselves produce corrective movements. The point is, the presence of equilibrium reactions indicate the integration of the cerebral cortex into the reflex machinery. These reactions do not usually appear until 6 months of age and along with optical righting allow the infant to assume bipedal locomotion. Any stimulus that acts to accelerate the body or head out of alignment is detected by the semicircular canals, otoliths, and retina. Appropriate postural adjustments are then made for correction of alignment of protection from falling. These adjustments include the following:

1. Rotation of the head and eyes in the direction of the acceleration
2. Activation of those trunk muscles needed to pull the body away from the fall
3. Extension and abduction of the limbs to act either as counterweights for balance or shock absorbers for protection

Eliciting these reactions in some patients with CNS disorders can be a useful treatment technique for activating weak, nonfunctioning muscles.

Centrally programmed movements

Centrally programmed movements are rhythmic patterns of movement, such as walking, that can occur without any outside stimulus or conscious monitoring, yet can be altered voluntarily by the motor cortices in terms of rate and amplitude. The basis of these movement patterns is probably a complex self-reexciting chain of neurons that is able to maintain an alternating rhythm of excitation.

The probable location of these circuits has been attributed to the basal nuclei. This supposition is based on studies that have shown that a dog or monkey with its brainstem transected just above the midbrain cannot ambulate even though it can right itself. A decorticate animal (only the cerebral cortex is removed), however, is able to ambulate to some degree. It is also known that electrical stimulation of the subthalamus, caudate, and globus pallidus can produce crude forward progression movements.

The cerebral cortex is also important for the manifestation of these circuits, just as it is for the cerebellar and equilibrium reactions. Patients with basal nuclei lesions have difficulty initiating movements (Parkinson's disease) or have involuntary movements (athetoid). In both instances there is an apparent loss of these centrally programmed circuits. The motor cortices can overcome these to a degree and produce locomotion (unless there is severe involvement). However, lesions of the motor cortices or their descending projections often severely impair ambulation.

It is interesting that these programmed rhythmic circuits are partly inborn and partly learned. Most animals can walk after birth, and even human babies have reflex stepping movements. Much practice is necessary, however, to make these circuits efficient producers of movement.

Voluntary movements

The production of willed movements is complex. Although many areas of the cerebral cortex contribute to the production of movement, it is generally accepted that the motor cortices are the primary output source to the other motor centers and spinal motoneurons. Once a program of movement is initiated, it can be carried out by centrally programmed or reflex circuits involving these centers and directed to the spinal cord via the descending tracts. As mentioned, the motor cortices do not act in isolation. Most motor acts involve the attainment of some goal either in the present sensory field or incorporated into the individual's memory. Thus the initiation of movement involves input from many sources to the motor cortices. These sources include the following:

1. Sensory association areas for interpretation of ongoing sensory activity and recall of prior experiences
2. Frontal lobe for the reason as to why a particular movement should occur
3. Limbic system to add an emotional-motivational force to the movement
4. Thalamic projections that maintain a level of alertness to allow motor response to occur

Although voluntary movements can occur with lesions of the cerebral cortex, they are sluggish and often nonfunctional. It is through the motor cortices that we can attain the marvelous alternations in rate and amplitude of contractions.

Association movements

There are two kinds of association movements. The first are movements that normally occur as part of a total program of movement but are not necessary for the production of that movement (e.g., the arm swing during gait, raising the arms during a yawn, and facial movements associated with emotional expression).

Although these associated movements occur without conscious effort, they can willfully be inhibited and thus are not reflex responses. There is clinical evidence to support that their origin is probably in the basal nuclei. Patients with a damaged cortex from CVA often demonstrate flexion and abduction of the arms while yawning, yet are unable to produce these movements willfully. Patients with Parkinson's disease, however, frequently lack these associated movements. For example, they can voluntarily move their facial muscles but generally show little emotional reaction on their faces (deadpan faces).

The second type of associated movements involve an overflow of neural impulses within the spinal cord. An attempt to flex or extend a limb at one joint elicits the flexion or extension synergies in that same limb as well as exciting a synergy circuit on the contralateral side. This contralateral overflow generally behaves in the following manner:

1. Resisted flexion or abduction in an upper limb elicits the flexor synergy in the other upper limb. Resisted extension or adduction elicits the extensor synergy in the other upper limb.
2. Resisted flexion in the lower limb elicits the extension synergy in the other lower limb. Resisted extension elicits the flexion synergy in the other lower limb.
3. Resisted abduction of one lower limb elicits abduction in the other lower limb. The same is true for adduction. This is known as *Raimiste's sign*.
4. In homolateral synergies involving

both extremities on one side, resisted flexion of one extremity elicits flexion in the other extremity.

These associated movements can be seen in healthy people. Squeezing a dynamometer not only produces elbow flexion on the same side but usually causes the other hand to close. EMG studies have estimated this overflow of neural impulses to the contralateral synergy circuits to be approximately 15%. In fact, it is known that a casted extremity will have less atrophy if the uninvolved limb receives heavy resistive exercises.

These overflow-associated reactions are used by some therapists in the treatment of patients with CVA or head injury. Without the normal cortical inhibition of these circuits they sometimes can be elicited more easily; and while these movements are nonfunctional (i.e., it is hard to dress or feed oneself using synergies), they can be useful in providing tone to a flaccid muscle and as a motivational force so that the involved limb is not completely paralyzed.

Two points are important. The first is that the overflow synergies do not manifest themselves in terms of full range or with much force. Thus strong elbow flexion of the uninvolved arms of a patient with CVA may produce only slight shoulder abduction and arm flexion on the involved side. However, there will be facilatory impulses on the appropriate motoneurons for the entire flexor synergy. As with the tonic reflexes, these EPSPs can be useful additions in producing voluntary movements.

The second point is that sometimes these overflow movements appear with the wrong stimulus. For example, strong extension of the uninvolved arm of a patient with CVA sometimes elicits components of the flexor synergy in the other arm, while a strong quadriceps contraction of the uninvolved leg may elicit extension of the other leg. The reason for this is unknown.

ORGANIZATION OF MOTONEURONS AND PROJECTIONS TO THEM
Final common pathway

The alpha motoneuron and its axon are collectively known as the *final common pathway* and the *lower motoneuron*. This is an accurate description considering the following variety of influences that converge on the alpha motoneuron and determine its frequency of firing:

1. Spinal influence: gamma loop, Renshaw cells, FRAs, and pressure receptors for extensor thrust
2. Tonic and righting reflexes
3. Reticular formation and brainstem nuclei
4. Cerebellum, basal nuclei, and thalamus
5. Cerebral cortex

All of these influences affect the level of polarization of the resting membrane of the alpha motoneurons through the summation of EPSPs and IPSPs. The end result is the contraction or relaxation of somatic muscles. The role of the therapist is to adjust this level of polarization on the appropriate alpha motoneurons to excite those muscles needed for functional movement and to inhibit those muscles that interfere with functional movements.

Descending tracts

The converging data from the higher centers on the alpha motoneurons are primarily delivered via five descending tracts that collectively represent the upper motoneurons. The axons in these tracts synapse either directly on the motoneurons or on internuncials, which in turn synapse on the motoneurons. These tracts are diagrammed in Figs. 9-15 and 9-16 and described below; see also Fig. 6-1 for the position of these tracts in the spinal cord.

Corticobulbospinal tract. This tract is often referred to as the corticospinal or pyramidal tract. About 40% of these axons origi-

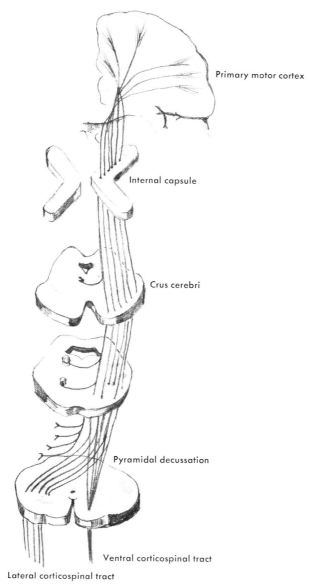

Primary motor cortex

Internal capsule

Crus cerebri

Pyramidal decussation

Ventral corticospinal tract

Lateral corticospinal tract

Fig. 9-15. Descending corticobulbospinal tract. Note that fibers for leg (medial surface) assume more dorsal position in internal capsule, while those for face (lateral surface) assume anterior position. In medulla, note that most fibers decussate to form lateral corticospinal tract. Keep in mind that other cortical areas contribute fibers to this tract. It should be mentioned that technically term *bulbar* refers to structures derived from rhombencephalon. However, cortical fibers to cranial nerves III and IV are often called corticobulbar fibers also. (Adapted from Netter, F.: The CIBA collection of medical illustrations. I. The nervous system, Summit, N.J., 1972, CIBA Chemical Co.)

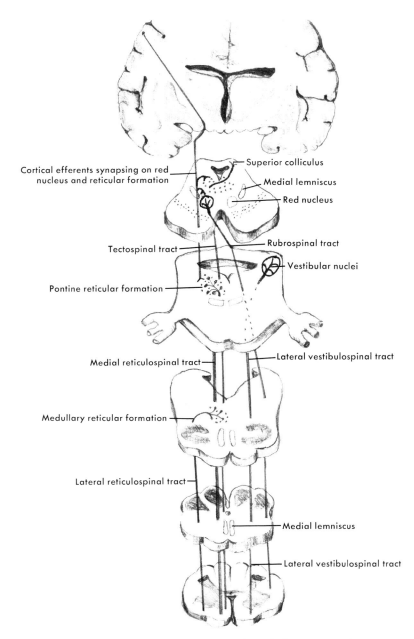

Cortical efferents synapsing on red nucleus and reticular formation

Superior colliculus

Medial lemniscus

Red nucleus

Tectospinal tract

Rubrospinal tract

Vestibular nuclei

Pontine reticular formation

Medial reticulospinal tract

Lateral vestibulospinal tract

Medullary reticular formation

Lateral reticulospinal tract

Medial lemniscus

Lateral vestibulospinal tract

Fig. 9-16. Major extrapyramidal pathways to spinal cord. Interconnecting synapses within brainstem have been omitted along with medial longitudinal fasciculus and its spinal component, medial vestibulospinal tract. Note that rubrospinal tract is located near lateral corticospinal tract.

nate in areas 4 and 6, 20% from areas 3, 1, and 2, and the remaining 40% from area 8 and the sensory association areas. Obviously the primary motor cortex is not the sole producer of movement. In fact, the giant Betz cells in area 4, which were once thought to be the primary source of voluntary movement, actually contribute only 3% or 4% of the 1 million fibers in this tract. As axons leave their respective cortical areas they descend through the white matter of the coronal radiata and converge to form the genu and anterior portion of the posterior limb of the internal capsule. As they continue to descend through the brainstem the bulbar fibers leave the tract and cross to the opposite side to innervate the motor nuclei of the cranial nerves. In the medulla this tract contains primarily corticospinal fibers, which then form the pyramids. In the caudal medulla 85% to 90% of the corticospinal fibers decussate to form the *lateral corticospinal tract.* This tract descends in the lateral funiculus of the spinal cord, giving off axons to innervate appropriate internuncials and motoneurons in the ventral horn. Note in Fig. 9-15 that the corticobulbospinal fibers are arranged in a very organized fashion (e.g., the bulbar fibers are in the genu of the internal capsule and the corticospinal fibers for the leg are lateral to those for the arm). This arrangement is important, as certain lesions such as tumors may involve only a portion of this tract, creating an unexpected set of symptoms.

The remaining 10% to 15% of the corticospinal fibers descend uncrossed in the ventral funiculus as the *ventral corticospinal tract.* These axons may terminate ipsilaterally or cross in the commissures to the opposite side. Of interest, many patients with CVA show very mild motor deficit on the "uninvolved" side. However, this may also result from bilateral hemispheric control.

The function of this tract is to produce voluntary movement. The bulbar fibers are involved with the motor nuclei associated with

cranial nerves III, IV, V, VI, VII, IX, XI, and XII. Conscious movements of the eyes, face, jaw, and neck are handled by these fibers. The spinal fibers are primarily concerned with skilled movements of the distal muscles. Although all somatic muscles are innervated by these fibers, the largest percentage is to the distal flexor muscles (lamina IX).

The remaining efferent cortical fibers involved with the motor system descend through the internal capsule to the other suprasegmental motor centers. Since these fibers do not pass through the pyramids of the medulla, they are considered part of the extrapyramidal motor system. These fibers are basically involved with complex feedback circuits involving the basal nuclei, cerebellum, red nucleus, reticular formation, and thalamus.

Reticulospinal tracts. There are two tracts in each half of the brainstem. The *lateral reticulospinal tracts* originate in the medulla and descend in the cord in the ventral funiculus. When stimulated this tract inhibits reflexes, extensor tone, and cortically induced movements. The *medial tracts* originate in the pons and descend in the cord in the ventral funiculus. When stimulated the effect is the opposite of that seen with the lateral tract. The importance of these tracts is apparent, considering the large afferent input from areas 4 and 6, and the cerebellum, and the sensory systems to the RF. Thus a good part of the motor programming is delivered to motoneurons via the reticulospinal tracts. This is especially true for proximal and trunk muscles. In fact, experimental pyramidotomy on monkeys produces only a loss of rapid, agile, and isolated movements. Stereotyped useful movement patterns are retained, but they appear to require more deliberation and attention. Also of importance is that these reticulospinal tracts seem to be a major regulator of gamma motoneuron activity and thus of muscle spindle biasing. This becomes apparent in lesions that interrupt the cortical suppression to the midbrain and pontine reticular formation. This results in spasticity of the physiologic extensors.

It is important to mention that the reticulospinal tracts are also involved with the autonomic nervous system.

Vestibulospinal tracts. There are two pairs of vestibulospinal tracts. The *lateral vestibulospinal tract* arises primarily from the lateral vestibular nucleus (Deiters' nucleus) and descend in the anterolateral part of the spinal cord. This tract receives both vestibular and cerebellar input and is facilatory to extensor muscles for the maintenance of posture and equilibrium. The influence of this tract can be demonstrated with experimental lesioning that results in marked reduction of extensor spasticity in animals with previous intercollicular transection. The *medial vestibulospinal tract* is actually the spinal part of a brainstem reflex tract known as the *medial longitudinal fasciculus (MLF)*. All of the vestibular nuclei contribute axons to this tract, which projects rostrally to the midbrain and caudally to the cervical cord in the anterior funiculus. This tract along with the tectospinal tracts is involved with reflex head, eye, and probably arm movements associated with visual and auditory stimuli. The MLF is also associated with lateral gaze movements, which require coordinated input to cranial nerves III, IV, and VI.

Tectospinal tracts. These tracts originate in the superior colliculi and descend to the cervical cord adjacent to the MLF. Their function has been mentioned above.

Rubrospinal tracts. These fibers originate in the red nucleus, which receives its input from the cerebellum and cerebral cortex. The fibers cross and descend the entire length of the cord in the lateral funiculus adjacent to the lateral corticospinal tract. In spite of its impressive afferent connections, the red nucleus does not seem to be a major component of the motor systems in humans. Stimulation of it or the rubrospinal tract ex-

cites flexor muscles, which may be one of the reasons why many patients with lesions above the red nucleus have spasticity in the arm flexors. Recall that midbrain lesions produce spasticity in all the extensors.

Hierarchical order of motor system

The maturation of the motor system from birth can be thought of as an ascending dominance of one over another. This dominance frequently involves an inhibitory effect on the neural circuits of the lower center. For example, a spinal animal will demonstrate spasticity and hyperreflexia as a result of loss of inhibitory influences from the reticular formation. Evidence of this can be seen in an animal with a transection at the pontomedullary junction. The medullary RF is primarily inhibitory to spinal circuits. This results in flaccidity and hyporeflexia. If the transection is at the intercollicular level the RAS and vestibular nuclei become the dominant motor centers. This causes marked spasticity of the physiologic extensors. The more rostral motor centers operate in a similar fashion. It is important to keep in mind that the neural circuits of each motor center are retained throughout life. When a higher center is lesioned there can be a release of regulatory control, allowing the next lower center to dominate.

Experiments utilizing transections or ablations of different levels of the neuraxis have revealed some interesting functional aspects of the motor system. The basic patterns of movement of forward progression (alternating flexion and extension of diagonally paired limbs) are built into the spinal cord and its propriospinal pathway. Stimulation of either lateral reticulospinal tract can set these patterns in motion. The righting reflexes, which orient the body, head, and eyes to an upright quadriped position, require an intact brainstem up to the diencephalon. These reflexes produce a posture that can utilize the forward progress movements. Stability in progression is handled by the subthalamus and basal nuclei, which provide the necessary responses to correct for displacements of the center of gravity. A decorticate monkey, for example, will right itself from a supine position and move about its cage, although somewhat stiffly. Protection from loss of balance involving extension-abduction of the extremities requires an intact cerebral cortex and its pathways involving the vestibular system. The visual system is not essential to righting or protective reactions; blind persons have these reactions.

The role of the motor cortex seems to concern skilled movements of one extremity. The extraphyramidal component (area 6) influences the basal nuclei to make the appropriate postural adjustment, while the pyramidal system (area 4) controls the fine movements of the distal muscles. Lesions of area 6 result in the dystonic posturing seen in patients with CVA. Lesions of area 4 produce surprisingly mild deficits.

To place the above section into another perspective, the supraspinal motor systems can be organized into the following categories.

Corticobulbospinal system. This system is primarily concerned with skilled activities. There is a greater density innervation of the distal muscles compared with the proximal muscles. Also there is a higher percentage of monosynaptic activity on alpha motoneurons than with the other three systems, accounting for the fine rapid movements produced by this system. Electrical stimulation of area 4 can produce responses ranging from quick flick movements of the thumb or flexion of an entire limb, depending on the site and intensity of stimulation. When the cortical area of an extensor muscle is stimulated a brief hyperpolarization period follows on those alpha motoneurons. With flexor stimulation a brief depolarization period follows. The rationale is probably that a skill such as playing the piano requires the distal flexors

to be constantly ready to contract. Bilateral effects from unilateral stimulation can be seen in the upper facial muscles, the jaw, the swallowing mechanism, and trunk muscles. Lesions of this sytem produce contralateral hemiparesis with loss of fine distal movements. Spasticity usually is present, and there may be alterations of tone. Mass movement patterns of the proximal muscles remains.

Neocerebellar-thalamic-cortical system. This sytem provides a continuous monitoring of the corticobulbospinal system. Proprioceptive vetibular, and somatic sensory data are constantly fed into the neocerebellum to provide corrections for cortically induced movements. Lesions of the neocerebellum produce an inappropriate modulation of these movements, resulting in lack of synergy (ataxia).

Basal nuclei-thalamic-cortical system. This system is poorly understood but seems to be involved with complex repetitive automatic movements such as walking and chewing. Lesions of the basal nuclei produce involuntary movements or rigidity and resting tremor.

Brainstem nuclei system. This system is involved with the regulation of the muscle tone and spinal reflexes, adjustment in posture, and equilibrium. It provides the foundation on which the other three systems can operate. Lesions of the brainstem nuclei have various effects, depending on the severity and location.

Cranial nerve motor components

The emphasis so far in this chapter has been on the regulation of the spinal motoneurons by the suprasegmental motor centers. The motor components of the cranial nerves, however, deserve equal time because a large number of patients with CVA and other neurologic disorders have brainstem involvement. The same suprasegmental motor centers, with the exception of the red nucleus, can also regulate these motor components. It is important to mention, however, that the movements of the eyes, jaw, face, tongue, and head are largely reflex in nature and that the neural pathways are complex.

Nerves III, IV, and VI are responsible for moving the eyeball in its socket, focusing the lens, constricting the pupil, and partly raising the eyelid. The innervation to these nerves is from the frontal and occipital lobes, vestibular nuclei, midbrain tectum and pretectum, and reticular formation. Except when the eyes converge (both medial recti muscles contract) for near vision, they always move in the same direction, such as in watching a Ping-Pong game or someone on a trampoline. This conjugate activity requires very precise simultaneous innervation of all six eye-moving nerves. For example, when looking to the right the right abducent and left oculomotor nerves are active, the right oculomotor and left abducent nerves are inactive, and both trochlear nerves are active. Damage to any of these nerves or their nuclei or a muscle weakness interferes with conjugate gaze and usually produces a constant deviation of one or both eyes from the normal forward gaze position. Such a deviation is known as *strabismus.*

The control center for conjugate gaze is the frontal eye fields (area 8). Impulses from the ipsilateral visual cortex are sent to area 8 and are then projected to the contralateral pons (lateral gaze center) and to the pretectum (vertical gaze center). The impulses from these centers are then relayed to the appropriate cranial nerves via the MLF, which produces a fusion of images from each eye, fixation on a moving object, and maintenance of a horizontal relationship with the visual fields.

Stimulation of area 8 on one side produces contralateral deviation of the eyes (the side the visual cortex sees). Therefore a lesion of area 8 produces deviation of the eyes toward the side of the lesion because of the unopposed action of the other frontal eye field. A unilateral lesion of the dorsomedial tegmental area of the pons involving the lateral gaze center or of the MLF produces an inability to look to the side of the lesion. Bilateral stimulation of the frontal eye fields produces upward gaze. Thus some patients with head injury may be unable to look upward. A similar loss of upward gaze can be found in pineal tumors that compress the pretectal area.

The center for convergence is the visual cortex. Impulses from the retina that reach the visual cortex "out of focus" cause an immediate reflex response to the superior colliculi and then to both oculomotor nerves. The result is accommodation of the lens and constriction of the pupil to sharpen the image and medial deviation of both eyes to retain a fused image. A lesion of the oculomotor nerve or its nucleus not only produces external strabismus but also double vision for close work and contralateral gaze, a constantly dilated pupil, and ptosis as a result of loss of parasympathetic impulses to the levator palpebrae (Weber's syndrome).

Eye movements are also regulated by the vestibular mechanism and proprioceptive data from the eye muscles. This input is primarily involved with keeping the eyes fixed on an object when either the object or the person is moving. When riding in a car and watching a row of telephone poles go by, the eyes will track each pole back as far as it can and then jump forward to follow the next one. This rapid oscillation back and forth is known as *nystagmus.* Nystagmus also occurs normally in any rotatory movement of the head such a falling or spinning around a center point. The uneven stimulation of the semicircular canals produces impulses that are transmitted via the MLF to cranial nerves III and VI. The result is a rapid deviation of the eyes forward alternating with a slower return component. Except under the above circumstances nystagmus is normally absent because the input from each set of semicircular canals is balanced. Any unilateral lesion of the labyrinth, vestibular nerve, vestibular nuclei, or MLF can produce nystagmus. Nystagmus may be present with the eyes at rest or only noticed with lateral or upward gaze. In the clinical assessment of nystagmus the quick component is to the side opposite the lesion.

The facial and hypoglossal nerves provide interesting patterns of innervation. The motor nucleus for the facial nerve is divided into two components. One part innervates the upper half of the face, and the other part innervates the lower half. Fibers from both halves compose the facial nerve, and any lesion of the nerve will produce ipsilateral facial paralysis (Bell's palsy). The cortical innervation to each half, however, is different. The part for the lower facial muscles receives contralateral cortical input, while the part for the upper facial muscles receives bilateral cortical input. Thus any unilateral involvement of the motor cortex or its descending fibers to the facial nucleus results in paralysis of the lower facial muscles only.

The hypoglossal nerve presents an interesting "contralateral" situation. Each nerve when activated alone causes contralateral tongue deviation. Thus the left hemisphere innervates the right nerve and causes the tongue to deviate to the left. Bilateral stimulation of the nerves causes the tongue to stick out straight. However, a lesion of one nerve or its nucleus produces tongue deviation to the side of the lesion, while a lesion of the motor cortex or its descending projections causes the tongue to deviate to the side opposite the lesion. This is usually transient.

The motor control of the jaw is quite simple. Each motor nucleus for the trigeminal nerve receives bilateral cortical input. Thus a lesion of one motor cortex or its projections produces only mild weakness. A unilateral lesion of the trigeminal nerve or its motor nucleus weakens the ability to chew on one

side, although the person will have no major difficulty eating.

Movements of the head are performed by various groups of muscles that have contralateral and ipsilateral effects. The spinal accessory nerve supplies the upper trapezius and sternocleidomastoid, while the other muscles are supplied by the upper cervical nerves. A unilateral lesion of a spinal accessory nerve produces weakness in contralateral head rotation and a tendency for the head to deviate to the side of the lesion at rest. There is also shoulder depression and weakness in arm elevation on the lesioned side. A unilateral lesion of the motor cortex or its projections can produce a contralateral deviation of the head, contralateral shoulder depression, and weakness rotating to the side of the lesion.

TEST QUESTIONS

1. In a patient with spinal cord injury the flexor withdrawal response:
   a. Can be elicited by touching the foot.
   b. Can produce movement in the ankle, knee, and hip.
   c. May be inhibited somewhat by firm maintained contact.
   d. May be useful in ambulation training.
2. Which one(s) of the brainstem nuclei is involved with protective reflexes?
   a. Vestibular nuclei.
   b. Tectum.
   c. Red nucleus.
   d. Substantia nigra.
3. Providing proximal joint stability to allow the distal part of the extremity is a function of:
   a. Cerebellum.
   b. Red nucleus.
   c. Reticular formation.
   d. Basal nuclei.
4. In Parkinson's disease there is:
   a. Bradykinesia, dysmetria, and cogwheel rigidity.
   b. Subthalamic lesion.
   c. Loss of posterior column sensations.
   d. Benefit from isometric exercises.
5. The resting tremor associated with lesions of the basal nuclei:

a. Can be produced with a low dopamine level in the globus pallidus.
b. Will disappear with a lesion of the crus cerebri on the same side as the tremor.
c. Is generally manifested distally.
d. Is referred to as a pill-rolling tremor.

6. A lesion of the neocerebellum on the left produces the following symptom(s) ipsilaterally:
   a. Intention tremor.
   b. Loss of kinesthesia.
   c. Muscle rigidity.
   d. Dysmetria.

7. A transection of the brainstem at the midpons with ablation of the vestibular nuclei produces:
   a. Constant state of sleep.
   b. Hypotonia and flaccidity.
   c. Complete loss of all sensations.
   d. Ataxic gait.

8. In a decorticate patient (the cerebral cortex has been lesioned, but the thalamus and rest of the CNS are intact) there is:
   a. Complete anesthesia.
   b. Marked ataxia and intention tremor.
   c. Maintenance of standing balance.
   d. Homonymous hemianopia.

9. The electroencephalogram:
   a. Is altered in lesions of the internal capsule.
   b. Is altered in lesions of the cerebellum.
   c. Is a function of synaptic activity between cortical neurons.
   d. Varies with age.

10. An early tumor of the falx cerebri in the region of the central sulcus involves:
    a. Area 44 for verbal expression.
    b. Areas 5 and 7 for body awareness.
    c. Area 4 for motor fibers to the arm and face.
    d. Areas 41 and 42 for hearing.

11. In a patient with a head injury to the anterior part of the frontal and temporal lobes you would expect:
    a. Decreased ability to block out interfering extraneous stimuli.
    b. Marked difficulty interpreting somatic sensations.
    c. Damage to the entire hippocampus.
    d. Difficulty remembering instructions.

12. Which influence(s) is involved with muscle tonus?
    a. Renshaw cells,

b. Gamma motoneurons.
c. Cerebral cortex.
d. Head position.

13. The basic synergies of the spinal cord:
    a. Are used in normal activities such as walking.
    b. Are the same movement patterns as the flexor withdrawal and extensor thrust reflexes.
    c. Cannot be elicited in a patient with T6 spinal cord injury.
    d. Are present in patients with a basal nuclei lesion.

14. To increase the tone in the left triceps you could:
    a. Turn the head to the right.
    b. Lie supine.
    c. Lean to the left from a sitting position.
    d. Look toward the ceiling.

15. A child with cerebral palsy dominated by the tonic reflexes:
    a. Is able to assume a sitting position but requires support.
    b. Has difficulty bringing objects into the mouth.
    c. Is able to crawl normally.
    d. Is able to walk because of the tonic lumbar reflex.

16. In a patient with severe head injury with opisthotonus there would be:
    a. Reduced response to FRA stimulation.
    b. Crude walking movements.
    c. Decrease in extensor tone in the prone position.
    d. Basal ganglia symptoms of cogwheel rigidity and resting tremor.

17. A tumor pressing on the lateral side of the cervical cord:
    a. Reduces facilitation to flexor muscles.
    b. Disrupts vestibulospinal and tectospinal influences.
    c. Initially involves the lower extremities.
    d. Disrupts 85% of the corticospinal fibers.

18. A lesion of the right side of the pons produces:
    a. Left hemiparesis and hemianesthesia.
    b. Deviation of the eyes to the right.
    c. Bilateral internal strabismus.
    d. Nystagmus.

19. A lesion of the left internal capsule results in:
    a. Contralateral hemiplegia.

b. Contralateral paralysis of the lower facial muscles.
c. Deviation of the tongue to the right.
d. Rotation of the head to the left.

SUGGESTED READINGS

Barr, M.: The human nervous system, New York, 1972, Harper & Row, Publishers.

Carpenter, M.: Core text of neuroanatomy, ed. 2, Baltimore, 1978, The Williams & Wilkins Co.

Chusid, J. G.: Correlative neuroanatomy and functional neurology, Los Altos, Calif., 1976, Lange Medical Publications.

Netter, F.: The CIBA collection of medical illustrations. I. The nervous system, Summit, N. J., 1972, CIBA Chemical Co.

Noback, C., and Demarest, W.: The human nervous system, ed. 2, New York, 1975, McGraw-Hill Book Co.

Willis, W. D., Jr., and Grossman, R. G.: Medical neurobiology: neuroanatomical and neurophysiological principles basic to clinical neuroscience, ed., 2, St. Louis, 1977, The C. V. Mosby Co.

Yahr, M., and Purpura, D., editors: Neurophysiological basis of normal and abnormal motor activities, New York, 1967, Raven Press.

CHAPTER 10

Neurology

In this chapter material presented in the preceding chapters is related to patients with neurologic disorders. The emphasis of this chapter is *not* a discussion of the etiology, pathology, and so on, of all of the known neurologic disorders. This not only would be repetitious of the many well-established neurology texts but would also require the addition of many hundreds of pages.

What this chapter presents is the process of a neurologic evaluation of a patient, specific neurologic tests used in the diagnosis, and an analysis of the more common neurologic disorders. This analysis emphasizes the anatomic site of the lesion, anticipated symptoms, and the related findings from the specific neurologic tests.

Highlighting this chapter are 15 case studies that will require you to determine the site of the lesion from a given set of symptoms. The ability to perform this task will improve tremendously with experience. By knowing the site of a lesion (and in essence the symptoms to expect), the therapist can be more effective in the evaluation and treatment of patients with neurologic disorders.

Restated, the areas of emphasis in this chapter are the following:

1. Components that are assessed in the nine subdivisions of the neurologic examination
2. Site of lesion from a given set of symptoms
3. Expected set of symptoms for a given lesioned area of the nervous system
4. Probable results of neurologic tests for a given neurologic disorder

METHOD OF NEUROLOGIC DIAGNOSIS

The data collected during the evaluation of a neurological disorder are vital not only to the diagnosis but to determining the specific area of involvement. This knowledge can then be used to select an appropriate and effective program of treatment. The physician is the primary initiator and collector of this medical information. On initial contact with a new patient the physician (1) records the patient's history, (2) performs a general physical examination, and (3) performs a specific examination of the components of the nervous system. The specialized skills of the physical, occupational, and speech therapists, nursing personnel, and social worker may be called on by the physician to assist in this data collection. To better visualize the therapist's role in this information-gathering process a detailed look at the history, physical examination, and neurologic examination is presented.

History

With the overall goal of determining the diagnosis and site of involvement the first step is generally to identify the chief complaint of the patient. In other words, what is causing the patient to seek medical help. Some chief complaints are obvious, such as paralysis of the facial muscles on one side of the face. Since the area of involvement is limited to the facial nerve or its motor nucleus, a diagnosis of Bell's palsy can quickly be made an an effective treatment program initiated. Other complaints such as numbness of an arm or decreased deep-tendon reflexes are less specific as to the site of involvement, and further examination is needed.

With the chief complaint noted, a series of questions is asked to determine the characteristics of the present illness. For example:

How long has the patient had this problem?

Was the onset sudden or insidious?

Are the symptoms progressing?

Are the symptoms present all of the time or only part of the time?

The final part of the history taking deals with the patient's family and social background and previous medical history:

Has any other member of the family had similar or related symptoms?

Has there been any serious disease such as cancer, diabetes, or CVA in the family history?

What is the patient's racial and cultural background?

This information from the patient's history can provide important clues as to the nature of the disorder. For example, some disorders such as tumors have a slow onset, while many vascular lesions can have a sudden onset. Diseases such as Friedreich's ataxia have a definite genetic linkage, while others such as cerebrovascular accidents (CVA) can be found in other family members or the patient may be the first to have one. Multiple sclerosis is often linked culturally to some groups such as those of Northern European background whose diet has a high percentage of saturated fats.

Physical examination

Like the history taking, a general examination of the body can provide further clues as to the nature of the disorder. An external examination of the skull, neck, spine, and extremities is performed to reveal any areas of laceration, swelling, tenderness or muscle spasm and general mobility. The color and temperature of the skin are noted as well as any areas of loss of hair. Internally the cardiac, respiratory, digestive, and genitourinary systems are evaluated. Detailed analysis of blood and urine samples is usually necessary. In some instances pulmonary function and cardiac stress tests may be indicated. The results from the physical examination often are a determining factor in deciding whether to perform a specific neurologic examination. For example, a chief complaint of malaise and melancholy could indicate a problem with the liver or pancreas or with the frontal lobes or limbic system.

Neurologic examinations

If the results from the history and physical examination indicate a neurologic disorder or if the results are insufficient to indicate anything, a neurologic examination is performed.

The neurologic examination is usually broken down into nine sections that encompass most of the functional activities of the nervous system (see box on p. 298):

1. Cerebral function
2. Cranial nerves
3. Cerebellum
4. Motor systems
5. Sensory systems
6. Reflexes
7. Gait
8. Autonomic nervous system
9. Miscellaneous findings

The examination can begin with any of the sections, and the organization may be rearranged to suit the needs of individual examiners. For example, reflexes may be included

NEUROLOGIC EXAMINATION

1. Cerebral function
   a. General behavior and cooperation
   b. Level of consciousness
   c. Intellectual functioning (orientation, memory, insight)
   d. Emotional reactions
   e. Sensory integration (agnosia)
   f. Motor integration (apraxia)
   g. Language functioning (aphasia, alexia, dysarthria)
2. Cranial nerves (right, left)
   a. Olfactory
   b. Optic
   c. Oculomotor, trochlear, abducent
      (1) Strabismus
      (2) Ptosis
      (3) Pupils
      (4) Convergence
      (5) Conjugate movements
      (6) Nystagmus
   d. Trigeminal
      (1) Motor
      (2) Sensory
   e. Facial
      (1) Motor
      (2) Sensory
   f. Vestibulocochlear
      (1) Hearing
      (2) Balance
   g. Glossopharyngeal, vagus
      (1) Swallowing
      (2) Phonation
      (3) Gag reflex
   h. Spinal accessory
   i. Hypoglossal
3. Cerebellar function (extremities and trunk)
   a. Dysmetria
   b. Adiadochokinesia
   c. Intention tremor
   d. Rebound
4. Motor function (extremities, trunk, neck)
   a. Muscle status
      (1) Tone
      (2) Atrophy
      (3) Fasciculations
      (4) Tenderness
   b. Strength
      (1) Paralysis
      (2) Paresis
      (3) Synergies
   c. Gross motor
      (1) Transfers
      (2) ADL
      (3) Self-mobilization
      (4) Contractures
   d. Fine motor
      (1) Writing
      (2) Handicrafts, dexterity
5. Sensory function (body and face, right and left)
   a. Somatic function
      (1) Light tough
      (2) Pain/temperature
      (3) Discriminative touch, kinesthesia, vibration
   b. Paresthesias
6. Reflexes
   a. Myotatic (right, left)
      (1) Biceps (C5-C6)
      (2) Triceps (C7-C8)
      (3) Quadriceps (L2-L4)
      (4) Achilles (L5-S2)
      (5) Jaw (V)
   b. Resistance to stretch
      (1) Flaccid
      (2) Spastic
      (3) Rigid
   c. Superficial
      (1) Corneal
      (2) Gag
      (3) Abdominal
      (4) Plantar
7. Gait
   a. Ataxic
   b. Synergy
   c. Uneven stride
   d. Uneven timing
   e. Assistive devices
8. Autonomic system function
   a. Light reflex
   b. Skin temperature and color
   c. Tropic changes
      (1) Nails
      (2) Hair loss
   d. Bowel and bladder
9. Miscellaneous findings
   a. Babinski's sign
   b. Hoffman's sign
   c. Clonus
   d. Kernig's sign
   e. Brudzinski's sign
   f. Romberg's sign
   g. Resting tremor
   h. Cogwheel rigidity
   i. Opisthotonus

in the motor or sensory systems. Some sections may be omitted. For example, a patient may demonstrate normal mental abilities during the history taking and physical examination, or observation of the face may indicate there is no need to test the cranial nerves. If a complete examination is indicated, however, assessing the level of cerebral functioning is often performed first.

Much of the data in this first section will indicate the functional ability of the different cortical area. The major cerebral functions to assess include the following:

1. General behavior. Is the patient aggressive, hostile, cooperative, nervous, etc.?
2. Level of consciousness. There are four basic levels of consciousness:
   a. Normal: the person is alert, eyes are open, response to verbal and painful stimuli, normal voluntary and reflex movements.
   b. Lethargy: the person is drowsy, eyes may be closed, decreased response to stimuli, decreased speed and coordination of movement.
   c. Stupor: the person appears asleep, eyes generally are closed, no response to verbal stimuli, may vocalize in response to painful stimuli, no voluntary movements but mass flexor withdrawal to pain, decreased spontaneous eye movements.
   d. Coma: no vocalization, decerebrate posturing to pain or no response, eyes closed, no spontaneous eye movements.
      It is important to remember that a patient's level of consciousness may alter during the day and from day to day.
3. Intellectual functioning.
   a. Is the patient oriented as to time and place?
   b. Can the patient recall a series of five or six numbers (recent memory) and several major historical events (long-term memory)?
   c. Can the patient show insight into statements such as "a stitch in time saves nine"?
4. Emotional reactions.
   a. Does the patient show outbursts of emotions inappropriately such as rage, crying, laughing?
   b. Is there a lack of normal emotions such as anger or depression?

5. Sensory integration. Can the patient identify an object by touch (with the eyes closed), by vision alone, and by hearing the sound an object makes? Ability to detect that a stimulus is present but inability to identify it or determine its usefulness is known as agnosia. For example, inability to name an object from its picture is known as verbal (visual) agnosia.
6. Motor integration. Can the patient perform a simple skill such as tying a knot or opening a sealed envelope? Inability to do this is called apraxia. Apraxia is not paralysis or weakness of a muscle; rather it indicates difficulty in knowing how to use a muscle group in a functional manner.
7. Language skills.
   a. Does the patient speak clearly and at the proper speed?
   b. Are the words in the proper order and appropriate in meaning?
   c. Can the patient understand spoken words? Difficulty in verbal expression or comprehension is known as asphasia. Difficulty in articulating words is known as dysarthria.
   d. Can the patient comprehend written material and mate an object to its written name: Difficulty in this area is known as alexia.

The second section of the examination deals with the cranial nerves. This area should receive special attention whenever a brainstem lesion is suspected.

1. Olfactory. Can the patient detect aromas such as coffee or tobacco? Loss of the sense of smell is known as anosmia and can occur with lesions of the medioinferior region of the frontal lobe or the anteriomedial part of the temporal lobe (limbic system).
2. Optic. Is there complete vision in each eye, or are parts of the visual field missing? Recall that each optic tract contains fibers conveying data from the same half of the visual field of each eye. A lesion of the optic tract can produce a symptom known as contralateral homonymous hemianopia.
3. Occulomotor, trochlear, abducent.
   a. Do the eyes point straight forward at rest? A deviation from this position is known as strabismus.
   b. Is there drooping (ptosis) of the eyelids (III)?

c. Do the pupils react to light and during close vision (III)?

d. Do both eyes converge for close vision (III)?

e. Are there conjugate eye movements in all directions?

f. Is nystagmus (rhythmic oscillation) present at rest or on movement (medial longitudinal fasciculus)?

4. Trigeminal.
   a. Are there sensory deficits in the face or scalp (three divisions)?
   b. Does the eye blink when the cornea is touched with a wisp of cotton (corneal reflex)?
   c. Does the jaw deviate to one side when protruded?
   d. Is there sensitivity to loud noises?

5. Facial.
   a. Are the muscles of facial expression weak?
   b. Can the patient blow up a balloon or drink through a straw?
   c. Does the corner of the mouth droop?
   d. Can the patient wrinkle his or her forehead?
   e. Is there loss of taste in the anterior two thirds of the tongue?
   f. Is the corneal reflex present?
   g. Is the patient sensitive to loud noises?

6. Vestibulocochlear.
   a. Are there complaints of vertigo (dizziness) or nausea?
   b. Is there nystagmus or difficulty with balance?
   c. Can faint sounds be heard, such as a ticking watch held at least 4 inches from the ear?
   d. Are there complaints of ringing or buzzing in the ear?

7. Glossopharyngeal, vagus.
   a. Does the patient have difficulty swallowing?
   b. Is the patient's voice nasal or hoarse?
   c. Does the uvula deviate during phonation?
   d. Is the gag reflex present?

8. Spinal accessory.
   a. Can the patient turn his or her head side to side? Can the patient shrug his or her shoulders?

9. Hypoglossal.
   a. Can the patient enunciate correctly?
   b. Does the tongue deviate when protruded?

The third section of the examination concerns the cerebellum. In general, lesions of the cerebellum produce noticeable difficulties with coordination manifested in the extremities or the trunk. Lack of coordination is known as asynergy or ataxia. It is important to mention that most cerebellar symptoms become obvious only with the initiation of movement.

1. Can the patient reach for objects, bring the two index fingers together with arms raised, or touch the tip of his or her nose with the index fingers? Inability to do this is known as dysmetria and often results in overshooting (pastpointing) the desired goal.

2. Can the patient rapidly pronate and supinate the forearm? Inability to perform rapidly alternating movements is known as adiadochokinesia.

3. Are there tremors present when a movement is initiated? Such tremors are known as intention tremors and often become more pronounced toward the end of the range of movement.

4. If resistance is applied to a contracting muscle, can the patient relax the contraction as soon as the resistance is released? For example, the patient attempts to flex an elbow while the examiner applies strong resistance. When the examiner lets go, an intact cerebellum will sense the release of tension and prevent further elbow flexion by contracting the triceps. Inability to do this is known as rebound phenomenon.

5. Is the patient's gait awkward or staggering, with a broad base of support? This is ataxic gait.

The fourth section concerns the functioning of the motor system. Information from this section will help determine the integrity of the motoneurons and the descending motor tracts.

1. Muscle status. What is the general condition of the major muscle groups in the extremities, trunk, and neck? Are the muscles hypotonic or hypertonic on palpation? Are there any areas of atrophy? Can fasciculations be seen or palpated? Is there any tenderness?

2. Muscle strength.
   a. Do some muscles produce no movement (paralysis)?
   b. Is there marked weakness (paresis) in some muscles?
   c. Can isolated joint movements such as knee extension or elbow flexion be performed? Patients who have movement throughout a limb while attempting to move only one joint are said to be using a synergy pattern. For example, a patient attempting to extend a knee may also produce hip extension and ankle plantar flexion, which are part of the extension synergy for the lower extremity.
3. Gross motor.
   a. Can the patient transfer from bed to chair, chair to toilet, and so on?
   b. Can the patient perform all the normal activities associated with daily living?
   c. Can the patient move about independently if using crutches, cane, or wheelchair? Are there any contractures that limit the patient's ability to function?
4. Fine motor.
   a. Can the patient write, string beads, cut with a pair of scissors? Lack of fine motor control often indicates a problem with the corticospinal (pyramidal) system.

In assessing the motor system, it is important to keep in mind the distribution of any abnormalities. For example, weakness and hypertonus in the antigravity muscles on one side of the body are often found in patients with CVA. On the other hand, paralysis of part of an extremity such as the hypothenar muscles could indicate a peripheral nerve, plexus, or ventral root problem.

The fifth section of the examination concerns the integrity of the somatic sensory system. Like in the motor system, the distribution of sensory disturbances is important. This is especially true in patients with spinal cord injury where the area of sensory loss often indicates the level of lesion. In performing a sensory examination a strong knowledge of dermatomes and the fiber tracts is essential. Keep in mind that sensory testing must be done with the patient's eyes closed.

1. Can the patient detect the touch of a pen or pencil tip?
2. Can pain be felt from the sharp point of a pin?
3. Can hot and cold water be perceived?
4. Can objects such as a paper clip or key be identified (stereognosis)?
5. Are there areas of numbness?
6. Can the patient determine whether a joint is flexed or extended?
7. If the examiner places a joint in a position, can the patient match that position with the other extremity?
8. Can the patient feel the vibration of a tuning fork?
9. Are there tingling or burning sensations (paresthesias)?
10. Can two points held close together (about 1 cm) be detected on the fingertips?

The sixth section measures the deep-tendon (myotatic) reflexes and the presence or absence of the superficial reflexes. The grading of the deep-tendon reflexes ranges from zero (0) to ++++, with ++ the grade for a normal muscle. The procedure involves a rapid firm tap of the tendon with a reflex hammer, which in turn applies a quick stretch to the muscle spindle. The tendons usually assessed are the biceps, triceps, quadriceps (patellar tendon), and Achilles tendons. It is important to test both sides because it is possible for a normal person to have grades of + or +++ related to emotional or psychologic factors. The jaw and wrist may also be tested.

The myotatic reflexes not only determine the integrity of the peripheral nerve, dorsal and ventral roots, and spinal cord but also the excitatory-inhibitory influences acting on the motoneurons. Therefore a large variety of neurologic lesions produce alterations of the myotatic response.

The resistance to stretching of a muscle is directly related to the response to a tendon tap, since both excite the muscle spindle. A tendon tap grade of 0 offers no resistance to stretch and is known as flaccidity. Grades of

+++ and ++++ offer moderate to strong resistance to stretch and are known as spasticity or rigidity. Rigidity is generally associated with lesions of the cerebellum and basal nuclei, while spasticity is usually found with lesions of the cerebral cortex, internal capsule, parts of the brainstem, and the spinal cord. Flaccidity is usually found with lesions of the peripheral nervous system, although some CNS lesions can cause flaccidity. (The characteristics of spasticity and rigidity are discussed in Chapter 9.)

The superficial reflexes are different from the deep-tendon reflexes in that cutaneous receptors initiate the response. Superficial reflexes are graded as either present (+) or absent (0). The corneal reflex is absent only with lesions of the fifth and seventh cranial nerves or of the spinal tract of the trigeminal nerve. The gag reflex indicates the integrity of the ninth and tenth cranial nerves. There are about eight other superficial reflexes, but the two most commonly tested are the abdominal and plantar. Both are tested with a pointed but not sharp object, such as the tip of a pen. In the abdominal reflex stroking (scratching) of the abdomen normally causes the umbilicus to deviate toward the stimulus. In the plantar reflex stroking the lateral border of the sole and across the metatarsals normally causes the toes to curl (plantar flexion). In lesions of the pyramidal motor system the abdominal reflex is diminished or absent, while the plantar response is extension and fanning of the toes (especially the big toe). This plantar response is known as Babinski's sign. The rationale as to why the abdominal reflex diminishes and the plantar reflex changes is uncertain.

The seventh section concerns an analysis of gait. Since normal gait requires an intact sensorimotor system, many lesions of the peripheral and central nervous system can produce disturbances in gait.

1. Does the patient have a wide base of support and seem to stagger and lurch forward? These findings indicate an ataxic gait.

2. Is there a limp? If so, determine whether it is from pain, weakness, or contracture.

3. Does one or both legs operate in a synergy pattern of movement? Usually the appearance of synergy indicates involvement of the pyramidal system.

4. Are the stride lengths and the amount of time spent on each foot equal? Deviations could indicate pain, weakness, contracture, sensory deficit, or synergy patterns.

5. What assistive devices are used? Can the patient walk without them?

The eighth section concerns the autonomic nervous system (ANS). In spite of its many functions, the ANS can be assessed in several ways:

1. Do the pupils constrict to light and for near vision?

2. How does the patient's skin feel? Is it warm or cold? Are there areas of excess sweating or dryness? What is the color?

3. Are the nails brittle?

4. Is there uneven hair loss?

5. Are there bowel or bladder complications such as incontinence?

Disturbances of ANS function are most commonly seen in peripheral nerve lesions, although parts of the CNS may also be involved.

The final section concerns a variety of miscellaneous signs and symptoms associated with specific CNS disorders. This list is by no means all-inclusive, but it does represent the more commonly tested items.

1. Babinski's sign. Stroking the lateral part of the sole and across the metatarsals with a pointed but not sharp object normally causes the toes to curl (plantar reflex). Lesions of the pyramidal system cause the toes to dorsiflex and abduct, and the ankle may even dorsiflex. The Babinski response is actually part of the flexor withdrawal reflex and is normally regulated by the pyramidal motor system. Infants and a few normal adults demonstrate a Babinski response.

2. Hoffman's sign. Rapid flicking of the patient's index finger (causing a quick stretch of the

flexor digitorum superficialis and profundus) produces flexion in all the fingers and thumb. Hoffman's sign is not present in a normal nervous system. When present, it indicates damage to the pyramidal system.

3. Clonus. Rapid dorsiflexion causes an oscillation of the ankle in plantar and dorsiflexion. The basis of clonus is a hyperactive stretch reflex of both the plantar and dorsiflexors such that quick stretch of the plantar flexors causes them to contract and stretch the dorsiflexors, which in turn causes the dorsiflexors to contract, and so on. Clonus can also be found in the knee and wrist and can range in intensity from mild to severe.

4. Kernig's sign. Flexion of the hip with the knee straight (the patient is supine) causes knee flexion as a result of an involuntary spasm of the hamstring muscles. The presence of Kernig's sign usually is associated with an irritation of the meninges such as from meningitis and a herniated intervertebral disc.

5. Brudzinski's sign. Forcible neck flexion on the chest causes flexion of both legs and thighs. Brudzinski's sign indicates irritation of the meninges.

6. Romberg's sign. Romberg's sign is positive when the patient can stand with the eyes open but loses balance when the eyes are closed. The presence of Romberg's sign indicates dorsal column damage.

7. Resting tremor is oscillation of agonist and antagonist at a general frequency of 3 to 5 times per second. A resting tremor is usually found in the fingers and wrist, and sometimes in the neck creating a head tremor. A resting tremor disappears or diminishes with muscle contraction and disappears with sleep. It is characteristic of basal nuclei disorders. Specifically, lesions of the substantia nigra pathways to the globus pallidus, as in Parkinson's disease, produce a resting tremor.

8. Cogwheel ridigity. In cogwheel rigidity the resistance to stretch is not constant as in regular rigidity but alternately increases and decreases. This "cogwheel" effect is felt as a series of releases of tension during the stretching. Cogwheel rigidity is associated with Parkinson's disease and in some respects can be thought of as a combination of rigidity and resting tremor. The cogwheel effect is not apparent during movement.

9. Opisthotonus. Opisthotonus is a condition of strong spasticity in the neck and back extensors, causing marked arching of the entire spine. This posturing is associated with decerebrate rigidity in which the extensors of the extremities are also markedly spastic. This spasticity of the physiologic extensors is generally caused by an interruption of inhibitory fibers descending to the brainstem. The unregulated reticular formation and vestibular nuclei send continuous excitatory impulses to the extensor motoneurons. The extensor mucles have the characteristics of spasticity. Thus the term decerebrate rigidity is somewhat incorrect. This condition can be seen clinically in patients with severe head or brainstem lesions.

Specialized neurologic tests

Despite all of the data from the history, physical, and neurologic examinations, the physician may need further information to add clarity to the medical picture the patient is presenting. A variety of specialized neurologic tests can yield specific data as to the precise location of a lesion and can differentiate the lesion from other similar conditions. These tests are discussed briefly to provide a background as to the general techniques involved, interpretation of results, and clinical usefulness. For greater detail about these procedures, current textbooks on neuroradiology or electrodiagnostic techniques are helpful.

Spinal tap (lumbar puncture). A spinal tap is performed whenever there is a need to analyze the cerebrospinal fluid (CSF). Numerous conditions such as infections, hemorrhage, hematomas, and tumors can present abnormal CSF findings (see Table 7-1).

The procedure involves having the patient lie on one side with the knees as close to the chest as possible. After administration of a local anesthetic a needle is inserted into the subarachnoid space between vertebrae L-4 and L-5. Since this is the area of the cauda equina, there is no danger of puncturing the spinal cord. If one of the spinal nerves is struck the patient may feel lightning-like pains, and the needle can be moved backward. When the needle is correctly in place the CSF pressure is measured and then several 2 to 3 ml samples are drawn off for analysis. The lumbar puncture can cause headache for up to 48 hours afterward. The headache is relieved when the patient lies down and is exacerbated with sitting. During this time therapy treatments may be discontinued.

Myelogram. The myelogram allows the examination of the structures of the spinal canal by introducing a contrast medium into the subarachnoid space. A gas such as air or oxygen or a radiopaque substance such as iophendylate (Pantopaque) can be used as the contrast medium. The procedure begins with a lumbar puncture into the subarachnoid space. After a certain amount of CSF is removed the contrast medium is injected into the subarachnoid space. Because of different densities the medium will "float" on the remaining CSF. Thus by tilting and rotating

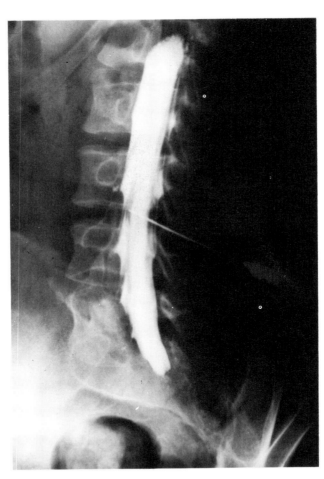

Fig. 10-1. Myelogram, using radiopaque contrast medium. Cauda equina nerves and lowest portion of spinal cord are coated with dye, showing normal filling of subarachnoid space. (From Willis, W. D., Jr., and Grossman, R. G.: Medical neurobiology, ed. 2, St. Louis, 1977, The C. V. Mosby Co.)

the patient the entire subarachnoid space around the spinal cord can be viewed. Any herniation of a disc, protrusion of bone, abscess, tumor, or adhesion that alters the shape or blocks the subarachnoid space can be detected (Fig. 10-1). The patient should remain flat for 24 to 48 hours after this procedure to prevent the headache associated with lumbar puncture. When a radiopaque dye such as iophendylate is used, as much as possible is removed after the procedure to reduce meningeal irritation. The gas can be left in as it will be absorbed into the bloodstream.

Pneumoencephalogram. The pneumoencephalogram is useful for studying the integrity of the subarachnoid space within the cranium and also the contours of the ventricular system. Since there is almost no spare room within the cranium, most space-occupying lesions such as tumors, hematomas, and aneurysms become obvious from this procedure as they cause either an improper filling of the subarachnoid space or a shift of the ventricles from midline.

The technique is similar to the myelogram in that a lumbar puncture is performed and measured quantities of CSF are replaced with a contrast medium (usually a gas). Unlike the myelogram, the patient sits in a special "somersaulting" chair. This chair can be rotated in any direction, thus allowing the exact positioning of the patient in any plane. In this way all subarachnoid and ventricular areas (such as the posterior horns of the lateral ventricles) can be filled (Fig. 10-2).

It is generally believed that the gas injected into the subarachnoid space enters the fourth ventricle via the foramen of Magendie. The CSF in the ventricles that is displaced by the gas exits via the foramen of Luschka. As in the myelogram, numerous x-ray or fluoroscopic films are taken.

The injected gas is absorbed into the bloodstream and the CSF replenished usually within 72 hours. The patient is kept flat in bed during this time to minimize headaches.

Ventriculogram. In some instances the injection of air into the subarachnoid space does not completely fill the ventricular system. This may be caused by an obstruction or increased intraventricular pressure. In these cases direct injection of a contrast medium into the ventricular system may be necessary to determine whether the symmetry or midline shift of the ventricles has been altered. This procedure is performed with the patient under light anesthetic. A small burr hole is made through the skull 1 to 3 cm lateral to midline and 1 cm in front of the coronal suture or 3 cm lateral to midline and 7 cm above the external occipital protuberance. A needle can then be inserted into one of the lateral ventricles and measured amounts of CSF can be replaced with a contrast medium. Although there will be some focal damage to neural tissue from the needle, there are usually no clinical manifestations.

Although this procedure sounds somewhat drastic, the complications are generally considered no more serious than for the pneumoencephalogram, and the data obtained are usually vital to the life of the patient.

Angiogram. Another area that may require special analysis is the vascular circulation of the brain. A variety of conditions can interfere with the normal filling of the vascular system. Tumors and hematomas can displace or block blood vessels. An aneurysm or a highly vascular tumor can indicate an area of excess blood. A thrombus or embolus can prevent the filling of vessels distal to the clot.

The angiographic procedure involves the injection of a contrast medium directly into the arterial system of the brain. The common carotid artery is used to study the anterior circulation of the brain. Since the vertebral artery is small and harder to palpate, the subclavian artery is often punctured instead to study the posterior circulation. The patient is

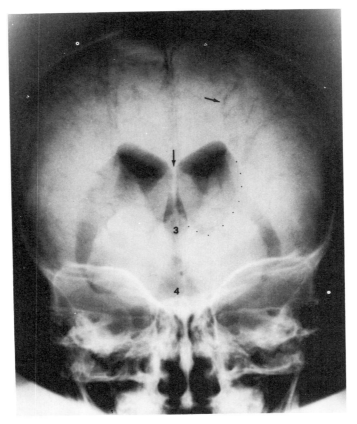

Fig. 10-2. Pneumoencephalogram showing normal filling of lateral, third, and fourth ventricles in anteroposterior view. Note symmetry of lateral ventricles and midline position of third and fourth ventricles. A space-occupying lesion would alter these characteristics. Also note septum pellucidum *(midline arrow)* in its normal midline position. *Upper arrow* points to air in subarachnoid space of sulcus. (From Willis, W. D., Jr., and Grossman, R. G.: Medical neurobiology, ed. 2, St. Louis, 1977, The C. V. Mosby Co.)

in a supine position and lightly sedated for comfort and to reduce excess movements. As soon as the contrast substance has been injected, a series of ten pictures are taken within 10 seconds (Fig. 10-3). The contrast medium then passes into the venous and dural sinus system and additional pictures may be taken at this time. By comparing the pictures with the normal distribution of vessels, those that are displaced or did not fill or areas of excess blood can readily be pinpointed.

Brain scan (radioisotopic encephalography). The brain scan involves an intravenous injection of a short-lived radioactive substance. Emitted gamma rays can then be detected with a scanner (e.g., a Geiger counter) or a specialized camera as the substance travels through the cerebral circulation. In lesions such as tumors (especially me-

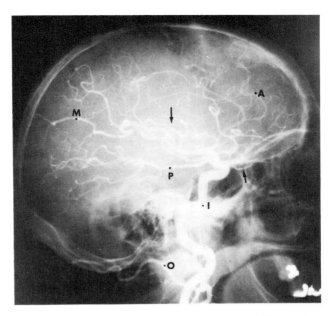

Fig. 10-3. Right carotid angiogram, lateral view, showing normal filling of internal carotid *(I)*, anterior cerebral *(A)*, middle cerebral *(M)*, and posterior cerebral *(P)* arteries. *Upper arrow* points to vessels that form Sylvian triangle. *Lower arrow* points to ophthalmic artery. (From Willis, W. D., Jr., and Grossman, R. G.: Medical neurobiology, ed. 2, St. Louis, 1977, The C. V. Mosby Co.)

ningiomas), hematomas, abscesses, and infarctions where there is an increased vascular supply or pooling of blood there will be an increased uptake of the substance, which can then be detected. There are several drawbacks to this procedure. One is that if the lesion is near a normal vascular area it may not "stand out" from the background. The other is that the brain scan offers few clues as to the nature of the lesion. Although the results can be 85% to 90% accurate in indicating the presence of a lesion, they do not tell whether it is neoplastic, vascular, inflammatory, or traumatic.

An alternate brain scan procedure is to perform a lumbar puncture and inject the radioisotopic compound into the subarachnoid space. This procedure is called radioisotopic cisternography (a cistern is actually a large subarachnoid space). It is used to obtain data about the flow and absorption of CSF. It normally takes from 24 to 48 hours for the compound to reach the superior sagittal sinus. During this time the scan can be repeated at intervals. Normally the compound will flow over the hemispheres of the cerebrum and not enter the ventricular system. Any obstruction of the dural sinuses or of the subarachnoid space can prevent the flow of the isotope over the hemispheres as it moves into the ventricles (Fig. 10-4).

The above procedures can provide specific valuable detail about the integrity of the CNS. They all, however, involve invasion of the sterile field of the body and in most cases the introduction of a foreign substance into the bloodstream or ventriculomeningeal system. Complications can result, and permanent residual deficits such as hemiparesis have been reported with some of these pro-

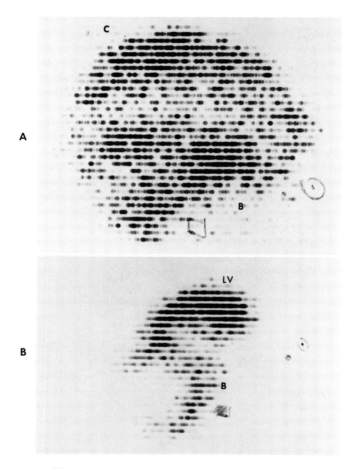

Fig. 10-4. A, Normal brain scan, right lateral view, after injection of isotope into lumbar sub-arachnoid space. **B,** Right lateral scan of head showing presence of isotope in ventricular system, *LV*, and lack of filling of subarachnoid space over hemisphere. Basilar cistern, *B,* in front of brainstem has filled normally. *C,* area around superior sagittal sinus. (From Willis, W. D., Jr., and Grossman, R. G.: Medical neurobiology, ed. 2, St. Louis, 1977, The C. V. Mosby Co.)

cedures. The following techniques have the advantage of being noninvasive.

Echoencephalogram. The echoencephalogram is used to detect space-occupying lesions such as tumors and hematomas. The procedure is simple and there are no adverse reactions or complications to the patient. The technique involves the use of an ultrasound generator and a receiver displaying the sound echoes on an oscilloscope. The patient is awake and can be sitting or lying down. The transducer (1 to 10 MHz/sec) is applied to the skull (which has been wetted with water). It first transmits a sonic beam and then records that portion of the beam reflected back by the different structures. Since bone,

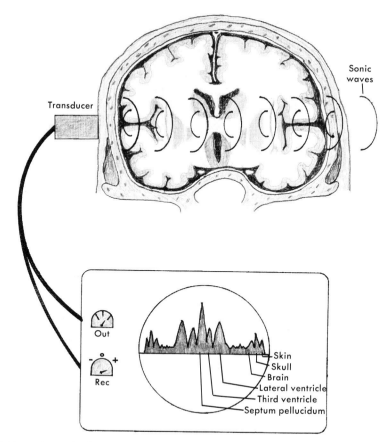

Fig. 10-5. Recording of echoencephalogram. Transducer acts as sender of sonic waves and receiver of reflected waves. Different densities of skin, bone, brain tissue, and CSF reflect varying amounts of sound waves.

neural tissue, cerebrospinal fluid, and dura mater have different densities and positions, varying degrees of sound are reflected back at different intervals. Of importance are the septum pellucidum, lateral ventricles, and third ventricles, which are midline structures. Any displacement of these by a lesion can readily be detected (Fig. 10-5).

Electroencephalogram. The electroencephalogram involves the recording of the electrical activity from the neurons of the cerebral cortex. It is one of the oldest and best known of the neurologic diagnostic procedures. The procedure is harmless to the patient and relatively simple to perform. Sixteen electrodes are placed on the scalp. The head does not have to be shaved, since a jellylike electrode paste can be used to facilitate contact. The position of these electrodes is standardized to aid the reliability of the readings and to allow the recording of the entire convexity of the cortex. The patient is either recumbent or seated with the eyes closed in a grounded, wire-shielded room while recordings are made from the individual electrodes in relation to each other (see Fig. 9-8). Readings are taken for about 20 minutes so that there is time for any abnormal activity to be manifested. To facilitate the emergence of any abnormal findings, the patient is asked to hyperventilate (respiratory alkalosis increases neuron excitability) and is exposed to a 2-minute period of rhythmic light flashing (1 to 30/sec). In some cases the EEG pattern is taken during sleep and after the injection of a weak seizure-inducing drug.

Interpretation of the results is complex because there are many variations with no apparent clinical significance. The recorded potentials are interpreted in terms of their frequency, amplitude, and structure. In general the appearance of large-amplitude slow-frequency potentials in a conscious patient is cause for concern. The EEG can provide useful information for localizing brain traumas, hematomas, abscesses, tumors, vascular disorders, and infections and in epilepsy. The value of the EEG in nonorganic brain disorders such as headache is questionable.

Roentgenography. X-ray films of the skull and spine are probably the most common diagnostic tool used for neurologic conditions. Films are routinely taken for a variety of conditions such as head injury, tumors, hematomas, vascular anomalies such as aneurysms, and hydrocephalus. The value of x-ray views of the spine and skull is that these bony coverings are fairly standard in shape; thus any deviations can be detected. For example, some tumors erode an area or become calcified, cerebral arteriosclerosis can calcify some arteries, and most bone fractures can be spotted. Also of importance is the pineal gland, which calcifies after puberty in about 55% of the population. Since it is a midline structure, a space-occupying lesion can cause it to shift. With the advent of portable x-ray machines and Polaroid, x-ray films can be processed quickly and easily.

Computer-assisted tomography (CAT scan). The CAT scan is one of the newest and most precise methods of localizing a lesion. It uses an x-ray machine, the EMI scanner, that produces a very thin x-ray beam. This beam is passed through the head, usually at an angle of 20° to 25° in relation to Reid's baseline at the level of the eye (Fig. 10-6, A). The silver iodide crystal detectors then record the photon transmission through the head at that level. The scanning unit is then rotated 360° around the head at 1° intervals. A computer then correlates the absorption value and displays this slice of the brain on an oscilloscope. The scanning unit is then moved up a specified distance and the procedure repeated. The results from the CAT scan are a series of eight pictures of x-ray slices through the head (Fig. 10-6, B). These pictures show the different structures of the brain as related to their different densities. Thus it is possible to "see" white and gray matter, the convolutions, the ventricles, the subarachnoid space, and the skull. Any abnormal mass such as a tumor, calcification, vascularity, atrophy, edema, or dilation of the ventricles becomes visible, and its exact location can be identified.

The entire procedure takes about 30 minutes. During most of this time while the machine is scanning the patient must remain motionless to prevent artifacts from appearing. In some cases where enchanced accuracy is needed to localize a lesion a contrast medium can be injected intravenously.

Electromyography and nerve conduction. Electromyography (EMG) and nerve conduction can be discussed together because they are often performed together and with the same equipment. The procedure for electromyography involves the insertion of a small needle into a muscle. The needle acts as a recording electrode and is able to detect the muscle action potentials. Surface electrodes can be used, but they provide inferior results. The electrical impulses picked up by the needle are amplified and projected

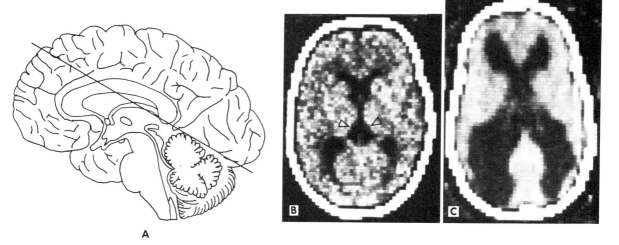

Fig. 10-6. CAT scan. **A,** Midsagittal view indicating level at which both scans were taken. **B,** Normal scan showing lateral and third ventricles and upper portion of superior cistern (dark) and skull (outer white ring). Note subarachnoid space (thin dark space between skull and brain tissue). **C,** Moderate hydrocephalus showing dilation of ventricles and compression of brain. (**B** and **C** from Harwood-Nash, D. C.: Neuroradiology in infants and children, vol. 2, St. Louis, 1976, The C. V. Mosby Co.)

through an oscilloscope or a polygraph machine for a permanent written record. Polaroid cameras and tape recorders can also be used to make permanent recordings. The electrical activity of a muscle is examined during insertion of the needle, at rest, and during contraction. A normal muscle is fairly standard in terms of frequency, amplitude, and of interest the sound of its action potentials. At rest there is little activity and the oscilloscope baseline is only sporadically broken. With minimal contraction individual motor unit action potentials can be seen and heard as a thumping sound (most EMG machines have a speaker in addition to the oscilloscope). With a strong contraction the baseline is obliterated and there is a sound of low-pitched static.

The primary usefulness of the EMG is its ability to detect the fibrillation potentials associated with denervation. Any involvement of the peripheral nerve, ventral root, or anterior horn cell such as from trauma can produce fibrillation potentials. These potentials, which represent the hypersensitive response of one muscle cell, are very prominent on the oscilloscope with the muscle at rest. It is usually important to perform a series of EMGs over time to determine whether the lesion is progressing, as in a nerve tumor, or whether there is recovery, as in a regenerating nerve. In the latter case the fibrillation potentials are replaced by small polyphasic motor unit potentials (nascent units), which indicate that regeneration has occurred (Fig. 10-7).

The EMG can also be used to detect certain myopathies such as the myotonias and the muscular dystrophies, in disorders of the neuromuscular junction such as myesthenia gravis, and in some hysterical conditions.

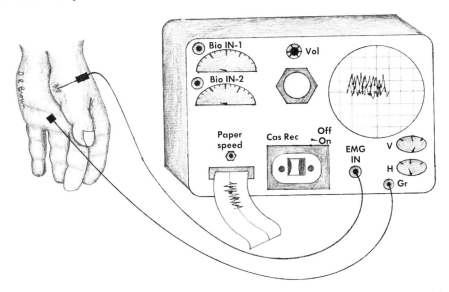

Fig. 10-7. Electromyography. This procedure requires no outside stimulation; it simply records electrical output from portion of muscle around inserted needle. This output can be displayed on oscilloscope, recorded on tape or paper, and made audible through speaker. Biofeedback units are not in use.

The procedure for recording the conduction velocity of a nerve and the usefulness of this technique are discussed in Chapter 5.

Electrodiagnostic tests. These tests include the strength-duration curve, the galvanic-faradic (twitch-tetanus) response, and the polarity formula. Their purpose is to determine the presence of denervation and possible reinnervation. The techniques are discussed in Chapter 5.

SYMPTOMS OF NERVOUS SYSTEM LESIONS

In this section are defined the various symptoms associated with peripheral and central nervous system lesions. No attempt is made to construct clusters of these symptoms into syndromes, because the list would be endless. In the box on p. 313 many of these symptoms are grouped into two categories:

those associated with lower motoneuron lesions and those associated with upper motoneuron lesions.

Muscle tone

Muscle tone is best determined by palpation of a muscle at rest (relaxed). The firmness or consistency that a muscle has is the result of the inherent viscosity of the muscle tissue and the asynchronous discharge of motor units from activity of the alpha motoneurons. Muscle tone varies from muscle to muscle and with the arousal/emotional state of the individual. Muscle tone should be assessed with the muscle at a standard length so that any influence from the muscle spindle is constant.

hypotonus Muscle feels soft or mushy.
hypertonus Muscle feels firm or taut.

COMPARISON OF SYMPTOMS ASSOCIATED WITH LOWER AND UPPER MOTONEURON LESIONS

Lower motoneuron	Upper motoneuron
Flaccid paralysis, decreased DTRs, hypotonus	Spastic or rigid paresis, increased DTRs, hypertonus (flaccidity, hypotonus, and hyporeflexia are possible)
Marked atrophy (denervation)	Mild atrophy (disuse)
No involuntary movements	Possible involuntary movements
No asynergistic movements	Possible asynergistic movements
Fibrillations and tropic responses	None
Fasciculations	None
No abdominal or cremaster reflexes	Same
No clonus or Hoffman's or Babinski's signs	Clonus and Hoffman's and Babinski's signs

Deep-tendon reflexes

The myotatic reflexes commonly assessed are the patellar, Achilles, biceps brachii, wrist extensor, and temporomandibular. A normal amount of movement is graded as ++ from a firm tap of the tendon or lower jaw with a reflex hammer.

hyporeflexia Decreased response to tendon tap (+ or 0).
hyperreflexia Increased response to tendon tap (+++ or ++++).

In addition to a tendon tap the application of a quick stretch can also determine the integrity of the involved pathways by the amount of resistance offered by the muscle.

flaccidity No response (resistance) to quick stretch.
spasticity Increased resistance to quick stretch. Spasticity is the result of increased facilitation (overactive) gamma motoneurons. The rate of sensitivity is the result of overbiasing of the nuclear bag (the faster the stretch, the greater the response). Spasticity is abolished by dorsal root section, is found predominantly in the antigrav-

ity muscles, and is associated with the clasp-knife response.
rigidity Also an increased resistance to quick stretch. True rigidity, however, represents an increased facilitation of alpha motoneurons directly. There is little sensitivity to the rate of stretch. Rigidity is only moderately affected by dorsal root section, and is generally found on both sides of the joint.

Decreased voluntary movement

These symptoms involve the amount of voluntary movement produced.

paresis Weakness or reduced voluntary movements. Since the velocity is often slowed, coordinated movements are affected.
paralysis Absence of voluntary movements.

Involuntary movements

Involuntary movements occur without control of the part of the individual. They generally disappear with sleep.

resting tremor Rapid oscillation (usually 3 to 5 times per second) of agonist and antagonist. Usually seen in the hands and wrist (pill-rolling tremor) and the head.

athetosis Slow, writhing, wormlike movements of the extremities (more common in the proximal muscles), face, and tongue.

choreiform Brisk, graceful, almost purposeful movements of the extremities (more common distally), face, and tongue.

ballismus Violent flinging movements of the extremities.

Posturing (dystonia)

These symptoms involve a fixed posturing of one or more extremity. The trunk, neck, and eyes can also be involved.

decerebrate rigidity Actually a marked spasticity of the physiologic extensors and adductors.

opisthotonus Marked spasticity of the neck and back extensors, resulting in an arched posture.

decorticate posturing Marked spasticity of the antigravity muscles, resulting in flexion, adduction, and internal rotation of the upper limb and extension, adduction, and internal rotation of the lower limb.

Asynergistic movements

Asynergistic movements show a lack of coordination.

dysmetria Inability to judge distances when reaching for something, resulting in past-pointing.

adiadochokinesia Inability to perform rapidly alternating movements such as pronation-supination of the wrist.

ataxia Bizarre forceful distortion of the basic movement patterns in proximal muscles. An ataxic gait is a staggering, lurching, pattern with the legs set far apart.

scanning speech Ataxia of the diaphragm, resulting in speech patterns that lack normal rhythm and emphasis. Words may be pronounced with excess force, and pauses may be too long.

intention tremor Tremor (usually of the distal muscles) of the extremity frequently seen toward the end of a movement, resulting in a rhythmic oscillation of the hand as it approaches the object.

Abnormal reflexes and signs

Abnormal reflexes and signs involve abnormal superficial and deep-tendon responses.

Babinski's sign Stroking along the lateral border and metarsals of the foot with a sharp (but not pain-producing) object causes the toes to extend and fan out.

Hoffmann's sign Flicking of the index finger produces clawing movements of the other fingers and thumb.

clonus Rapid dorsiflexion of the ankle, resulting in rhythmic plantar and dorsiflexion of the ankle. Clonus is best seen if the examiner attempts to hold the ankle in dorsiflexion. Clonus may be sustained or unsustained and may also occur at the wrist and knee.

abdominal reflexes Normal response is a twitch-like contraction of the abdominal muscles from a light stroking stimulus. Lack of this response can indicate a lesion.

cremester reflex Normal response is contraction of the scrotum from stroking of the medial side of the thigh. Lack of this response can indicate a lesion.

Atrophy

A muscle that is not used will atrophy and show loss of contractile proteins.

denervation Loss of lower motoneuron control results in severe loss of muscle bulk.

disuse Immobilization of a muscle causes severe atrophy. Disruption of the motor centers causes mild to moderate atrophy.

Sensory disturbances

Alterations in perceived sensations can vary tremendously depending on the extent and area of involvement.

anesthesia Complete loss of all sensations.

hyperesthesia Increased sensitivity (lowered threshold) to one or more sensations. Individual hyperesthesias such as increased sensitivity to pain (hyperalgesia) can occur.

hypesthesia Decreased sensitivity to sensations. Individual hypesthesias can occur.

paresthesia Tingling or buzzing sensation.

astereognosis Inability to tell the shape of an object by touch alone.

agnosia Inability to determine the meaning of or judge the value of a sensation. Agnosias can occur with the visual, auditory, or tactual systems.

thalamic syndrome Hypersensitivity to noxious

and nonnoxious stimuli often resulting in painful or unpleasant sensations to simple touch.

tinnitus Ringing or buzzing in the ear.

Trophic changes

These symptoms are the result of denervation.

fibrillation potentials Hypersensitive response to individual muscle cells.

skin and nail changes Skin becomes dry and inelastic, and the nails become brittle.

Miscellaneous symptoms

fasciculations Contraction of a single motor unit that can be seen and felt through the skin.

nystagmus Rhythmic oscillation of the eyes (usually side to side). Usually occurs in both eyes.

Romberg's sign Inability to maintain balance with the eyes closed.

Kernig's sign Inability of a supine person to extend the knee with the hip flexed.

Brudzinski's sign Strong flexion of a supine person's neck causes flexion of both legs and thighs.

cogwheel rigidity Oscillatory rigidity in which the resistance to stretch is released momentarily every few seconds.

lead-pipe rigidity Maintained resistance to stretch.

clasp-knife reflex Sudden release of resistance to stretch of a spastic muscle.

vertigo Feeling of dizziness or whirling motion.

syncope Fainting spell.

Anatomic relationships

Symptoms and diseases of the nervous system can be broken down anatomically several ways. In the boxes on pp. 316-319 specific areas are listed along with the symptoms to be anticipated from lesions of those areas. Because of the complexity of the nervous system and the variations in a disease process, the information given may differ slightly from what is seen clinically.

One method is to relate symptoms to those associated with upper motoneuron and lower motoneuron lesions. An upper motoneuron lesion involves any CNS center or fiber tract that regulates the activity of alpha motoneurons. A lower motoneuron lesion involves the ventral horn cells or the motor axons.

Upper motoneuron symptoms generally include increased tone, paresis, hyperactive DTRs, little atrophy, absent superficial reflexes, and the presence of pathologic reflexes and signs such as clonus and Babinski's sign.

Lower motoneuron symptoms generally include decreased tone, flaccidity, paralysis, decreased or absent DTRs, marked atrophy, fasciculations, fibrillations, the reactions to denervation such as elevated chronaxie, no pathologic reflexes or signs, and tropic symptoms (box, p. 313).

For both upper and lower motoneuron lesions (1) severity of the symptoms depends on the extent of the lesion, (2) variations are possible, and (3) sensory symptoms may accompany the motor symptoms.

Another anatomic breakdown of neurologic symptoms concerns the pyramidal and extrapyramidal systems. The pyramidal system is often referred to as the corticospinal pathway. It includes the fibers that pass through the pyramids of the medulla and from the corticospinal tracts. These fibers originate from many areas of the cerebral cortex and generally are concerned with isolated movements and skilled movements of the distal muscles. The extrapyramidal system includes parts of the cerebral cortex and the CNS centers such as the basal nuclei, reticular formation, and certain brainstem nuclei that do not project their axons through the pyramids. This system is more concerned with muscle tone and postural movements.

Both systems are in close proximity anatomically (especially in the internal capsule), and some areas such as the motor cortex are involved with both systems. It is often useless to try to separate the two clinically as many lesions involve both systems. To generalize, however, pyramidal system lesions produce flaccidity, hyporeflexia, hypotonus, a deficit in isolated voluntary or fine movements, loss of superficial reflexes, and the appearance of Babinski's and Hoffmann's signs. These symptoms were observed in monkeys

DIFFERENTIATION OF ANATOMIC LEVELS OF LESIONS BASED ON CLINICAL SYMPTOMS

Muscle and neuromuscular junction
  Decreased or absent DTRs
  Marked atrophy
  Weakness/paralysis, hypotonus
  Hypertonus, hypertrophy
  Pain, tenderness
Peripheral nerves
  Weakness/paralysis
  Flaccidity, decreased or absent DTRs
  Marked atrophy
  Fibrillations, increased chronaxie
  Decreased conduction velocity
  Anesthesia, hyperesthesia, paresthesia
  Pain
  Tropic, vasomotor changes
Dorsal (sensory) root
  Radicular pain (Lasègue's sign)
  Paresthesia, numbness
  Decreased/increased sensibility
  Decreased or absent DTRs
  Marked atrophy, hypotonia
  Bowel/bladder retention if S2 to S4
    involved
  Ataxic gait, Romberg's sign
  No signs of denervation
Ventral (motor) root
  Weakness/paralysis
  Decreased or absent DTRs
  Marked atrophy, flaccidity
  Fasciculations (early), fibrillations and
    increased chronaxie (later)
  Tropic and vasomotor changes
  Incontinence if S2 to S4 involved
Sympathetic ganglia
  Horner's syndrome
  Tropic and vasomotor changes
Complete cord section
  Anesthesia and paralysis
  Increased DTRs and other spinal
    reflexes
  Clonus
  Babinski's sign
  Atrophy
  Bowel/bladder incontinence
  Excess sweating
  Loss of vasomotor tone

Brainstem
  Dorsolateral lesions
    Involvement of cranial nerves V, VII,
      VIII, IX, X, and XI
    Nystagmus, vertigo
    Cerebellar signs
    Horner's syndrome
    Possible loss of pain and temperature
      for extremities and trunk
  Centromedial lesions
    Involvement of cranial nerves III, IV,
      VI, and XII
    Contralateral spastic paresis with
      upper motoneuron signs in
      extremities
    Contralateral dorsal column sensory
      loss
    Internuclear ophthalmoplegia
    Nystagmus
Cerebellum
  Ataxic gait
  Dysmetria
  Adiadochokinesia
  Rebound phenomenon
  Intention tremor
  Nystagmus
  Staccato (scanning) speech, dysarthria
  Hypotonia
Cerebrum
  Internal capsule/thalamus
    Contralateral loss of dorsal column
      sensations
    Contralateral spastic paresis of
      extremities with upper motoneuron
      signs
    Contralateral lower facial weakness
    Dysarthria
    Thalamic syndrome
    Contralateral homonymous
      hemianopsia
  Basal nuclei
    Involuntary movements (chorea,
      athetoid, ballistic)
    Parkinson's syndrome

DIFFERENTIATION OF ANATOMIC LEVELS OF LESIONS BASED ON CLINICAL SYMPTOMS—cont'd

Dorsal funiculus
  Paresthesia
  Loss of dorsal column sensations
  Ataxic gait with Romberg's sign
  Dysmetria (at night)
  Charcot joints
Lateral funiculus
  Spastic paralysis
  Increased DTRs, clonus
  Mild atrophy
  Babinski's and Hoffmann's signs
  Loss of abdominal and cremaster signs
  Loss of pain and temperature
    (contralateral)
Ventral funiculus/ventral horn, central cord
  Weakness, decreased DTRs, and
    hypotonia
Intramedullary cord/ventral horn
  Loss of pain and temperature bilaterally
  Loss of crossed reflexes
  Lower motoneuron signs
  Horner's syndrome

Cerebral cortex
  Contralateral loss of dorsal column
    sensations (usually upper and lower
    limb)
  Contralateral spastic paresis with
    upper motoneuron signs (usually
    upper and lower limb)
  Contralateral lower facial weakness
  Communications disorders (dominant
    hemisphere)
  Apraxia, agnosia
  Personality changes, emotional
    instability
  Contralateral homonymous
    hemianopsia
  Deviation of both eyes to side of
    lesion (transient)
  Poor attention span, poor ability to
    learn or remember, disorientation,
    difficulty with abstract thinking
  Poor body awareness, hemiattention

with lesions of area 4 (suppressor zone intact) or one pyramid. Extrapyramidal system lesions produce hypertonus, spasticity or rigidity hyperreflexia, clonus, involuntary movements, and some gross motor paralysis.

COMMON NEUROLOGIC SYNDROMES

The common neurologic syndromes associated with disorders of the various levels of the nervous system are considered on the following pages, beginning distally with the muscle and neuromuscular junction and continuing proximally to the cerebral cortex. Following this section is a series of case studies, from which the site of the lesion can be identified from a given set of symptoms. The box on p. 318 provides a generalized relationship between a specific area of the nervous system and the kinds of symptoms to be expected. Keep in mind that this table is not all-inclusive as there can be exceptions. The boxes on pp. 316 and 317 present the same information in a different perspective.

Muscle and neuromuscular junction

Although not directly connected to the nervous system, involvement of this area can produce some interesting symptoms.

Muscular dystrophies and polymyositis. In these conditions there are varying degrees of loss of muscle cells. The resulting symp-

SITES OF INVOLVEMENT ASSOCIATED WITH MORE COMMON NEUROLOGIC SYMPTOMS

Symptoms	Anatomic site
Anesthesia	Peripheral nerve, dorsal root, cranial nerve V, midbrain tegmentum
Flaccidity, hypotonus, decreased DTRs	Ventral horn, ventral or dorsal root, peripheral nerve, cranial nerve, precentral gyrus, pyramids
Fibrillations and tropic changes	Ventral horn, ventral root, peripheral nerve
Burning pain	Peripheral nerve
Hyperesthesias	Dorsal root, thalamus, peripheral nerve
Radiating pain, lightning pain	
Dissociated sensory loss (able to perceive some sensations)	Dorsal root, cranial nerve V
Ataxic gait, difficulty with balance	Spinal cord, medulla, pons, internal capsule, postcentral gyrus
	Dorsal columns, cerebellum and its peduncles, cranial nerve VIII, vestibular nuclei
Difficulty swallowing	Cranial nerves IX and X, tegmentum of medulla (nucleus ambiguus)
Dysarthria	Cranial nerves V, VII, XII, cerebellum and its peduncles
Diplopia	Cranial nerves III, IV, and VI
Nystagmus	Cranial nerve VIII, vestibular nuclei, cerebellum, and dorsomedial part of pons (medial longitudinal fasciculus)
Loss of taste	Cranial nerves VII, IX, tegmentum of medulla (solitary tract)
Ptosis	Cranial nerve III, lateral part of medulla and spinal cord (descending reticulospinal fibers to sympathetic preganglionics), T1 to T4, ventral roots, upper three ganglia of sympathetic chain
Spasticity, hypertonus, increased DTRs, clonus, Babinski's and Hoffmann's signs	Premotor area of cortex, internal capsule, most lesions of brainstem and spinal cord
Dysmetria, adiadochokinesia, scanning speech, rebound phenomenon, intention tremor	Cerebellum, cerebellar peduncle
Rigidity, bradykinesia, resting tremor, athetosis, chorea, ballismus	Basal nuclei
Decorticate posturing	Premotor area of cortex, internal capsule
Homonymous hemianopsia	Optic tract, some midbrain lesions, lateral geniculate body, posterior limb of internal capsule, occipital lobe
Agnosia, apraxia, aphasia	Cerebral cortex

LEVEL OF CLINICAL PARALYSIS OR SENSORY LOSS RELATED TO ANATOMIC SITE

Level of involvement	Anatomic site
Lower extremity	Spinal cord below T1, cauda equina, parasagittal lesion of cervical or thoracic cord, lumbosacral plexus
Upper extremity	Early intramedullary lesion of cervical cord, brachial plexus
All four extremities	Cervical cord, anterior lesion of medulla
Upper and lower extremity and head on same side	Contralateral cerebral cortex, contralateral internal capsule
Upper and lower extremity on one side and head on other side	Unilateral lesion of midbrain, pons, and medulla; head (cranial nerve) symptoms ipsilateral to lesion, extremity symptoms contralateral

toms are weakness and atrophy leading to paralysis, flaccidity, and decreased or absent deep-tendon reflexes (DTRs). An EMG reveals a decrease in amplitude and frequency of motor unit potentials. There are, however, no reactions of degeneration or decrease in conduction velocity, since the peripheral nerves are not involved.

Myasthenia gravis. In myasthenia gravis there is either insufficient acetylcholine or too much cholinesterease. The effect is to reduce the number of muscle cell action potentials, resulting in symptoms similar to those listed above. Since the muscle cells are not involved, these symptoms vary in severity depending on the condition of the junction. Thus in the morning after a night's rest there can be less weakness and flaccidity because of a buildup of acetylcholine. With repeated movements or exertion fatigue rapidly advances and the symptoms become more prominent. This fatiguability can be demonstrated on an EMG by applying a strong faradic stimulus to a peripheral nerve and recording from an innervated muscle. There

will be a pronounced decline in the frequency of motor unit potentials. Excitability may return after a rest period. This procedure is known as Jolly's myasthenic reaction.

Myotonic disorders. In myotonia there is a slowness to relax after voluntary contraction, probably the result of an abnormal muscle cell membrane. The patient can have difficulty performing rapidly alternating movements (adiadochokinesia), hypertrophy, and hypertonus. This slowness to relax should not be thought of as spasticity, since there is no resistance to passive stretch when the patient is at rest. An EMG will be normal except for a prolonged motor unit discharge to a single stimulus and an excess amount of insertion activity from the needle.

It is important to point out the lack of sensory symptoms in the above disorders. The myositis disorders are generally the only ones that produce pain or tenderness.

Peripheral nerves

The peripheral nerves are subject to a variety of disorders including hereditary, vascular, infectious, metabolic, tumor, and trauma. Because they contain motor and sensory fibers for the somatic and autonomic nervous systems, lesions of peripheral nerves can produce a wide variety of symptoms depending on the extent of involvement.

Damage to the somatic motor fibers results in varying degrees of weakness or paralysis, flaccidity, decreased or absent DTRs, atrophy, fibrillation potentials, decreased conduction velocities, and increased chronaxie values. Damage to the somatic sensory fibers results in degrees of anesthesia or decreased sensation and decreased or absent DTRs. Irritative sensory fiber involvement such as from a neuroma can cause burning sensations, pain, paresthesias (tingling), or hyperesthesias. Keep in mind that the sensory disturbances associated with peripheral nerve involvement do not follow dermatomal patterns (see Fig. 5-1). With certain polyneuropathies, such as those associated with chronic diabetes mellitus and alcoholism, there can be "glove" and "stocking" anesthesias of the hands and feet. These disturbances represent symmetric involvement of all the distal nerve fibers. They differ from the "glove" and "stocking" anesthesias of hysterical patients in that there is no sharp boundary between areas of sensation and areas of loss.

Damage to the autonomic fibers results in an initial loss of sweating (anhidrosis) and vasomotor tone, resulting in a warm and dry area of skin. Later effects include loss of hair, brittle nails, dry inelastic skin, and cold cyanotic skin as a result of vasomotor hypersensitivity.

Peripheral nerve involvement can be put into three general categories. Traumatic injuries usually sever or crush all the fibers in one or more nerves of one limb. All of the above symptoms can be found in the distribution of the involved nerve(s). Nerve compression usually involves only one nerve and initially may involve only the outer fibers

of the nerve. Since the larger type A fibers are located peripherally, symptoms can include weakness, decreased DTRs, slower conduction velocities, and decreased sensibility. Pain, temperature, and touch sensations may still be perceived. With increased compression pain and paresthesias will be felt. Eventually all of the fibers can be compressed, resulting in the symptoms of a severed nerve. The final category is the neuropathies. These include hereditary, infectious, nutritional, metabolic, toxic, and vascular conditions that affect the integrity of the axon, myelin, or connective tissue coverings of the nerve. The distinguishing feature of the neuropathies is the symmetric involvement of all four extremities.

A brief description of the more common types of peripheral nerve involvement seen by allied health therapists is presented below.

Traumatic conditions

DIRECT INJURY. Direct injury can result from the penetration of the skin by a bullet, knife, or other object.

ERB'S PALSY. The roots of the brachial plexus can be torn during birth if excess traction is applied to the arm or to the head. Usually roots C5 to C6 are involved. The loss of function of the infraspinatus, deltoid, biceps, brachialis, and brachioradialis results in a typical posture of adduction, internal rotation of the arm, elbow extension, and forearm pronation (waiter's tip). There are sensory deficits over the deltoid and radial side of the forearm.

Compression injuries

SATURDAY NIGHT PARALYSIS, CRUTCH PARALYSIS. In both of these conditions the radial nerve is compressed as it passes through the axilla. Saturday night paralysis is usually associated with an intoxicated person who falls asleep with an arm over the back of a chair. Crutch paralysis results from a person bearing weight on the top of the crutches instead of the handgrips. In both conditions the patient looses elbow, wrist, and finger extension, grip strength is greatly weakened, and supination is weakened. There are sensory deficits over the dorsum of the arm and forearm and the radial side of the dorsum of the hand. Since the maintained compression causes axonal damage, the electrical changes associated with denervation are present.

THORACIC OUTLET (CERVICAL RIB, SCALENUS ANTICUS). The lower roots or cord of the brachial plexus or the subclavian blood vessels can be compressed as they pass through the first rib and clavicle (thoracic outlet). The compression is intermittent and is accentuated by lifting or repetitive arm movements. The symptoms also are intermittent and include deep aching pain or paresthesia in the ulnar portion of the hand and forearm, weakness and clumsiness of the small hand muscles but only minimal atrophy, cyanosis, numbness, and tropic changes. There is generally only minimal axonal damage in this condition.

CARPAL TUNNEL SYNDROME. The medium nerve can become compressed at the wrist as it passes from the flexor retinaculum and the carpal bones. The symptoms are confined to the radial side of the hand and include pain, paresthesia, and numbness. The symptoms can be intermittent. Untreated cases can show symptoms of denervation.

TUMORS. The connective tissue covering of a peripheral nerve can develop a tumor. Usually the larger fibers are compressed first. Any of the peripheral nerves are subject to tumors.

Neuropathies

PERONEAL MUSCULAR ATROPHY (CHARCOT-MARIE-TOOTH DISEASE). This is a hereditary disease that results in degeneration of the outer fibers of the distal nerves or the motoneurons. Initial symptoms appear in the muscles of the feet and later spread to the muscles of the leg, hands, and forearm as more proximal portions of the nerves become involved. The primary symptoms are weakness or paralysis, marked atrophy, decreased DTRs, denervation signs, loss of kinesthesia, and a steppage gait. Sensibility is often unaffected.

HEAVY-METAL POISONING. In heavy-metal poisoning (e.g., from lead, mercury, arsenic) the distal nerves undergo a demyelinating and axonal degeneration process resulting in pain (often severe), glove and stocking anesthesias, foot and wrist drop, paresthesias, decreased or absent DTRs, and signs of denervation. These symptoms appear in several days to 2 weeks after ingestion of the metal. Excess dosage can lead to death as the result of cardiac and renal complications.

ACUTE POSTINFECTIVE POLYNEURITIS (GUILLAIN-BARRÉ SYNDROME). Acute postinfective polyneuritis is an inflammatory reaction primarily of the spinal roots and also the peripheral nerves resulting in edema and degeneration, especially in motor nerves. The symptoms include a rather sudden onset of marked flaccid paralysis of all four limbs that usually starts in the feet and ascends to the shoulders. The paralysis can affect the trunk, neck, and facial muscles, causing respiratory and swallowing difficulties. The urinary and rectal sphincters are rarely involved. With the paralysis is a loss of DTRs, a slowing of nerve conduction velocity, marked tenderness to touch, and paresthesias. These can be signs of denervation. This disorder is characterized by a preliminary upper respiratory infection, symmetry of symptoms, and a very high recovery rate.

ALCOHOLIC POLYNEURITIS. The toxic effects of chronic excessive alcohol consumption can produce degeneration of the distal nerves in all four limbs. The symptoms are similar to those associated with polyneuritis from heavy-metal toxicity with the addition of autonomic signs in the hands and feet.

VITAMIN DEFICIENCIES. Conditions of malnutrition or pernicious anemia with insufficient levels of B vitamins can cause serious degenerative changes in peripheral nerves.

The symptoms are those typical of polyneuritis and can involve the somatic motor fibers to the respiratory muscles as well. CNS symptoms can also occur.

DIABETIC POLYNEUROPATHY. Chronic diabetes, especially when uncontrolled, can cause ischemia to nerve trunks as a result of excess deposits of lipids in the blood vessels supplying them. This produces typical polyneuritis symptoms, especially in the feet. The sensory loss coupled with the generalized poor circulation often lead to nonhealable ulcers on the feet. This unfortunately often requires amputation of the involved part to prevent gangrene.

Dorsal and ventral roots

The dorsal and ventral roots (and dorsal root ganglia) are not commonly involved in isolated lesions; instead they are usually involved with spinal cord disorders and polyneuropathies. If the dorsal root or ganglia is involved symptoms can include radiating (radicular) pain, paresthesia, numbness, or hyperesthesia in the dermatomal distribution; decreased or absent DTRs; atrophy and hypotonia of the muscles supplied by that cord segment; and disturbances of bowel and bladder if S2 to S4 dorsal roots are involved. If the lumbar roots are involved the person is unable to stand with the eyes closed (Romberg's sign) and the gait is ataxic. There are no signs of denervation.

If the ventral roots are involved the symptoms include weakness or paralysis, decreased or absent DTRs, marked atrophy, flaccidity, fasciculations, signs of denervation, tropic and vasomotor changes if T1 to L2 roots are involved, and incontinence of bowel and bladder if S2 to S4 roots are involved.

Only a few disorders primarily involve the spinal roots.

Ruptured intervertebral disc. Most herniations of a disc cause the nucleus pulposa to spread posteriorly into the spinal canal, compressing the dorsal and ventral roots. The most prominent symptoms are radiating pain, paresthesia, mild hypoesthesia, weakness, hypotonia, and decreased DTRs. The most common discs involved are those between C5 and C6, C6 and C7, L4 and L5, and L5 and S1. When the lumbar discs are ruptured the term sciatica is often used because of the involvement of the component of the sciatic nerve. Lasègue's sign, in which pain is experienced when the patient tries to extend the knee with the hip flexed, is present with sciatica.

Herpes zoster (shingles). Herpes zoster is an acute infection of the first-order (sensory) neuron involving primarily the dorsal root ganglia and sensory ganglia associated with the cranial nerves. The entire course of the neuron can show inflammatory changes. The symptoms include burning or shooting pains, hyperalgesia, and only minimal sensory impairment. Several days after the onset of pain the skin supplied by the involved ganglia erupts into small vesicles and becomes erythematous.

Sympathetic ganglia

No common disorders directly involve the sympathetic ganglia except for traumatic injuries to the back. The main syndrome is associated with interruption of autonomic fibers in Horner's syndrome. Signs are ipsilateral and include a flushed face, mild ptosis, constricted pupil, and a sunken eyeball (enophthalmos). Horner's syndrome occurs when the superior cervical ganglia, the preganglionic (T1 to T4) fibers going to the ganglia, or the descending autonomic fibers to those preganglionics are damaged.

Sympathectomy of other parts of the chain is a method of treatment for chronic circulatory insufficiency; however, the accompanying tropic changes and hypersensitive vasomotor reactions limit the effectiveness of this treatment.

Spinal cord

The spinal cord is subject to a variety of disorders, which can be categorized into traumatic, infectious, toxic-metabolic, congenital, vascular, and degenerative conditions. Because different areas of the white and gray matters are concerned with different functions, the site of the lesion within the cord can produce a great variety of symptoms. Also important is the level of disorder and the extent (number of segments) of involvement. Keep in mind that damage to the spinal cord can produce signs of lower motoneuron involvement (ventral horn motoneurons) as well as signs of upper motoneuron involvement (descending motor tracts).

A complete transection of the cord results in paralysis and anesthesia below the level of involvement, clonus, increased DTRs and other spinal reflexes, atrophy, Babinski's sign, loss of superficial reflex, disturbances in sexual function, and bowel and bladder incontinence. In injuries involving an entire segment areflexia and signs of denervation occur as a result of dorsal and ventral horn neurons.

Lesions of the dorsal funiculus can produce loss of joint position sense, pressure, vibration, and discriminating touch, an ataxic gait, Romberg's sign, and dysmetria (upper extremity ataxia). Paresthesia, numbness, hypotonia, and decreased DTRs can occur if the dorsal horn is involved.

Lesions of the lateral funiculus can produce spastic paralysis with clonus, increased DTRs, Babinski's and Hoffmann's signs, and disuse atrophy. Because the motor fibers for the leg are located laterally in the lateral corticospinal tract, a parasagittal tumor of the cervical cord can affect the legs first. Other symptoms include loss of pain and temperature contralateral to and one to two levels below the lesion and loss of the superficial reflexes. It is surprising that there are few or no symptoms from the loss of the spinocerebellar tracts.

Lesions of the ventral funiculus can cause weakness, decreased DTRs, and hypotonia predominantly in the proximal muscles as a result of loss of the reticulospinal and vestibulospinal tracts. The ventral horns may be involved.

Intramedullary lesions can cause loss of pain and temperature bilaterally and loss of crossed reflexes as a result of interruption of the crossing axons. If the ventral horn is also involved loss of the motoneurons produces lower motoneuron symptoms (flaccid paralysis, decreased DTRs, marked atrophy, and signs of denervation). It is possible to have Horner's syndrome if the intermediolateral horn at T1 to T4 is involved.

The common spinal cord disorders are briefly discussed below.

Traumatic lesions

DIRECT INJURY. The spinal cord can be damaged anywhere along its course, and any portion or the entire cord may be involved. Most cord injuries initially produce a sudden onset of flaccid paralysis, bowel or bladder incontinence, areflexia, and marked sensory loss below the level of injury. This period is known as spinal shock and is primarily a result of the edema and hemorrhage associated with the trauma. Spinal shock can last for several hours or for several weeks. After this period clonus, spasticity, hyperactive spinal reflexes, and Babinski's sign appear. Varying degrees of motor ability and sensation can return with incomplete lesions. Some degree of bowel or bladder control and sexual functioning can return if the lumbosacral cord is preserved. Interruption of higher center regulation of the preganglionic sympathetic neurons can result in warm skin and edema as a result of loss of vasomotor tone and intermittent periods of excess sweating. The loss of vasomotor tone can cause the patient to feel faint if he or she stands quickly because of pooling of blood in the legs (orthostatic hypotension). Keep in mind that at the level of lesion the dorsal and ventral horns, spinal roots, and spinal nerves can be involved, producing dermatomal zones of anesthesia and autonomic and lower motoneuron involvement. Severe pain and paresthesias can also occur with dorsal root involvement. Also remember that the spinal roots and nerves angle downward (see Fig. 2-1). Thus a cord injury at T_{12} may also involve spinal nerves T_9 to T_{11}. Autonomic dysreflexia may also occur (see Chapter 5).

COMPRESSION OF CORD. Collapse of the vertebral column as a result of metastatic carcinoma, tuberculosis, forward movement of the vertebrae (spondylolisthesis), or degeneration of the vertebrae (spondylolysis), spinal cord tumors, ruptured discs, hemorrhage, and meningitis can create a compression of the cord interferring with nerve conduction. Symptoms usually develop gradually and can involve any area of the cord as well as the spinal nerves and their roots. Early symptoms are often severe radiating pain followed by paresthesia, numbness, coldness, and tightness of the limbs. Motor symptoms usually develop later and begin with weakness, stiffness, and clumsiness of the limbs. Spasticity and lower motoneuron signs can develop later. Loss of sympathetic preganglionics or the higher center regulatory fibers to them can produce the autonomic symptoms mentioned. Keep in mind that the topographic arrangement of the ventral horns and fiber tracts may cause unusual symptom-site lesion relationships. For example, lateral compression of the lateral corticospinal tracts in the cervical cord can produce more initial involvement of the leg (see Fig. 6-1).

Infections. Although the CNS is well protected from infections by the surrounding tissues and the blood-brain barrier, the various bacteria, viruses, and other organisms can invade the CNS through the blood, from penetrating wounds, and via retrograde passage along a nerve trunk. Also the blood-brain barrier hinders the passage of the body's defense mechanisms into the CNS. Onset of symptoms is usually gradual and involves diffuse areas. The inflammatory reaction to the infecting agent causes edema of CNS tissue, thrombosis of arteries, and resultant degeneration of neurons. Only a few CNS infections are generally confined to the spinal cord.

TABES DORSALIS. Tabes dorsalis is a form of neurosyphilis of the lumbosacral cord. The dorsal columns and dorsal roots are primarily involved.

EPIDURAL ABSCESS. Epidural abscess usually begins over the dorsum of the thoracic cord. The initial symptom is radiating pain. The entire cord along many segments can become compressed if untreated.

TETANUS. Tetanus is a unique infection that involves the inhibitory internuncials to anterior horn cells. This results in tonic spasm of skeletal muscles.

HERPESZOSTER. See p. 322.

POLIOMYELITIS. The poliomyelitis virus shows a preference for diffuse infiltration of motoneurons in the spinal cord and brainstem. Initially the meninges are inflamed, causing fever, neck stiffness, Kernig's sign, and muscle tenderness. Later the motoneurons become involved, initially showing fasciculations and later flaccid paralysis and the signs of denervation. The paralysis may be widespread or localized; patchy asymmetric involvement is common.

TRANSVERSE MYELITIS. Transverse myelitis is an inflammation of the gray and white matters and is limited longitudinally to a few segments. There is degeneration of the myelin sheath, of axon cylinders of the fiber tracts, and of the central gray matter. Involvement may be partial or complete. The symptoms resemble those of a spinal cord injury.

Toxic-metabolic conditions. These conditions include overdoses of drugs, chemicals, heavy metal and poisons, toxic effects of infectious agents, water and electrolyte imbalance, carbohydrate and amino acid imbalance, vitamin deficiencies, and failure of the liver, pancreas, and kidneys. Most of these conditions primarily involve the brain because of the specialized nature of those cells. Those commonly demonstrating spinal cord involvement are as follows.

LEAD POISONING. Among other neurologic complications excess lead intake can cause degeneration of anterior horn cells.

COMBINED SYSTEM DISEASE (POSTEROLATERAL SCLEROSIS). This condition is associated with pernicious anemia. The insufficient levels of vitamin B_{12} cause degeneration of the white matter in the dorsal and lateral funiculi.

Congenital disorders. Congenital disorders can be related to genetic factors, the nutritional-metabolic environment of the mother, and extrinsic factors such as infectious diseases. The effects of these factors on the developing embryo can be extremely varied. Those involving the spinal cord are as follows.

FRIEDREICH'S ATAXIA. Friedreich's ataxia is a degenerative disorder of the cerebellum and dorsal spinal cord that does not manifest itself until middle or late childhood (5 to 15 years). Initially the fasciculus gracilis is involved, followed by the lateral corticospinal and posterior spinocerebellar tracts and the cerebellum. The earliest clinical sign is an ataxic gait. Symptoms progress to involve the upper extremities and can vary with the extent of corticospinal versus dorsal column involvement. Cerebellar signs such as nystagmus, dysarthria, and intention tremor can also be present.

PROGRESSIVE SPINAL MUSCULAR ATROPHY (WERDNIG-HOFFMANN SYNDROME). Progressive spinal muscular atrophy is a degenerative disorder of the ventral horn cells and motoneurons of the brainstem. The symptoms of lower motoneuron involvement may be present at birth or may develop within a few months after birth.

SYRINGOMYELIA. Syringomyelia usually results from an incomplete closure of the neural tube, which leads to gliosis (scar formation) and cavitation of the core of the spinal cord and medulla. Symptoms usually appear between 25 to 40 years of age and initially interfere with the descussating sensory fibers of the lower cervical and upper thoracic cord. This results in bilateral loss of pain and temperature in the hands, arms, and upper chest. Later the ventral horns, sympathetic preganglionics, corticospinal tracts, spinothalamic tracts, and dorsal columns can become involved producing additional symptoms. The process extends to the medulla to involve the hypoglossal nuclei, spinal tract of the trigeminal nerve, and eventually the vital centers.

SPINA BIFIDA. Spina bifida results from a failure of a vertebra to close properly. The defect commonly occurs in the lumbosacral area. In the mildest form, spina bifida occulta, symptoms can be relatively mild or may not appear at all. The symptoms can include bladder or bowel dysfunction, lower motoneuron signs in the muscles below the knee, and sensory impairment in the sacral dermatomes. Associated lipomas, adhesions, bony spicules, or malformation of the cord may be the cause of these symptoms. The more involved forms of spina bifida are meningocele and myelomeningocele. With meningocele there is a cystlike sac, usually over the sacrum, that contains dura and arachnoid matter and CSF. The cyst may compress the cord and lumbosacral roots or become infected, producing symptoms similar to those mentioned above. In myelomeningocele there is a similar sacral sac that contains not only meninges but also nerve roots, spinal cord, or both. Symptoms are more severe because of direct involvement of neural tissue and usually include bowel or bladder incontinence, sexual impotence, sensory loss, and lower motoneuron involvement. In both cases the sac is repaired early in life to prevent possible infection. Hydrocephalus is a postsurgical complication.

Vascular conditions. Although the blood supply to the cord is not commonly interrupted, infarctions and hemorrhages can occur with very severe symptoms.

OCCLUSION OF THE ANTERIOR SPINAL ARTERY. The most common site of occlusion of the anterior spinal artery seems to be in the cervical region. The symptoms are basically lower motoneuron involvement at the level of infarction as a result of loss of the ventral horns, upper motoneuron involvement below the level as a result of loss of descending motor tracts, and loss of pain and temperature below the level. Bowel, bladder, and sexual functions are disturbed. Dorsal column sensations are preserved. Onset is usually very sudden.

HEMATOMYELIA. In hematomyelia there is hemorrhaging within the spinal cord. The commonest causes are falls or blows on the spine, which may cause no direct cord damage but rupture a blood vessel, and congenital intramedullary angioma. In both cases the commonest site of hemorrhaging is in the central gray matter of the cervical cord. Symptoms are usually rapidly progressing. Initially there can be severe radiating pain in the neck and arms. Later symptoms develop that are similar to those of syringomyelia. The extent of involvement varies with the amount of hemorrhage.

Degenerative or demyelinating conditions. The destruction of CNS neurons and myelin can occur from a variety of causes. In fact, many of the diseases mentioned earlier, such as Friedreich's ataxia, involve degeneration of neurons or loss of myelin as a result of some infectious, toxic or metabolic, congenital, or vascular influence. Two diseases continue to have unproved causes and are thus often placed in this category.

MULTIPLE SCLEROSIS. The myelin sheath and to some extent the axon cylinder are destroyed and replaced with a sclerotic plaque (scar). The white matter of the spinal cord and the brainstem, cerebellum, and cerebrum are primarily affected, although the cerebral cortex, cranial nerves, and spinal roots can also be involved. The clinical picture in terms of age of onset, speed of onset, and symptoms is varied. The motor symptoms commonly seen are ataxic gait, spastic paralysis of the legs, cerebellar signs of the upper extremities, dysarthria, and scanning speech. Ocular symptoms include dimness of vision in one eye (retrobulbar neuritis of the optic nerve), nystagmus, and internuclear ophthalmoplegia. The latter symptom involves diplopia on lateral gaze with nystagmus of the abducting eye. Sensory symptoms include paresthesia and loss of dorsal column sensations, and anesthesia can occur in the lower limbs. Other symptoms include reduction of vision, nystagmus, loss of conjugate gaze, and euphoria (although marked depression can occur). Multiple sclerosis is a slowly progressive disease often characterized by periods of remission and exacerbation.

AMYOTROPHIC LATERAL SCLEROSIS. Amyotrophic lateral sclerosis (motoneuron disease, progressive muscular atrophy) is a slowly progressive disease that involves the anterior horn cells, motor nuclei of the medulla, and corticospinal tracts in the upper cord. The symptoms are generally lower motoneuron involvement of the upper extremities and tongue and upper motoneuron involvement of the lower extremities. Fasciculations are common in the hands and tongue, as is marked atrophy. Bowel, bladder, and sexual functions can be impaired. There is no sensory impairment.

Cranial nerves and brainstem

Lesions that directly involve the cranial nerves and brainstem are primarily concerned with tumors and infarctions. Trauma to the brainstem is generally fatal because of involvement of the vital respiratory and circulatory centers. Congenital defects of these areas are rare and usually not compatible with life. The major infectious and degenerative diseases involve other areas of the CNS and either have been mentioned or will be in the section on the cerebrum.

The nature of the blood supply and tumor growth is such that most brainstem or cranial nerve lesions are either dorsolateral or centromedial. Lateral lesions involve the cerebellar peduncles and cranial nerves V, VII, VIII, IX, and X and their nuclei. Centromedial lesions involve cranial nerves III, IV, VI, and XII and their nuclei, the medial lemniscus, the corticobulbospinal tract descending in the crus cerebri, the pons, and the pyramids. Involvement of the other motor and sensory tracts and other structures varies with the extent of the lesion. An important clue that often points to a brainstem lesion is ipsilateral cranial nerve involvment with contralateral extremity symptoms. A review of the location of the cranial nerves is advisable at this point.

The common brainstem or cranial nerve disorders are as follows.

Weber's syndrome (oculomotor alternating hemiplegia). In Weber's syndrome the anteromedial portion of the midbrain is involved, lesioning the oculomotor nerve and crus cerebri. Ipsilateral symptoms include lateral strabismus, partial ptosis, diplopia, and dilated pupil. Contralateral symptoms are weakness of lower facial muscles and tongue (the tongue deviates from the side of the lesion), spastic weakness of the extremities with upper motoneuron signs, and loss of lateral gaze.

Millard-Gubler syndrome. Millard-Guber syndrome (abducent or facial alternating hemiplegia) is caused by a lesion of the anteromedial portion of the pons, producing ipsilateral facial nerve involvement (see Bell's palsy below) and internal strabismus and contralateral spastic weakness of the extremities. The tongue deviates from the side of the lesion.

Pontocerebellar angle tumor syndrome (acoustic neuroma). This syndrome usually involves a tumor of the eighth nerve as it enters the lower pons. There is often compression of the dorsolateral portion of the pons involving cranial nerves V and VII as well. Symptoms of eighth nerve involvement includes tinnitus, deafness, vertigo, and nystagmus. Other symptoms are ipsilateral Bell's palsy and anesthesia and loss of the corneal reflex and tearing.

Tic douloureux (trigeminal neuralgia). Tic douloureux is a disorder characterized by a brief attack of severe pain in the distribution of one or more divisions of one trigeminal nerve. There is no apparent evidence of organic disease, and the attack often occurs without any stimulus.

Bell's palsy. Bell's palsy is an inflammation involving the seventh nerve within the stylomastoid foramen. The inflammation may involve the nerve, bone, or periosteum. In any case edema leads to compression of the nerve fibers. There is no apparent cause for the inflammation, but exposure to a cold draft frequently is part of the patient's history. The symptoms are unilateral and appear suddenly. They include paralysis of the muscles of facial expression with lower motoneuron signs, loss of the corneal reflex, inability to close the eye, loss of taste on the anterior two thirds of the ipsilateral half of the tongue, reduced salivation, and hyperacusis.

Wallenberg's syndrome. Wallenberg's syndrome is caused by a lesion of the dorsolateral portion of the upper medulla, occasionally the result of involvement of the posterior inferior cerebellar artery. Structures involved are the inferior cerebellar peduncle, spinocerebellar tracts, lateral spinothalamic tract, cranial nerves IX and X, spinal tract of V, and part of the descending sympathetic fibers. The symptoms are classic and include ipsilateral ataxia, intention tremor, dysmetria, contralateral loss of pain and temperature over the body, dysarthria, bilateral loss of pain and temperatrue over the face, and Horner's syndrome.

Hypoglossal alternating hemiplegia. Hypoglossal alternating hemiplegia is involvement of the anteromedial portion of the lower medulla, producing ipsilateral paralysis of the tongue and contralateral spastic weakness of the extremities with upper motoneuron signs. The tongue deviates toward the side of the lesion.

Cerebellum

In general cerebellar lesions involve either the midline (vermis) and flocculonodular area (archicerebellum) or rather large areas of the hemispheres (primarily neocerebellum). In the former there are primarily disturbances in posture and equilibrium. Symptoms include loss of balance when standing or walking (truncal ataxia) with a tendency to fall backward, an ataxic broad-based gait, and hypotonia. There is little or no ataxia of the extremities. Involvement of the neocerebellum produces symptoms such as intention tremor, dysmetria, anataxic broad-based gait, dysarthria scanning speech, nystagmus, pendular tendon reflexes, adiadochokinesia, rebound phenomenon, and hypotonia. If only one hemisphere is involved the extremity symptoms are ipsilateral. These symptoms can also occur in some brainstem lesions that disrupt the afferent or efferent fibers of the cerebellum.

The primary cerebellar lesions include the following.

Multiple sclerosis. The demyelinating process associated with this disease varies in intensity and area of involvement. When the cerebellum is involved all of the above symptoms can occur with the exception of hypotonia and pendular reflexes.

Friedreich's ataxia. The degenerative process can involve the cerebellar cortex, deep nuclei, and afferent and efferent projections.

Alcoholic cerebellar degeneration. Chronic consumption of alcohol can result in degeneration of the archicerebellum.

Cerebellar abcess. Infections of the CNS can lodge in the cerebellum. The primary symptoms are those related to neocerebellar involvement.

Encephalitis. Certain viral infections such as mumps, measles, rabies, and those transmitted by tick or mosquito can involve the cerebellum. Symptoms are usually mild. They may be transient if the patient recovers or masked by basal nuclei symptoms.

Vascular disorders. The collateral circulation is well developed in the cerebellum. Isolated infarctions are not common, and when they do occur the prognosis is good. Cerebellar symptoms can occur with infarctions of the posterior inferior and the superior cerebellar arteries, since these arteries supply the inferior and superior cerebellar peduncles respectively.

Tumors. Cerebellar tumors are primarily seen in children. They usually arise in the midline but can involve the hemispheres as well. Extension of the tumor into the fourth ventricle, causing secondary hydrocephalus, is a major complication. Tumors of other parts of the brain or brainstem may compress the cerebellum or its peduncles, producing cerebellar symptoms.

Direct injury. Trauma to the back of the skull can cause extensive cerebellar damage. If the deep nuclei are not damaged remarkable recovery can occur over several months.

Cerebrum

It is ironic that this part of the nervous system, the most specialized and vital to functioning, is also the most susceptible to disease and injury. Lesions of the cerebrum can be divided into two main categories: cortical lesions, which involve gyri of the different lobes, and subcortical lesions, which involve the internal capsule, thalamus, and basal nuclei.

Subcortical lesions of the internal capsule often affect an entire one half of the body because of the close proximity of descending corticobulbospinal fibers and the ascending sensory fibers from the thalamus. The motor symptoms are usually spastic paresis with upper motoneuron signs, lower facial muscle weakness, and mild dysarthria. Patients are often able to produce reflex association movements such as abduction and external rotation of the involved shoulder during a yawn, yet when asked to do so voluntarily they cannot. The reason for this is that other motor centers such as the basal nuclei have not been lesioned. These patients can also demonstrate some of the tonic and righting reflexes, since the lesion is above the midbrain. Spasticity, however, may interfere with the manifestation of these reflexes and with association movements. Sensory symptoms are primarily the loss of dorsal column sensations (kinesthesia, discriminating touch, pressure, and vibration). The loss of these sensations, especially kinesthesia, is a major deterrent to the development of fine motor skills such as dressing and writing and of gross motor skills such as ambulation. If the thalamus is not involved in the lesion spinothalamic sensations can still be perceived although not well localized. If the most posterior portion of the internal capsule is involved contralateral homonymous hemianopia can occur. Although auditory data travel in the internal capsule, a lesion on one side does not produce deafness, since both ears project to each hemisphere.

Involvement of the thalamus can cause contralateral hemianesthesia, although more commonly there is an overreaction to sensation. This overreaction is known as the *thalamic syndrome* and may be the result of interruption of cortical inhibitory fibers. The loss of this inhibition on the thalamic nuclei that normally perceive pain, temperature, and touch may result in the unpleasant, disagreeable feelings associated with this syndrome (i.e., nonnoxious stimuli such as touch or joint motion are perceived as unpleasant). Other thalamic symptoms can include mild ataxia, and choreoathetoid movements if the ventrolateral nuclei are involved. These symptoms occur on the opposite side of the body and can be transitory.

Lesions of the basal nuclei produce three kinds of symptoms: involuntary movements, rigidity, and resting tremor. The involuntary movements include athetosis, chorea, dystonia, and ballismus. They are associated with lesions of the caudate, putamen, and subthalamus. As mentioned, they occur at rest and involve abnormal posturing movements or positions of the extremities. Rigidity and resting tremor are often seen together clinically and probably represent damage to the substantia nigra cells. Remember that the substantia nigra acts to inhibit the globus pallidus. Basal nuclei symptoms are generally seen bilaterally. When only one side is lesioned symptoms appear contralaterally.

Lesions of the cerebral cortex can be extremely varied depending on the area involved and the extent of the lesion. Lesions of the frontal lobe can produce changes in personality such as temper tantrums or euphoria; emotional instability such as inappropriate laughter or crying; poor attention span; disorientation; difficulty with abstract thinking, learning, and memory; deviation of the eyes to the side of the lesion (area 8); lower facial muscle paralysis; spastic extremity paresis with upper motoneuron signs; motor aphasia; and apraxia. It is possible to have flaccid paralysis if area 4 is lesioned with the suppressor zone (the anterior portion of area 4) intact. Motor symptoms are contralateral, with the greatest deficit seen in the distal flexor. Keep in mind, however, that each hemisphere is involved with some degree of bilateral control.

Lesions of the parietal lobe can cause loss of dorsal column sensations, tactual agnosia, apraxia, and poor body awareness such as forgetting to wash or dress one side. Keep in mind that the inability to perceive discriminating touch and joint position sense interferes with the learning of skilled activities.

Lesions of the temporal lobe can produce personality and mood changes (memory and learning difficulties) and receptive aphasia (verbal agnosia). As with the internal capsule, deafness is not a symptom unless both temporal lobes are damaged.

Lesions of the occipital lobe can produce contralateral homonymous hemianopia and visual agnosia. Blindness can occur if both lobes are involved as in a traumatic accident.

Lesions of the parieto-temporal-occipital junction in the dominant hemisphere can produce disorders of reading (dyslexia), writing (dysgraphia), and calculating (dyscalcula).

Other symptoms associated with cerebral cortex lesions are seizures, headache, nausea, vomiting, and dilated pupils. These symptoms are not necessarily specific to any lobe and are frequently related to increased intracranial pressure.

An interesting feature of cortical or internal capsule lesions involving the motor areas is that association movements are often not affected. A patient with CVA, for example, may have lower facial paralysis but still be able to move the involved side of the mouth for crying or smiling. Also the involved arm may move when the patient yawns. These movements are possible because extrapyramidal centers (probably the basal nuclei) have fiber pathways that do not run with the corticospinal (pyramidal) fibers.

The major disorders of the cerebrum are as follows.

Traumatic injuries. Any area of the cerebrum can be damaged in a trumatic accident, although the cortex is usually more involved than the basal nuclei. The damage can occur from several sources. Direct penetration by an object such as a bullet or piece of skull bone can cause focal or extensive tissue damage. Rupture of a blood vessel by shearing forces or a penetrating object can cause blood to hemorrhage into the dural or arachnoid spaces or intracerebrally. This can damage cerebral structures by compression. Finally, any accident that interferes with respiration or cardiac output can cause anoxic damage of cerebral neurons.

Initial symptoms usually involve a loss of consciousness, which may last only a few minutes or may persist for months. During prolonged periods of stupor or coma the patient may show decorticate or decerebrate (or a combination) posturing. This excess resistance to movement can make range of motion quite difficult. After recovery any of the symptoms mentioned above can occur, depending on the site or extent of involvement.

Of special interest are patients who sustain a subdural hematoma from a head injury. Since most of these are of venous origin, the intracranial pressure builds up slowly. The patient may lose consciousness for only a few moments and then recover with no apparent symptoms but several hours or days later may lapse into a coma. For this reason all persons who receive trauma to the head, regardless of whether they lose consciousness, should be closely watched for the signs of increased intracranial pressure for 72 hours.

One final point is that of *contracoup*. In contracoup the cerebrum (usually the cortex) opposite the side of injury can be damaged on a sudden impact. For example, in a head-on collision in which a person's forehead smashes the windshield, the occipital lobes can also be damaged.

Infectious disorders. The cerebrum can be infected by agents that are airborne or blood-borne, that enter from direct penetration, from retrograde passage along nerve trunks, and from direct infection of the CSF.

BRAIN ABSCESS. Infection may come directly from a penetrating wound or secondarily from infections of the lung, heart, sinuses or ear. The commonest sites of brain abscess are the frontal and parietal lobes. The white matter is more involved than the gray matter. Early symptoms are those of generalized infection and increased intracranial pressure, followed by symptoms specific to the area of involvement.

MENINGITIS. A wide variety of organisms, including those that cause syphilis and tuberculosis, can cross the blood-brain and blood-CSF barriers to invade the brain, or the CSF can become infected by direct penetration. CSF is an excellent growth medium, and its circulation allows rapid spread of the infection. The inflammatory response to the infection can cause fever, headache, stiffness of the neck (nuchal rigidity), drowsiness, confusion, possible loss of consciousness, and cranial nerve involvement. Inflammation of cerebral blood vessels can cause symptoms of pyramidal, cerebellar, or basal nuclei involvement.

ENCEPHALITIS. The term encephalitis is applied to infections of the CNS and meninges that are due to virus organisms. Viral infections of the brain can be grouped into two categories: aseptic meningitis and encephalitis. In aseptic meningitis only the meninges are involved, producing symptoms similar to meningitis. In encephalitis there is damage to neurons. The main symptoms are weakness, rigidity, tremor, headache, and mental dullness.

Symptoms of encephalitis and aseptic meningitis can be found with measles, mumps, mononucleosis, and rabies. Fortunately these conditions are not frequent.

Toxic or metabolic disorders. The neurons of the CNS and the neuroglial cells are very sensitive to imbalances of many substances. Excesses or deficiencies in water, oxygen, sodium, potassium, calcium, magnesium, glucose, and the B vitamins can produce serious neurologic complications. Since the cerebral cortex is more highly developed, many early symptoms show involvement of this area. These excesses or deficiencies can result from improper intravenous fluids, poor dietary habits, excess vomiting or diarrhea, or malfunction of the pituitary, thyroid, parathyroid, pancreas, liver, or kidneys. The cerebrum can also suffer from the toxic effects of heavy-metal poisoning.

Of interest among these disorders is *Wilson's disease,* or hepatolenticular degenera-

tion. In this disease there is an abnormal copper metabolism resulting in high serum copper levels. The toxic effects primarily damage the liver, kidneys, and lenticular nuclei (putamen and globus pallidus).

Congenital disorders. The developing fetus requires an extremely delicate environment for proper development. Exposure to drugs, disease, excess x-ray examinations, anoxia, malnutrition, genetic abnormalities, toxic poisoning, and metabolic factors can interfere with the development or growth of the neural tube. Also trauma to the mother or during the birth process can damage neural structures.

The most common congenital disorder is cerebral palsy, which basically results from maldevelopment or trauma to the motor centers of their connections. Symptoms can vary and may be present at birth or may not appear for several months. The following are other common congenital disorders.

DOWN'S SYNDROME (MONGOLISM). Down's syndrome is a chromosomal disorder that affects the development of the whole body. The primary neurologic sign is severe mental retardation.

HYDROCEPHALUS. Hydrocephalus is caused by an increase in intracranial pressure as a result of blockage in the flow of CSF, usually in the cerebral aqueduct or foramen of Magendie. The pressure buildup causes the ventricles to dilate, the brain tissue to be compressed, and the head to enlarge. Because of the softness of the skull at birth, damage to brain tissue and resulting symptoms may not occur for several months after birth. Often associated with hydrocephalus is the *Arnold-Chiari malformation* in which there is downward displacement of the pons, medulla, and cerebellar vermis. This displacement may be the result of "anchoring" of the filum terminal.

PHENYLKETONURIA. Phenylketonuria (PKU) is caused by inability to properly metabolize the amino acid phenylalanine. High levels of serum phenylalanine may have toxic effects on neurons, their transmitter substances, or myelin synthesis. The main symptoms are mental retardation and delayed development.

Demyelinating or degenerating disorders. The destruction of the myelin sheath, neuron cell body, or its processes can occur from a variety of infectious, toxic, metabolic, and genetic factors. The process is usually slowly progressive, and the damage is often irreversible.

TAY-SACHS DISEASE. Tay-Sachs disease is a genetic metabolic disease primarily of Jewish people in which an enzymatic deficiency leads to an accumulation of gangliosides in CNS neurons. Symptoms include pyramidal tract signs, cranial nerve signs, dementia, seizures, and cerebellar signs.

MULTIPLE SCLEROSIS. The white matter and bordering gray matter of the cerebrum (especially in the frontal lobe) can be involved in multiple sclerosis. The main symptoms can be euphoria and inappropriate laughter, depression, poor judgment and memory, and mild dementia.

ALZHEIMER'S DISEASE. In Alzheimer's disease there is premature destruction of cortical neurons, probably the result of an enzymatic or metabolic deficiency. The destruction usually begins after 30 years of age and results in general dementia. The symptoms consist of inappropriate behavior, irritability, euphoria, disorientation, poor judgment and memory, paranoia, apraxia, and seizures.

HUNTINGTON'S CHOREA. Huntington's chorea is a genetic deficiency that results in reduction of a neurotransmitter in the basal nuclei. Other amino acid metabolism is interrupted, resulting in cortical atrophy, especially in the frontal lobes and cerebellum. Symptoms include athetoid and choreiform movements, ataxic gait, dementia with depression, slurred speech, and rigidity. The disease usually does not manifest itself until after 30 years of age.

PARKINSON'S DISEASE (PARALYSIS AGITANS). Parkinson's disease is the most common basal nuclei disorder and is characterized by a syndrome of three symptoms: resting pill-rolling tremor, rigidity, and bradykinesia. Other related symptoms include a masklike face, shuffling (festinating) gait, flexed posture, and loss of associated movements such as arm swing during gait. The cause of this disorder is believed to be a decreased level of the neurotransmitter dopamine resulting from a loss of substantia nigra neurons. The effects of low levels of dopamine cause the globus pallidus to produce an imbalance in the activity of the alpha and gamma motor systems. There appears to be a depression of gamma activity, which contributes to the bradykinesia, and an increase in alpha activity contributing to the rigidity. The decreased sensitivity of the muscle spindle combined with the heightened activation of the alpha motoneurons produces an oscillation of agonist and antagonist at rest.

Neoplasms. About 10% of the tumors that arise in the body are found in the CNS. Of those about 80% are found within the skull. Brain tumors can be roughly classified into those involving neuroglial cells (gliomas), the meninges (meningiomas), the blood vessels (angiomas), tumors of the skull, and tumors of the pituitary gland and infundibulum. Since any area of the cerebrum, brainstem, and cerebellum can be involved, the symptoms can be quite varied. However, as the tumor progresses certain symptoms indicate that there is increased intracranial pressure from the space-occupying lesion. These symptoms include headache, nausea, vomiting, seizures, papilledema, drowsiness, and signs of dementia. In addition to focal damage and increasing intracranial pressure a growing tumor can also cause (1) compression of blood vessels, (2) blockage of the circulation of CSF, and (3) compression of the brainstem.

Of all the varieties of tumors, astrocytomas are the most common and can be found anywhere in the brain. In children tumors seem to show a preference for the cerebellum.

CLINICAL CASE STUDIES

In this final section are presented descriptions of 15 disorders. For each case a brief history and a list of neurologic symptoms are presented. Questions are then asked about the site of involvement, the meaning of the symptoms, or which diagnostic test might be useful. Keep in mind that these cases were written to help relate neuroanatomy to clinical findings.

While they represent "traditional" pictures of these disorders, many variations can be seen clinically. At the end of this section is a discussion of each case.

Case 1. A 15-year-old boy was brought to the physician by his mother, who stated that he walked funny and frequently fell when he went out at night. She had noticed these problems for the last 2 months, and they seemed to be getting worse. The examination revealed the following:

1. An ataxic broad-based gait
2. Loss of vibration, pressure, and two-point descrimination of both lower limbs
3. Weakness, hypotonia, and decreased DTRs of the lower limbs
4. Positive Romberg's sign
5. Bilateral Babinski's sign

Symptoms 1 to 4 and symptom 5 indicate involvement of what area(s)?

Why would the boy tend to fall at night?

Case 2. A 50-year-old man went to his physician with a chief complaint of weakness and numbness of both legs. He noticed these symptoms about 6 months ago and is now finding it difficult to walk. Examination revealed the following:

1. Spastic paralysis of both legs
2. Increased Achilles and patellar reflexes
3. Babinski's sign
4. Loss of two-point discrimination and position sense below the knee, but pain touch crudely perceived.

What do symptoms 1-3 indicate?

Would you expect to find clonus?

Where in the CNS can a lesion involve only the lower limbs?

What does the slow onset indicate?

Case 3. A 58-year-old lawyer collapsed in the courtroom in the middle of a heated debate. She was rushed unconscious to a local hospital. She regained consciousness in a few hours. Several days later her physician found the following:

1. Spastic paralysis of the right limbs
2. Increased DTRs on the right
3. Ankle clonus
4. Babinski's and Hoffman's signs
5. Absent abdominal and cremester reflexes
6. Loss of dorsal column sensations on the right
7. Deviation of the tongue to the left when protruded
8. Dysarthria

Why are pain and temperature sensations retained?

What does the sudden onset indicate?

Case 4. A 48-year-old truck driver went to the physician complaining of weakness of both hands. He complained of difficulty in writing and dressing and was unable to drive long distances because his "arms gave out." The symptoms started 2 or 3 months ago and seemed to be getting worse. Examination revealed the following:

1. Marked weakness and atrophy of the intrinsic hand muscles
2. Slight weakness and atrophy of the forearm muscles
3. Bilateral loss of pain and temperature in the C8 to T2 dermatomes
4. Slight bilateral ptosis, pupillary constriction, and enopthalmos
5. Flushing of the face

Why are there no symptoms in the legs?

What is causing the ocular and facial symptoms?

Case 5. A 28-year-old fashion designer went to her physician complaining of trembling in her arms and legs that has been steadily progressing. An examination revealed the following:

1. Choreiform movements of the limbs
2. Generalized muscular rigidity
3. Masked expression
4. Slurred speech
5. Adiadochokinesia
6. Pigmentation of the cornea

Why are there no sensory systems?

How can you explain the adiadochokinesia?

Case 6. A 5-year-old boy was admitted to the hospital with complaints of vomiting and vertigo. His mother had noted that for the past few months he was increasingly "awkward" and "clumsy" when he played and seemed to walk with his feet far apart. He had become irritable lately and fell frequently. Further examination revealed truncal ataxia and nystagmus.

What does the nature of the onset and the symptoms of irritability and vomiting point to?

What do the symptoms of vertigo, nystagmus, and loss of balance indicate?

Case 7. A 39-year-old bus driver went to his physician with a complaint of double vision that had prevented him from working for 2 weeks. He stated that he had been in a fight at a bar 2 weeks before and was knocked unconscious. He came to in a few minutes but had to be helped home because he "had trouble seeing." He also complained that his face seemed to sag and that he dribbled his food. Examination revealed the following:

1. Internal strabismus of the left eye
2. Drooping of the mouth on the left side and inability to wrinkle the left side of the forehead
3. Inability to close the left eye and absence of the corneal reflex on the left
4. Loss of pain and temperature sensation on the right side of the face and extremities
5. Inability to look to the left with either eye, although convergence for close vision was present

What nuclei or tracts are involved?

Symptoms 2 and 3 are associated with what specific structure?

What other symptoms should be included?

What does symptom 5 indicate?

Case 8. A middle-aged actress went to her physician with a chief complaint of difficulty in walking. This was first noticed about 6 months ago and has steadily worsened. In the last 2 months she has noted increasing weakness of the arms and hands. Examination revealed the following:

1. Spastic paresis of both lower limbs, more so on the left
2. Bilateral Babinski's sign
3. Marked atrophy in the upper limbs, more so on the left
4. Fasciculations of the shoulders and arms
5. No sensory loss
6. No cranial nerve involvement
7. On EMG, fibrillation potentials in the hand and forearm muscles, more so on the left

What kind of involvement do symptoms 3, 4, and 7 point to?

What possible cause(s) could produce these symptoms?

What other diagnostic tests might be considered?

Case 9. A 54-year-old concert pianist went to her physician complaining that for the past 6 months her arms seemed heavy and that she had periods of double vision. She also complained of feeling unsteady when on her feet a lot. Examination revealed the following:

1. Weakness of the right medial rectus when individual maintained lateral gaze to the left
2. Moderate ptosis on the right, questionable on the left
3. Drooping of the left side of the face but no noticeable facial weakness
4. Noticeable extremity weakness only in the left wrist extensors
5. Sensations and reflexes normal

What CNS area(s) could produce these symptoms?

What diagnostic tests should be considered?

Case 10. A 62-year-old automobile salesman was admitted to the emergency room with complaints of severe headache, nausea, and "clumsiness" of his left extremities. He told the admitting physician that he had experienced much difficulty in dressing that morning. When he was about to drive to work he seemed confused as to how to start the car. As the day progressed his headache became worse. He finally asked his wife to take him to the hospital. Examination revealed the following:

1. Normal cranial nerves but right pupil dilated
2. Mild left extremity weakness with increased DTRs
3. Babinski's sign on the left
4. Decreased position sense in the hands and feet on the left
5. An unsteady gait with a tendency to bump into objects
6. Ability to read and speak correctly but not to construct figures
7. Partial neglect of the left side

What is the probable nature of this lesion?

What does symptom 6 indicate?

Where is the lesion?

What diagnostic test might be useful?

Case 11. A 58-year-old man collapsed suddenly during a spirited basketball game. He remained unconscious for 30 hours. Examination 2 weeks later revealed the following:

1. Flaccid paralysis of right upper limb, spastic paralysis of right lower limb
2. Tongue protruded to the right, but there was no atrophy
3. Right corner of the mouth drooped
4. Loss of dorsal column sensations on the right but pain and touch perceived
5. Pupils reacted normally to light
6. Right homonymous hemianopia

Why is the arm flaccid and the leg spastic?

What other symptoms would be expected in the right leg?

Why can pain still be perceived?

Where is the lesion located?

What diagnostic test(s) might be considered?

Case 12. A 50-year-old former jockey went to his physician complaining of radiating pain in the legs. He also complained of trouble walking at night. Examination revealed the following:

1. Loss of discriminating touch, position sense, and vibratory sense in both lower limbs
2. Hypotonia in lower limb muscles with decreased DTRs
3. Positive Romberg's sign
4. No real weakness in the legs but gait wide based
5. Urine retention

What do symptoms 1, 3, and 4 indicate?

What does symptom 2 indicate?

What diagnostic test might be helpful?

Why is there urine retention?

Case 13. A 28-year-old interior decorator complained that she was seeing double and that her arm shook when she tried to write. Examination revealed the following:

1. Anesthesia on the left side of the body
2. Analgesia and thermal anesthesia on the left side of the face
3. Ptosis and external strabismus on the right
4. Right pupil dilated and unresponsive to light or the accommodation-convergence reaction
5. Left arm and leg intention tremor and dysmetria

Where is the lesion located?

Why are there cerebellar signs?

Case 14. A 40-year-old museum curator complained to her physician of dizziness, nausea, and vomiting for the past several days. Examination revealed the following:

1. Deafness in the right ear
2. Nystagmus
3. Tendency to fall to the right
4. Analgesia and thermal anesthesia on the left side of the body and right side of the face

5. Hoarseness (dysphonia) and difficulty in swallowing (dysphagia)
6. Horner's syndrome

What structures are involved?

Where is the lesion located?

Why is there Horner's syndrome?

Case 15. A 37-year-old plumber was admitted to the emergency room complaining of severe headache and blurred vision. The symptoms had started that morning with no apparent cause. There was no history of vascular or tumor disorders. His general health has been good except for a history of respiratory infections and a persistent cough. An examination several days later revealed the following:

1. Mild fever
2. Complete right homonymous hemianopia
3. No reading or writing problems
4. No motor or sensory disturbances

What area(s) might be involved?

DISCUSSION OF CASE STUDIES

Case 1. This is Friedreich's ataxia. Symptoms 1 to 4 point to dorsal column or dorsal root involvement of the lumbar cord. Symptom 5 indicates lateral corticospinal tract involvement. The increased tendency to fall at *night* is the result inability of the vestibular system to compensate for the loss of kinesthetic and visual righting mechanisms.

Case 2. The first three symptoms indicate involvement of the descending corticospinal system. Clonus is also a symptom of a lesion of this system. Most lesions of the corticospinal system are unilateral, involving an upper and a lower limb. The only places where isolated involvement of both legs is common are the lumbar cord and the medial surface of the pre- and postcentral gyri. Since some pain and touch can be perceived and the onset was insidious, the most probable diagnosis is a tumor of the falx cerebri (parasagittal meningioma) in the area of the central sulcus. Most cord lesions involving both legs tend to cause more sensory loss.

Case 3. Symptoms 1 to 5 indicate corticospinal involvement; symptoms 7 and 8 indicate hypoglossal nerve or nucleus involvement. Since these two sets of symptoms are alternating (right limbs and tongue deviate to the left), the only possible site for the lesion is the ventromedial portion of the medulla on the left side. This area is supplied by a branch of the anterior spinal artery. The nature of the onset indicates a rupture of this artery involving the pyramids above the decussation, the medial lemniscus, and the hypoglossal nucleus. The spinothalamic tract is spared.

Case 4. This condition is known as syringomyelia, which involves the central part of the lower cervical and upper thoracic cord. The ventral horn cells, the crossing spinothalamic fibers, and the preganglionic sympathetics are lesioned. Symptoms 4 and 5 are indicative of Horner's syndrome and in this case occur because of involvment of the intermediolateral gray column of T1 and T2. The motor tracts to the legs are spared.

Case 5. This is Wilson's disease. The first symptom indicates involvement of the corpus striatum (putamen and caudate). Symptoms 2 to 4 point to involvement of the globus pallidus. Adiadochokinesia is usually considered a cerebellar symptom but can also occur with rigidity because of the difficulty in initiating movements.

Case 6. The insidious onset, irritability, and vomiting point to a slowly growing tumor. The symptoms of loss of balance, broad-based gait, and clumsiness point to the cerebellum, dorsal columns, or vestibular nuclei. The presence of nystagmus and vertigo point to the vestibular nuclei. The truncal ataxia confirms the idea that this is a tumor of the fourth ventricle involving the vestibular nuclei and flocculonodular lobe.

Case 7. This lesion involves a branch of the basilar artery that supplies the dorsal and lateral portions of the lower pons on the left side. The structures involved are the abducent nucleus, the facial nucleus, the spinal tract and nucleus of V, and the medial longitudinal fasciculus. Symptoms 2 and 3 relate to facial nerve or nucleus involvement. Additional symptoms would include loss of taste over the anterior two thirds of the tongue and hyperacusia. Symptom 5 indicates a lesion of the medial longitudinal fasciculus.

Case 8. Symptoms 3, 4, and 7 indicate lower motoneuron involvement of the legs. The probable disorders of a slowly progressive nature are amyotrophic lateral sclerosis or compression of the ventral part of the cervical cord. In either case the ventral horns and lateral corticospinal tract are involved, more so on the left. Additional tests might include lumbar puncture and myelogram.

Case 9. Because the symptoms present an alternating picture the brainstem is naturally suspect. It would appear that a lesion of the right side of the midbrain involving the right third nerve and the descending corticospinal fibers would be the first possibility. However, the lack of reflex changes and any real paralysis, the ptosis, and the suggestion of a fatigue factor point to a defect in the neuromuscular junction. The performance of a Tensilon test, in which a short-acting anticholinesterase (edrophonium chloride) is injected intravenously, caused the relief of the ptosis and an increase in the strength of the wrist extensors. These symptoms are associated with myasthenia gravis.

Case 10. The sudden onset with headache, nausea, and dilated pupil usually indicates an intracranial hemorrhage or an embolus. The inability to draw figures while still being able to read and speak indicates involvement of the nondominant hemisphere. Symptoms 4, 5, nd 7 along with the motor apraxia (difficulty in dressing and confusion about starting a car) indicate involvement of the parietal lobe. The preferred tests might be a lumbar puncture to check for blood in the CSF and increased pressure, an angiogram of the right carotid artery, brain scan, or CAT scan to localize the lesion.

Case 11. This is a somewhat typical picture of a patient with CVA. The spasticity in the leg probably indicates some degree of recovery of the neurons or fibers. Other expected right leg symptoms would be increased DTRs, clonus, and Babinski's sign. Pain can still be felt because the lesion is above the thalamus, either in the internal capsule or in the internal carotid artery just before it splits into the anterior and middle cerebral arteries. The absence of any personality or emotional symptoms tends to point to the internal capsule. The diagnostic tests and probable causes are similar to those for case 10.

Case 12. Symptoms 1, 3, and 4 point to involvement of the dorsal columns in the lumbar cord. Symptom 2 indicates dorsal root, ventral horn, ventral root, or peripheral nerve involvement. An EMG would be useful to eliminate the latter three. This case is an example of tabes dorsalis. The involvement of dorsal roots S2 to S4 will interrupt afferents from the bladder.

Case 13. This lesion is located in the tegmentum of the right midbrain involving the oculomotor nerve, medial lemniscus, spinothalamic tracts, trigeminal lemniscus, and red nucleus. Since cerebellar afferents going to the motor cortex pass through the red nucleus, the dysmetria and intention tremor are expected symptoms.

Case 14. This is Wallenburg's syndrome, which involves cranial nerves VIII (symptoms 1 to 3), IX, and X (symptom 5), the spinothalamic tract and spinal tract of V (symptom 4), and descending autonomic fibers to sympathetic preganglionics (symptom 6). The area involved is the dorsolateral portion of the upper medulla.

Case 15. Homonymous hemianopia can occur with a lesion of the optic tract, the dorsal part of the thalamus, the dorsal part of the posterior limb of the internal capsule (optic radiations), or the occipital lobe. Since there are no other symptoms, the most likely site is the occipital lobe. The history of respiratory infections points to the possibility of a secondary brain abscess.

SUGGESTED READINGS

Brain, L., and Walton, J.: Brain's diseases of the nervous system, New York, 1969, Oxford University Press, Inc.

Chusid, J.: Correlative neuroanatomy and functional neurology, ed. 16, Los Altos, Calif., 1976, Lange Medical Publications.

Everett, N.: Functional neuroanatomy, ed. 6, Philadelphia, 1971, Lea & Febiger.

Gilroy, J., and Stirling Meyer, J.: Medical neurology, ed. 2, New York, 1975, Macmillan, Inc.

Harwood-Nash, D. C.: Neuroradiology in infants and children, St. Louis, 1976, The C. V. Mosby Co.

Netter, F.: The CIBA collection of medical illustrations. I. The nervous system, Summit, N.J., 1972, CIBA Chemical Co.

Toole, J., editor: Comtemporary neurology series. III. Special techniques for neurologic diagnosis, Philadelphia, 1969, F. A. Davis Co.

Truex, R., and Carpenter, M.: Human neuroanatomy, ed. 6, Baltimore, 1969, The Williams and Wilkins Co.

Van Allen, M.: Pictorial manual of neurologic tests, Chicago, 1969, Year Book Medical Publishers.

Willis, W. D., Jr., and Grossman, R. G.: Medical neurobiology: neuroanatomical and neurophysiological principles basic to clinical neuroscience, ed. 2, St. Louis, 1977, The C. V. Mosby Co.

Answers to test questions

Chapter 1

1. a, d
2. a, b
3. a, c, d
4. a
5. c
6. a, b, c
7. c
8. c
9. a, b, d
10. b, e
11. b, c, d
12. a, d
13. a, b, c, d
14. a, b, c, d

Chapter 2

1. a, b, c
2. b, c
3. a, b
4. b
5. a
6. b, c
7. b, d
8. a (indirectly), b, d
9. b, d
10. b, c, a, d
11. d, a, b, c
12. a, d
13. b, c (splenium)
14. b, c, d
15. b, c
16. a, b, d
17. c
18. c, d
19. a, b (give off some fibers), c
20. a, b, d

Chapter 3

1. b, c, d
2. c
3. a, d
4. a, c, d
5. b, c
6. b, c (distal)
7. b, c
8. b, d
9. c
10. None
11. a
12. a, b, c, d
13. c
14. b, c, d
15. a, b, c
16. a
17. b, c
18. a, c, d (possibly)

Chapter 4

1. a. 1, 2, 3 (maybe), 4, 8 (maybe)
   b. 1, 4, 5 (paciniform)
   c. 7, 8
   d. 1, 5
   e. 1, 5, 7
   f. 6, 7, 8
2. a, c, d
3. a. None
   b. 1
   c. 5
   d. 3
   e. 2
4. a, b, c
5. b, c, d
6. a, c
7. a, b, c, d

8. b, c (indirectly)
9. b, c, d
10. a, b, d, c
11. a, b, c
12. a, d
13. c, d
14. a, b, d

Chapter 5

1. a, b, c, d
2. c, d
3. b, c
4. c
5. a
6. a. IX, X, XII (push food to throat)
   b. VII, IX, X
   c. V
   d. VII, IX
   e. III, IV, VI, VIII (reflex), II (reflex)
   f. II (reflex), III
   g. II (reflex), III, VI, VIII (reflex)
   h. IX, X
   i. VII
7. d
8. a, b, d
9. c, a, d, b
10. c, d
11. a, b, c, d
12. a, b, c, d
13. b, c, d
14. b, c, d
15. a, c, d
16. b, c
17. a, b, c
18. a, c
19. c
20. a, c
21. a, b, c, d (depending on overlap)
22. b, c, d
23. c
24. b, d
25. a. 5, 6
    b. 4, 5
    c. 1
    d. 2, 3

Chapter 6

1. a, b
2. a. 1, 3, 5
   b. 4, 6
   c. 7

3. a
4. c, d
5. a, b, c
6. d
7. b, c
8. b, d
9. a, b, c
10. a, c, d
11. a, b, d (possible)
12. b, c

Chapter 7

1. a, d
2. b, d
3. a, b, d
4. a, c
5. a, b, c
6. b, c, d
7. a, b, c, d
8. a, c, d (possible)
9. c, d
10. c
11. b, c
12. a, b, d
13. b, a
14. a, b, c, d
15. b, c
16. b, c
17. none
18. b, c
19. b
20. c
21. a, b, c
22. a, b, c

Chapter 8

1. a, c, d
2. a, b, d
3. a, b, c (possible), d
4. c
5. a, c, d
6. a, b, c, d
7. d
8. a, b, d (indirectly)
9. a, b
10. a, c
11. b, c
12. a, b, d
13. b, c
14. b, c, d
15. a, b, d

Chapter 9
1. a (sometimes), b, c
2. a, b
3. c, d
4. None
5. a, c, d
6. a, d
7. a (possible), b
8. c
9. a, c (partly), d
10. None
11. a, d
12. a, b, c, d
13. a, d (components of)
14. b, c, d
15. b
16. a, c (minimal)
17. a, c, d (on one side)
18. a, d
19. a, b, c (transient)

APPENDIX B

Brief biographic data on major contributors to neuroanatomy*

Adamkiewicz, Albert (1850-1921)

Polish pathologist. Described the blood supply of the spinal cord of humans (an anterior radicular artery supplying the lumbar region of the cord is known as the artery of Adamkiewicz).

Argyll Robertson, Douglas Moray Cooper Lamb (1837-1909)

Scottish ophthalmologist. The Argyll Robertson pupil includes, among other signs, pupillary constriction with accomodation but not in response to light.

Babinski, Joseph Francois Felix (1857-1932)

French neurologist of Polish origin. The Babinski sign, which consists of upturning of the great toe and spreading of the toes on stroking the sole, is characteristic of an upper motoneuron lesion.

Beevor, Charles Edward (1854-1908)

English neurologist. Contributed to knowledge of neurology, especially with respect to localization of function in the cerebral cortex.

Bell, Sir Charles (1774-1842)

Scottish anatomist, neurologist, and surgeon. Bell's palsy is a form of facial paralysis caused by interruption of conduction by the facial nerve. The Bell-Magendie law states that dorsal spinal roots are sensory, while ventral roots are motor.

Bernard, Claude (1813-1878)

French physiologist and one of the great investigators of the nineteenth century. One of his outstanding contributions was demonstration of the vasomotor mechanism.

Bodian, David

Contemporary American anatomist who has made important contributions to neurology. Bodian's stain for nerve cells and fibers uses the organic silver compound protargol.

Breuer, Josef (1842-1925)

Austrian physician and psychologist. Contributed to our knowledge of reflexes controlling respiratory movements.

Broca, Pierre Paul (1824-1880)

French pathologist and anthropolist. Localized the cortical motor speech area in the inferior frontal gyrus; also described a band of nerve fibers (the diagonal band of Broca) in the anterior perforated substance on the basal surface of the cerebral hemisphere.

Brodmann, Korbinian (1868-1918)

German neuropsychiatrist. Brodmann's cytoarchitectural map of the cerebral cortex, although contested, is used frequently when referring to specific regions of the cortex.

Brown-Séquard, Charles Edouard (1817-1894)

British physiologist and neurologist. The Brown-Sequard syndrome consists of the sensory and motor abnormalities that follow hemisection of the spinal cord.

Cajal. See Ramon y Cajal.

Clarke, Jacob Augustus Lockhard (1817-1880)

English anatomist and neurologist. Among numerous contributions, described the nucleus dorsalis of the spinal cord, which is also known as Clarke's column.

Corti, Marchese Alfonso (1822-1888)

Italian histologist. Described the sensory epithelium of the cochlea (organ of Corti).

Cushing, Harvey (1869-1939)

Pioneer American neurosurgeon. Contributed to many basic aspects of neurology, including the function of the pituitary gland, pituitary tumors, tumors of the eighth cranial nerve, and classification of brain tumors.

Darkschewitsch, Liverij Osipovich (1858-1925)

Russian neurologist. Nucleus of Darkschewitsch in central gray matter of the midbrain.

Deiters, Otto Friedrich Karl (1834-1863)

German anatomist. The lateral vestibular nucleus,

---

*Adapted from Barr, M.: The human nervous system, New York, 1972, Harper & Row, Publishers.

which is the origin of the vestibulospinal tract, is known as Deiters' nucleus.

Edinger, Ludwig (1855-1918)

German neuroanatomist and neurologist. An outstanding teacher of functional neuroanatomy and a pioneer in comparative neuroanatomy. The Edinger-Westphal nucleus is the parasympathetic component of the oculomotor nucleus.

Foerster, Otfrid (1873-1941)

German neurologist and neurosurgeon. Made important contributions to the study of epilepsy, pain, the distribution of dermatomes, brain tumors, and the cytoarchitecture and functional localization of the cerebral cortex; introduced the chordotomy operation for intractable pain.

Forel, Auguste H. (1848-1931)

Swiss neuropsychiatrist. Described certain fiber bundles in the subthalamus, which are known as the fields of Forel. The ventral tegmental decussation of Forel in the midbrain consists of crossing rubrospinal and rubroreticular fibers. Forel also proposed the Neuron Theory on the basis of the response of nerve cells to injury.

Friedreich, Nikolaus (1826-1882)

German physician. Best known for his studies of neuromuscular disorders; hereditary spinal ataxia is known as Friedreich's ataxia.

Galen, Claudius (131-201)

Roman physician. Galen was the leading medical authority of the Christian world for over 1400 years. His name is attached to the great cerebral vein.

Golgi, Camillo (1843-1926)

Italian histologist. Introduced a silver staining method that provided the basis of numerous advances in neurohistology. Awarded the Nobel Prize in Physiology and Medicine in 1905 (with Santiago Ramon y Cajal). Golgi staining method, type I and type II neurons, tendon spindles, etc.

Grunbaum, Albert S. F. (later Leyton, A. S. F.) (1869-1921)

British bacteriologist and physiologist. Worked with Sir Charles Sherrington on functional localization of the cerebral cortex.

Gudden, Bernhard Aloys Von (1824-1866)

German neuropsychiatrist. Described the partial crossing of the optic nerve fibers in the optic chiasm, together with certain small commissural bundles adjacent to the chiasm,. Gudden also studied connections in the brain by observing changes subsequent to lesions made in the brains of young animals.

Head, Sir Henry (1861-1940)

English neurologist. Studied the dermatomes, cutaneous sensory physiology and especially sensory disturbances and aphasia following cerebral lesions.

Henle, Friedrich Gustav Jacob (1809-1885)

German anatomist and a pioneer in histology. The endoneurial sheath surrounding individual fibers of a peripheral nerve is known as sheath of Henle.

Hering, Heinrich ewald (1866-1948)

German physiologist. Best known for his study of the reflex that initiates expiration.

Heschl, Richard (1824-1881)

Austrian anatomist and pathologist. Described the anterior transverse temporal gyri (Heschl's convolutions), which serve as a landmark for the auditory area of the cerebral cortex.

His, Wilhelm (1831-1904)

Swiss anatomist and a founder of human embryology. Proposed the Neuron Theory on the basis of his embryologic studies of the development of nerve cells.

Horner, Johann Friedrich (1831-1886)

Swiss ophthalmologist. Horner's syndrome, caused by interruption of sympathetic innervation of the eye, includes pupillary constriction and ptosis of the upper eyelid.

Horsley, Sir Victor Alexander Haden (1857-1916)

A founder of neurosurgery in England. Studied the motor cortex and other parts of the brain by electrical stimulation; introduced the Horsley-Clarke stereotaxic apparatus.

Huntington, George Sumner (1850-1916)

American general medical practitioner. Described a hereditary form of chorea, resulting from neuronal degeneration in the corpus striatum and cerebral cortex.

Jackson, John Hughlings (1835-1911)

English neurologist and pioneer of modern neurology. Gave a thorough description of focal epilepsy (jacksonian seizures) resulting from local irritation of the motor cortex.

Kluver, Heinrich

Contemporary American psychologist. The Kluver-Bucy syndrome is caused by bilateral lesions of the temporal lobes.

Korsakoff, Sergei Sergeievich (1854-1900)

Russian psychiatrist. Korsakoff's psychosis or syndrome, which is usually a sequel of chronic alcoholism, includes a memory defect, fabrication of ideas, and polyneuritis.

Lanterman, A. J.

Nineteenth century American anatomist. Described the incisures of Schmidt-Lanterman in myelin sheaths of peripheral nerve fibers.

Lissauer, Heinrich (1861-1891)

German neurologist. Described the dorsolateral fasciculus of the spinal cord (Lissauer's tract or zone).

Luschka, Hubert Von (1820-1875)

German anatomist. Among other contributions to anatomy, described the lateral aperture of the fourth ventricle (foramen of Luschka).

Luys, Jules Bernard (1828-1895)

French neurologist. Described the subthalamic nucleus (nucleus of Luys), whose degeneration causes hemiballismus.

Magendie, Francois (1783-1855)

French physiologist and pioneer of experimental physiology. The sensory function of dorsal spinal nerve roots and motor function of ventral roots constitutes the Bell-Magendie law. Also described the median aperture of the fourth ventricle (foramen of Magendie).

Marchi, Vittorio (1851-1908)

Italian physician and histologist. The Marchi staining method for tracing the course of degenerating myelinated fibers continues to be an invaluable neuroanatomic technique.

Meissner, Georg (1829-1905)

German anatomist and physiologist. His name is attached to touch corpuscles in the dermis and the submucous nerve plexus of the gastrointestinal tract.

Merkel, Friedrich S. (1845-1919)

German anatomist. Described tactile endings in the epidermis, known as the Merkel's discs.

Meyer, Adolph (1866-1950)

Prominent American psychiatrist. Those fibers of the geniculocalcarine tract that loop far forward in the temporal lobe constitute Meyer's loop.

Monro, Alexander (1733-1817)

Scottish anatomist, also known as Alexander Monro (Secundus). Including tenure by his father (Primus) and son (Tertius), the Chair of Anatomy in the University of Edinburgh was occupied by Alexander Monros for over a century. The interventricular foramen between the lateral and third ventricles is known as the foramen of Monro.

Nauta, Walle Jetze Harinx

Contemporary American neuroanatomist. Has made numerous contributions to knowledge of the functional anatomy of the central nervous system. The Nauta silver staining method is used widely for tracing the course of nerve fibers.

Nissl, Franz (1860-1919)

German neuropsychiatrist. Made important contributions to neurohistology and neuropathology. Introduced a method of staining gray matter with basic aniline dyes to show the basophil material (Nissl bodies) of nerve cells.

Pacini, Filippo (1812-1883)

Italian anatomist and histologist. Described the sensory endings known as the corpuscles of Vater-Pacini.

Penfield, Wilder Graves

Contemporary Canadian neurosurgeon. Has made fundamental contributions to neurocytology and neurophysiology, including function of the cerebral cortex, speech mechanisms, and factors underlying epilepsy.

Purkinje, Johannes (Jan) Evangelist (1787-1869)

Bohemian physiologist, pioneer in histologic techniques, and accomplished histologist. Described the Purkinje cells of the cerebellar cortex, Purkinje fibers in the heart, etc.

Ramon y Cajal, Santiago (1852-1934)

Spanish histologist acknowledged as the foremost among neurohistologists. Awarded the Nobel Prize in Physiology and Medicine (with Camillo Golgi) in 1906. Among innumerable contributions, Cajal vigorously championed the Neuron Theory on the basis of his observations with silver staining methods.

Ranvier, Louis-Antoine (1835-1922)

French histologist and a founder of experiemental histology. Described the nodes of Ranvier in the myelin sheaths of nerve fibers.

Rasmussen, Grant Litster

Contemporary American neuroanatomist. Has made numerous contributions to neuroanatomy, including a description of the olivocochlear bundle.

Reil, Johann Christian (1759-1813)

German physician. The insula, lying in the depths of the lateral fissure of the cerebral hemisphere, is known as the island of Reil.

Renshaw, Birdsey (1911-1948)

American neurophysiologist. Certain intercalated neurons in the ventral gray horn of the spinal cord are called Renshaw cells.

Rexed, Bror

Contemporary Swedish neuroanatomist. Divided the gray matter of the spinal cord into regions (laminae of Rexed) on the basis of differences in cytoarchitecture.

Rolando, Luigi (1773-1831)

Italian anatomist. Among various contributions to neurology, described the central sulcus of the cerebral hemisphere and the substantia gelatinosa of the spinal cord.

Romberg, Moritz Heinrich (1795-1873)

German neurologist. Romberg's sign consists of unsteadiness or swaying of the body when standing with the feet close together and the eyes closed (present in tabes dorsalis).

Ruffini, Angelo (1864-1929)

Italian anatomist. Described sensory endings, especially those known as the end-bulbs of Ruffini.

Schmidt, Henry D. (1823-1888)

American anatomist and pathologist. Described the incisures of Schmidt-Lanterman in myelin sheaths of peripheral nerve fibers.

Schwann, Theodor (1810-1882)

German anatomist. Formulated the Cell Theory (with M. J. Schleiden) and described the neurolemmal cells (Schwann cells) of peripheral nerve fibers.

Sherrington, Sir Charles Scott (1856-1952)

English neurophysiologist. A foremost contributor to

basic knowledge of the function of the nervous system. His researches included studies of reflexes, decerebrate rigidity, reciprocal innervation, the synapse, and the concept of the integrative action of the nervous system.

Sydenham, Thomas (1624-1689)

English physician, known as the English Hippocrates. Described the form of chorea to which his name is attached.

Sylvius, Francis de la Boe (1614-1672)

French anatomist. Gave the first description of the lateral fissure of the cerebral hemisphere.

Sylvius, Jacobus (also known as Jackques Dubois) 1478-1555)

French anatomist. Described the cerebral aqueduct of the midbrain.

Vater, Abraham (1684-1751)

German anatomist. Among other contributions to anatomy, described the sensory endings known as the corpuscles of Vater-Pacini.

Wallenberg, Adolf (1862-1949)

German physician. Described the lateral medullary syndrome.

Waller, Augustus Volney (1816-1870)

English physician and physiologist. Described the degenerative changes in the distal portion of a sectioned peripheral nerve, now known as wallerian degeneration.

Weber, Sir Hermann David (1823-1918)

English physician. Described the midbrain lesion causing hemisparesis and ocular paralysis.

Weigert, Karl (1843-1905)

German pathologist. Introduced several staining methods, including a stain for myelin and therefore for white matter in sections of nervous tissue.

Wernicke, Carl (1848-1905)

German neuropsychiatrist. Made a special study of disorders in the use of language. Wernicke's sensory language area, Wernicke's aphasia.

Westphal, Karl Friedrich Otto (1833-1890)

German neurologist. Among other contributions to neurology, described the Edinger-Westphal nucleus in the oculomotor complex.

Willis, Thomas (1621-1675)

English physician. One of the dominating figures in English medicine of the seventeenth century and one of the founders of the Royal Society. Among numerous contributions to the anatomy of the brain, described the arterial circle that bears his name.

Wilson, Samuel Alexander Kinnier (1739-1808)

British neurologist. Described hepatolenticular degeneration, now known as Wilson's disease.

Index